Cosmoparticle Physics

Cosmoparticle Physics

Maxim Yu Khlopov
Centre for Cosmoparticle Physics, "Cosmion", Russia

World Scientific
Singapore • New Jersey • London • Hong Kong

Published by

World Scientific Publishing Co. Pte. Ltd.

P O Box 128, Farrer Road, Singapore 912805

USA office: Suite 1B, 1060 Main Street, River Edge, NJ 07661

UK office: 57 Shelton Street, Covent Garden, London WC2H 9HE

British Library Cataloguing-in-Publication Data
A catalogue record for this book is available from the British Library.

COSMOPARTICLE PHYSICS

ISBN 981-02-3188-1

This book is printed on acid-free paper.

Printed in Singapore by Uto-Print

To Yakov Zeldovich, Andrei Sakharov,
Viktor Shwartsman, David Schramm and
all my teachers, friends and colleagues,
who are not with me now
but are forever in my heart and memory.
Their legacy is in the annals
of cosmoparticle physics.

PREFACE

Starting from 1980-s the cross-disciplinary, multidimensional field of links between cosmology and particle physics is widely recognised by theorists, studying cosmology, particle and nuclear physics, gravity, as well as by astrophysicists, astronomers, space physicists, experimental particle and nuclear physicists, mathematicians and engineers.

It is well understood that only joint efforts may lead to progress in this field.

The relationship between cosmology and particle physics is now one of the important topics of discussion at any scientific meeting both on astrophysics and high energy physics. National and international institutes, laboratories and centres all over the world study specific problems of such a relationship.

On some historical reasons many important results, trends and directions of scientific research, undertaken in 80-90s in this field, in USSR and then in Russia are not widely known. The present book is aimed to fill in this gap.

Covering this loophole in the English scientific literature the book, based dominantly on the most recent results, also includes discussion of some papers dated 80-s or early 90-s. One may surprise how up-to-date are the results of these papers. One may find how useful they are for solution of problems, clearly understood and formulated in the English literature only now.

However, the very general approach that appeared, evolved and highly developed in Russia in 80-90s is even more important. Originated from the works of

A.D.Sakharov, Ya.B.Zeldovich, M.A.Markov and their scientific schools this approach in its development naturally embeds all the existing trends in studies of fundamental links between cosmology and particle physics.

It offers the natural framework for self-consistent analysis of nontrivial relationships between the theory of Universe and particle theory. It provides the system of studies of cosmology and particle physics in the complex network of their indirect astrophysical and physical effects. It implies the specific name for the multidimensional field of science studying the fundamental relationship between micro- and macro- worlds.

This name "Cosmoparticle physics" is put into the title of the book.

The reason we use "cosmoparticle physics" to denote the problems generally referred to "astroparticle physics" or "particle astrophysics" is in no case the ambition of terminological originality. It has much deeper reasons than the simple following the Russian tradition to call "cosmonauts" those Americans call "astronauts".

There are many particular problems such as massive neutrino as hot dark matter, invisible axions or neutralino as cold dark matter in which astroparticle and cosmoparticle physics practically coincide. But the proper choice of dark matter candidate or of the mixture of different forms of dark matter appears in astroparticle physics as the matter of taste. Cosmoparticle physics can make this choice the matter of test.

Thus cosmoparticle physics naturally contains astroparticle physics as important element but represents much wider field of scientific research. The research in this field implies nontrivial system of cross-disciplinary studies, which are reflected in the specific form of scientific organization.

In USSR first under charge of Ya. Zeldovich and then of A. Sakharov the complex program of joint studies in cosmoparticle physics was elaborated.

Founded in 1992, Centre for Cosmoparticle physics "Cosmion" initiated practical realisation of some ideas of this program in complicated conditions of rapidly changing Russia.

The first results of scientific activities of more than forty scientific groups from more than twenty Russian centres and institutions, joined in the Russian National project "Cosmoparticle physics", were reported at the 1 International conference on Cosmoparticle physics "Cosmion94", dedicated to 80 Anniversary of Ya.B.Zeldovich and 5 Memorial of A.D.Sakharov, December 4-14, 1994, Moscow.

The 2 International conference "Astronomy. Cosmoparticle Physics", dedicated to 75 Anniversary of A.D.Sakharov (COSMION-96), May-June 1996, Moscow, and the 3 International conference on Cosmoparticle physics "Cosmion 97", dedicated to 10 Memorial of Ya.B.Zeldovich, December 8-14, 1997, Moscow succeeded in the studies of various links between cosmology and particle physics and of various aspects of cosmoparticle research.

These various studies, being important by them, are within the project co-ordinated and started to be arranged into cross-disciplinary networks, putting together astronomers, space researchers and experimental physicists into the particular cells of proper cosmoparticle studies. In the context of cosmoparticle research theoretical studies play new important role gluing together experiments and observations in different fields of science.

These studies and their results presented at Cosmion conferences uncover the features of cosmoparticle physics as the new self-consistent field of scientific research.

The system of cosmoparticle research is open. It does not substitute for the well-established fields of studies in particle and nuclear physics, cosmology and astrophysics but it illuminates nontrivial links between them and offers new solutions for their old problems. It offers to each scientist the possibility to use all his scientific skill and experience in the joint study of the fundamental laws of our reality. It is the aim of this book to facilitate participation in these studies.

The book uses personal author's experience in the presentation of the problems of cosmoparticle physics to both physical and astronomical audience. It accounts for the author's course of lectures "Introduction into cosmoparticle physics", given to students of Moscow Engineering Physics Institute. It also takes into account the similar course, given to post- graduate students of the University of Trieste.

Based on this experience the book concentrates the readers attention to the transparent outlining of the principal ideas of the modern particle theory and cosmology, their mutual relationship and nontrivial correspondence of their physical and astrophysical effects.

The puzzles of particles and Universe are bound in a complicated knot of problems. I am grateful to my teachers Ya.B.Zeldovich and A.D.Sakharov for my involvement into the beauty of these puzzles.

I express my gratitude to co-authors of all my papers, especially, to Z.G. Berezhiani, S.I. Blinnikov, V.M. Chechetkin, A.G. Doroshkevich, D. Fargion, S.S. Filippov, S.S. Gershtein, A.A. Klypin, R.V. Konoplich, A.D. Linde, G. Piragino, A.G. Polnarev, S.G. Rubin, A.S. Sakharov, M.G. Sapozhnikov, E.V. Sedelnikov, A.A. Starobinsky, to collaborators and participants of National Project "Cosmoparticle physics", to collaborators of International projects "Astrodamus", "Coseth", Eurocos-AMS for cooperation in unbinding some ties of this knot.

I am grateful to CCPP "Cosmion", ETHZ (Switzerland), ECT*(Trent) and INFN (Trieste) for important support during the work on the book, to A.V. Berkov, E.V. Sedelnikov and A.S. Sakharov for help in preparation of the manuscript.

CONTENTS

PREFACE vii

CHAPTER 1. INTRODUCTION 1

CHAPTER 2. THE HIDDEN SECTOR OF PARTICLE THEORY

1. Particle Theory – the Standard Model 19

1.1. Quantum Electrodynamics 24

1.2. Gauge Symmetry of Quantum Electrodynamics 29

1.3. Symmetry of Fundamental Particles 32

1.4. Standard Model of Electroweak and Strong Interaction 39

2. Esthetical and Practical Grounds for the Extensions of the Standard Model 47

2.1. New Physics from Grand Unification 48

2.2. New Physics in the Modern Trends to the True Unification 51

3. Neutrino Mass and Invisible Axion 58

3.1. Physics of the Mass of Neutrino 59

3.2. Neutrino Instability 65

3.3. The Axion Solution for the Problem of Strong CP violation 74

3.4 The Models of Invisible Axion 79

CHAPTER 3. THE HIDDEN PARAMETERS OF THE MODERN COSMOLOGY

1. The Old Standard Big Bang Paradigm

1.1. The Expanding Universe 91

1.2. The Black Body Radiation Trace of the Big Bang 100

1.3. Physical Cosmology of Old Big Bang Scenario 106

2. Inflationary Universe with Baryosynthesis and Dark Matter 128

2.1. Magnetic Monopoles in Old Big Bang Model 130

2.2. Inflational Recovery of Old Big Bang Model 144

2.3. Baryosynthesis 155

2.4. Nonbaryonic Dark Matter 160

CHAPTER 4. COSMOARCHEOLOGY OF THE VERY EARLY UNIVERSE 165

1. Primordial Black Holes as the Cosmoarcheological Tool

1.1 Primordial Black Holes 166

1.2. PBH Signatures of Dust-Like Stages in the Early Universe 170

1.3. Direct PBH Formation at Dust-Like Stages 175

1.4. PBH Evaporation 181

2. Formation of Black Holes in the First Order Phase Transitions 184

2.1. Field Configurations in Bubble Wall Collisions 185

2.2. Gravitational Collapse of False Vacuum Bag and Black Hole Formation 188

2.3. The First-Order Phase Transitions in the Early Universe 196

CHAPTER 5. PRIMORDIAL PARTICLES IN THE PERIOD OF BIG BANG NUCLEOSYNTHESIS

1. New Particle Effects on Nucleosyntehsis 205

1.1 Restrictions on the Total Cosmological Density 214

1.2. The Shift of Equilibrium Rates of β-processes 224

1.3. Non-equilibrium Particles and ^4He abundance 233

2. Metastable Particles and the BBBR Spectrum

2.1 Distorsion of the Thermal Background Spectrum 240

2.2 Dominance of Metastable Particles in the Period of Nucleosynthesis 248

CHAPTER 6. ANTIPROTONS IN THE UNIVERSE AFTER THE BIG BANG NUCLEOSYNTHESIS 259

1. PBH Evaporation as a Source of Nucleon–Antinucleon pairs 260

2. Effects of Antiproton Annihilation with ^{4}He nuclei on D and ^{3}He abundance 274

3. Restriction on Antiproton Sources

3.1. Constraints on PBHs and Physics of Their Origin 282

3.2. The Problem of Primordial Gravitino 291

3.3. Constraints on Antimatter Domain Structure 298

CHAPTER 7. NON-EQUILIBRIUM EFFECTS AS THE PROBE FOR NEW PHYSICS 301

1. The Non-Equilibrium Cosmological Nucleosynthesis 302

1.1. Kinetic Equation for Non-Equilibrium Particles 303

1.2. Kinetic Equation for Weak Sources on RD Stage 307

1.3. Non-Equilibrium Nucleosynthesis. Special Cases 314

2. ASTROBELIX project.

2.1. AstroNuclear Experiment ASTROBELIX 324

2.2. Li and Be Abundance Constraint on Antiproton Annihilation on RD Stage 329

3. Cosmological Backgrounds of Non-Equilibrium Particles

3.1. High Energy Neutrino from the RD Stage 332

3.2. High-Energy Neutrino Background as the Probe for New Physics 337

3.3. DUMAND as the Tool for Cosmoparticle Physics 346

CHAPTER 8. THE NEW PHYSICS IN THE LARGE-SCALE STRUCTURE FORMATION

1. Problems of the Large-Scale Structure Formation

1.1. The Problem of Initial Fluctuations 351

1.2. The Problem of Self-Consistent Structure Formation and BBBR Anisotropy 355

1.3. The Problem of Physical Self-Consistency in the Theory of Large-Scale Structure Formation 360

2. The Neutrino Rest Mass and the Large-Scale Structure of the Universe

2.1. Massive Neutrino in the Big Bang Universe 362

2.2. Gravitational Instability of Massive Neutrino 369

2.3. The Structure Formation in the Neutrino-Dominated Universe 382

CHAPTER 9. PROBES FOR THE DARK MATTER PARTICLES

1. The Structure of the Universe as the Detector for the Dark Matter Particles

1.1. The Structure Formation Constraints on Unstable Particles 389

1.2. The Troubles of Simple Dark Matter Models 401

1.3. Arguments for the Existence of Unstable Dark Matter Particles 409

2. Dark Matter in the Galaxy

2.1. The Condensation of Dark Matter in Galaxies 416

2.2. WIMP Annihilation in the Galaxy 420

2.3. Direct and Indirect Searches of WIMPs in the Galaxy 423

CHAPTER 10. MIRROR WORLD IN THE UNIVERSE

1. Mirror Particles

1.1. The Equivalence of Left and Right 433

1.2. Fractons and Alice Strings 437

2. Mirror Particles in the Early Universe

2.1. Inflation and Constraints on the Domain Structure 444

2.2. The Inhomogeneous Baryosynthesis and the Island Distribution of Mirror Baryons 448

2.3. The Big Bang Nucleosynthesis and the Mirror World 450

3. Formation of Astronomical Objects of the Mirror Matter 451

3.1. The Separation of Ordinary and Mirror Matter 454

3.2. The Gas Accretion on the Astronomical Objects of the Other Mirrority 466

4. Observational Physics of the Mirror Matter 475

4.1. Galaxies and Galaxy Clusters of Definite Mirrority 476

4.2. The Mirror Matter at the Scale of Globular Clusters 482

4.3. The Effects of Mirror Matter at Stellar Scales 486

CHAPTER 11. COSMOPARTICLE PHYSICS OF HORIZONTAL UNIFICATION

1. Physical Grounds for Horizontal Unification

1.1. The Symmetry of Fermion Families 493

1.2. The Gauge Model of Family Symmetry Breaking 498

2. The Early Universe from Horizontal Unification

2.1. The Inflationary Dynamics and Energy Scale 507

2.2. The Formation of PBH in the MHU 509

2.3. Early "Horizontal" Phase Transitions 513

2.4. The Large-Scale Modulation in the Distribution of Primordial Axion Field 517

3. MHU Dark Matter Cosmology

3.1. The Dark Matter Unification in the Hidden Sector of Models of Horizontal Unification 525

3.2. DM Models from Horizontal Unification 528

3.3. The History of the Universe Based on MHU 539

REFERENCES 543

SUBJECT INDEX 569

CHAPTER 1

INTRODUCTION

COSMOPARTICLE PHYSICS is a natural step in the development of widely accepted relationship between COSMOlogy and PARTICLE PHYSICS. This development revives the tradition of Natural philosophy of universal knowledge, the tradition to consider the world in its universal completeness and unity.

Cosmoparticle physics represents the new level of this tradition and, being exact science, uses quantitatively definite methods in description of the world and its structures at macro- and micro- scales.

In order to trace the tendencies, that lead to the formation of Cosmoparticle physics, one should remind the main steps of internal development of particle theory and cosmology.

The experimentally proven part of the modern particle theory, referred to as the standard model of electoweak and strong interactions, is built as an extension of electrodynamics on particle transformations in weak and strong interaction processes.

The possibility of such an extension follows from the principal change of the notion of charge and current in quantum field theory. Instead of the internal properties of eternal charged particles, treated by classical electrodynamics, one considers in quantum electrodynamics the electromagnetic current as the locally conserved bilinear combination of one-particle-state annihilation and creation operators, acting on the physical vacuum.

For instance, the electron current represents the succession of annihilation and creation of the one-electron state. Initial and final states may differ, that corresponds to the current of a quantum electromagnetic transition.

The next logical step is to consider different particles in initial and final states and to generalise the notion of the current of transition to the current of particle transformation.

The electromagnetic current is the source of the electromagnetic field. In quantum electrodynamics, annihilation of electron in the initial state and its creation in the final state is a quantum process of emission or absorption of a quantum of electromagnetic field.

In general, one can say, that annihilation of one particle in the initial state and creation of another particle in the final state is the process of emission or absorption of a quantum of particle transformation, mediating the corresponding interaction.

So, in the process of weak interaction, annihilation of neutrino and creation of electron is the source of W-boson — the quantum of the field of weak interaction.

The generalisation assumes the symmetry between particles in initial and final states. Mathematically, the symmetry provides the generalisation of gauge invariance of electrodynamics to local gauge invariance, and it makes possible to introduce the similar forms for weak, strong and electromagnetic interactions. For particles of matter (fermions), the Lagrangian density of all these interactions has the same form

$$L_{int} = gJ_{\mu}V_{\mu}, \tag{1}$$

where g is the coupling constant, J_{μ} is the current and V_{μ} is the field of the corresponding interaction. The observed difference in the interactions of fundamental particles is ascribed both to the difference in the groups of gauge symmetry and to the mechanism of spontaneous symmetry breaking.

The standard model of interactions of particles, based on the SU(2)xU(1) local gauge symmetry of electroweak interaction and the $SU(3)_c$ symmetry of quantum chromodynamics (QCD, the gauge theory of strong interactions) does not find at present any direct contradiction with the experimental data.

However, the internal theoretical inconsistencies and esthetical challenges to unify all the fundamental forces of Nature forces the theory go beyond the standard model, losing on this way the possibility of direct experimental proof.

The practical theoretical necessity to extend the standard model follows from such internal problems of the standard model as quadratic divergence of loop radiative corrections to the mass of Higgs field or strong CP violation in QCD.

The solution of the former problem implies supersymmetry — symmetry between bosons and fermions, giving rise to cancellation of boson and fermion loop contributions into the Higgs mass due to the difference in Bose-Einstein and Fermi-Dirac statistics.

Supersymmetry should be broken, since we do not observe it in fermion and boson mass spectra, and the search for supersymmetric partners of known particles is one of the strongest challenges for the next generation of particle accelerators. But there is no hope in searching for gravitino (supersymmetric partner of graviton), predicted in local supersymmetric models, even in far future accelerators, due to its very weak, semigravitational coupling to other particles.

The solution of the problem of strong CP violation in QCD implies the existence of invisible axion, pseudo-Goldstone boson, the "smaller brother" of π^0 , with superweak interaction, being practically elusive for the direct search at accelerators.

The esthetical motivation to extending the standard model is connected with the attractive idea of unification of fundamental forces.

The similarity in description of electromagnetic, weak and strong interactions, achieved in the standard model, finds deeper grounds in grand unified theories (GUT), extending the fundamental gauge symmetry and putting $SU(2)\mathrm{x}U(1)\mathrm{x}SU(3)_c$ group of symmetry of the standard model into the unique group G.

Arranging the set of known particles into the representation of the group G, one finds "white spots" to be occupied to make the representation complete.

The larger G is, the more numerous is the amount of new particles and fields, which should be introduced to complete the basic particle and field content to the set, corresponding to the full symmetry.

Such particles and fields correspond to the "hidden sector" of the theory, since they are hidden from the direct experimental probes either due to their large mass, or owing to very weak interaction with the known particles.

In the both cases (of weakly interacting and of superheavy particles) one needs some indirect means to test the respective predictions.

The same is true for the parts of hidden sector, such as "invisible axion", gravitino, etc. invoked due to practical needs to make the standard model self-consistent.

There are very few indirect effects of superheavy and/or superweakly interacting particles and fields, feasible for laboratory probes.

They are the mass of neutrino, CP violation, lepton and baryon number nonconservation (reflected in neutrino oscillations, double neutrinoless beta decay, flavour changing neutral currents, proton decay, neutron-antineutron and hydrogen-antihydrogen oscillations).

Being rare, these effects can be discriminated due to manifest violation of conservation laws, included in the standard model. Relatively small manifestations of such effects appeal for additional probes of hidden sector of particle theory.

The problem of proper choice for the extensive hidden sector increases in the models of theory of everything (TOE), putting into unique theoretical framework the foundations for all the four fundamental forces of Nature, including gravity.

Such a framework may follow from successive extension of gauge symmetry, for instance, from the combination of local gauge models and supersymmetry, as it takes place in supergravity. Here the unification follows from the extension of internal symmetries on the symmetries of space-time.

The alternative approach is based on the extension of geometry of space-time to include the description of particle interactions. In geometrical approach, the fundamental forces are ascribed to the effects of additional compactified dimensions, and symmetries of space-time are extended including symmetries of elementary particles.

The both trends are accumulated in superstring theories based on principally new fundamental concepts of string theories.

In the heterotic string theory one combines the d=10 heterotic string theory with the $E_8 \times E_8'$ gauge symmetry. Its hidden sector should contain, in principle, the whole zoo of particles, fields and new phenomena, arising in various extensions of the standard model, whose direct experimental search is either very hard or virtually impossible.

That's why the Universe has drawn the most serious attention of particle physicists as the possible source of information on elementary particles.

Ya. Zeldovich called the Universe "the poor man's accelerator". As A.Linde followed, even the richest man can not build the accelerator, that can reach the energies of GUT or TOE physics, which are naturally released at earliest stages of cosmological evolution.

So, the internal development of particle physics has lead to the Big Bang Universe as a probe of its fundamental ideas.

The modern cosmology is based on two observational facts, i.e. that the Universe expands and that the Universe contains the background electromagnetic black body radiation.

Putting them together, one comes to the ideas of Big Bang Universe.

One inevitably comes to the conclusion, that at early stages of cosmological expansion the physical conditions in the Universe should have been dramatically different from those we observe now.

Extrapolating, or, more precisely, interpolating the law of cosmological expansion to the past, one finds that at earlier stages of cosmological expansion the energy density of radiation exceeded the density of matter, so that the radiation dominance stage should have existed.

One can easily check, that matter and radiation were in equilibrium, that there were no galaxies, stars, and the matter was in the state of nearly homogeneous plasma.

The old Big Bang scenario was a self-consistent combination of general relativity, thermodynamics and laws of atomic and nuclear physics, well proven in laboratory. The laws of physics were applied to the evolution of the Universe as a whole, under the assumption that it contains only baryonic matter and electromagnetic radiation (and neutrinos) (Zeldovich, Novikov, 1975).

According to this scenario, nuclear reactions should have taken place to the end of the first three minutes, that lead to the primordial chemical composition.

This picture found qualitative confirmation in the comparison of predictions of Big Bang nucleosynthesis with the observed abundance of light elements.

It gave the qualitative explanation to the observed structure of inhomogeneities as a result of gravitational instability in nearly homogeneous matter.

However, quantitative disagreements, that turned out to be more and more profound, made the whole picture controversial, unless some additional fundamental elements were added to the basis of the whole construction.

The formation of the observed large-scale structure of the Universe corresponded in the old Big Bang scenario to the expected anisotropy of thermal electromagnetic background radiations is inconsistent with the observations.

On the other hand, the low baryon density is needed in order to reproduce the observed abundance of light element as a result of Big Bang nucleosynthesis (for review, see Copi and Schramm, 1996). It is inconsistent with the much higher density one needs to explain the formation of the large-scale structure as a result of development of gravitational instability.

The both problems seemed to found their laboratory based solution in the framework of the old Big Bang scenario when it was claimed in 1980, that the electron neutrino has the mass about 30 eV. The neutrino thermal background, as abundant as the thermal photon background, was one of well established predictions of the old Big Bang scenario.

Multiplying this abundance by the value of the neutrino mass, claimed to be measured in the ITEP experiment, one can find that the modern density of massive neutrinos should exceed the density of the baryon matter by one-two orders of magnitude.

One came to scenario of neutrino-dominated Universe, in which massive neutrinos, weakly interacting with the matter and radiation, has driven the cosmological large-scale structure formation and dominate in the modern cosmological density. The anisotropy of the thermal background radiation, predicted in this picture, seemed to be consistent with the observational data.

However, successive experimental studies did not confirm that the mass of electron neutrino might be as high, as it was claimed by Lubimov et al. (1980).

On the other hand, the cosmological analysis proved the necessity for the existence of the dark matter, dominated in the Universe in the period of the large-scale structure formation. However, it found serious problems in the neutrino-dominated Universe scenario.

It leads the physics of dark matter outside the experimentally proven standard model of elementary particles.

It made necessary to modify the old Big Bang scenario by adding a new fundamental element — the dark matter, and finding physical grounds for it's existence in the hidden sector of the particle theory.

The problem of the true physical nature of the cosmological dark matter is accomplished by the following fact.

From the cosmological viewpoint, the models of the large-scale structure formation by the hot, cold, unstable dark matter, or more sophisticated models, implying cosmic strings plus hot dark matter, late phase transitions etc. are very different.

However, they are not alternatives from the viewpoint of particle physics. These different models find grounds in different and, in general, complementary parts of hidden sector of particle theory.

So, in principle, the mixture of all these models should be considered as the general case.

Another important initial condition for large-scale structure formation is the spectrum of initial fluctuations.

It could be easily checked that the statistical fluctuations alone were too small to form the structure of inhomogeneities in the expanding Universe. One has to assume the existence of small initial inhomogeneities, originated from very early stages of the cosmological evolution.

The old Big Bang scenario had no physical mechanism for their origin.

Moreover, the main questions: why the Universe expands? Why its initial conditions were so close to flat Universe? Why they were so similar in causally disconnected regions? Why it contains matter and no antimatter? had no answers in the old cosmological paradigm.

The first three questions, in principle, found their solution in the inflational cosmological models (Gliner, 1965, 1970; Guth, 1981), that assume the existence of a stage of the superluminous (in the simplest case, exponential) expansion of the very early Universe.

Such a stage can not be provided by the matter, radiation, or relativistic plasma dominance, but can be realised under some conditions as the cosmological consequence of the particle theory, for instance, in strong first-order phase transition or by slow rolling of scalar field down to its true vacuum state.

Simultaneously the inflational models offered the physical mechanisms for the generation of the spectrum of initial fluctuations.

Most of these effects are related to experimentally inaccessible parts of the particle theory, in particular, to the mechanisms of symmetry breaking at superhigh-energy scales.

One can also find, that different inflational models follow from different theoretical grounds and, in general, may co-exist in the complete cosmological scenario.

A.Sakharov (1967), and then V.Kuzmin (1970) were first, who related the observed baryon asymmetry of the Universe with the physical mechanism of the baryon excess generation. They found that the out-of-equilibrium CP-violating effects in hypothetical processes, violating the baryon number conservation, can generate baryon asymmetry in the very early stages of the initially baryon symmetric Universe.

The models of grand unification provided physical basis for these original ideas of baryogenesis, having among their predictions the existence of baryon nonconserving interactions.

The mechanisms of baryosynthesis found some other grounds in supersymmetric models. Based on these models, the exostence of the primordial condensate of scalar quarks is possible. The decay of scalar quarks in ordinary quarks results in the excess of baryons.

In the standard model of electroweak interactions, the baryon nonconservation can take place at very high temperatures. The electroweak baryosynthesis is possible provided that the standard model is extended by larger Higgs sector and/or includes the lepton number nonconservation. The latter can be related with the mechanism of generation of the neutrino Majorana mass.

So the modern cosmological paradigm reflects the fundamental changes in our understanding of the Big Bang cosmology.

From the self-consistent, but basically controversial and incomplete old Big Bang scenario we come to the picture of inflational cosmology with the baryosynthesis and the (multicomponent?) nonbaryonic dark matter.

Thus, directly or indirectly, the old Big Bang theory is supplemented in the modern standard Big Bang Universe by at least three necessary elements: inflation, baryosynthesis and nonbaryonic dark matter. All these three elements are based on the physical laws, predicted by particle theory but having no experimental proofs.

There is a wide spectrum of different physical mechanisms for inflation and baryosynthesis. There are various candidates on the role of the dark matter particles.

Unfortunately, the early Universe with inflation and baryosynthesis, as well as the dark matter can not be observed directly by astronomical means. One should elaborate the system of indirect methods in order to make the proper choice between variants, corresponding to underlying cosmological scenarios and particle models.

The problem is that the space of cosmological and physical parameters is, in general, multidimensional, because the physical grounds for different mechanisms of inflation, baryosynthesis and for different dark matter candidates follow from different physical motivations and in general are not alternative but complementary.

On the other hand, cosmological tests of particle models should, in general, account both for the particular realisation of inflation, baryosynthesis and dark matter and for the additional modifications of cosmological scenarios, corresponding to the chosen realisation.

One may conclude that the internal development of particle physics forces to find cosmological probes for its foundations. In turn, the foundations of the modern cosmology have physical grounds, lying beyond the possibilities of direct experimental tests.

Taking apart the hopes on capturing and storing some relics of superhigh-energy physics, such as primordial magnetic monopoles (t'Hooft, 1974: Polyakov, 1974; Zeldovich, Khlopov, 1978; Khlopov, 1979; Preskill, 1979; Rodionov, 1996), some nontrivial way to probe this physics should be found.

The clear understanding of mutual dependence and relationship between the foundations of macro- and microworlds and the practical absence of direct experimental and astronomical means to probe them has resulted in the formation of cosmoparticle physics, studying these foundations in the complex analysis of their indirect effects.

The necessity of such a combination of indirect methods follows from the central problem for the cosmology and the particle theory.

Being the two extremes of the fundamental physics in the largest and smallest scales, neither cosmology, nor the particle physics can separately study foundations, however feasible its experimental tools are.

Their foundations turn to be in such a close interrelation that virtually can not be separated. Thus, the extremities in frontiers of our knowledge at micro- and macro- levels converge. The mythical self-eating snake Urobohros symbolises the wrong circle of problems, to which leads the one-dimensional development of the fundamental physics.

The cosmoparticle physics offers the nontrivial way out of this circle, based on the fundamental relationship of cosmology and particle physics and on the principal possibility to study it in the set of indirect cosmological, astrophysical and microphysical effects.

In particle theory, cosmoparticle physics seems to open the practical way towards the true Theory of Everything, towards the Theory of Fundamental Natural Forces having, however, indirect but still testable grounds.

In the theory of the Universe, it opens the way to Physical Cosmology with the physically self-consistent description of the structure and the evolution of the Universe.

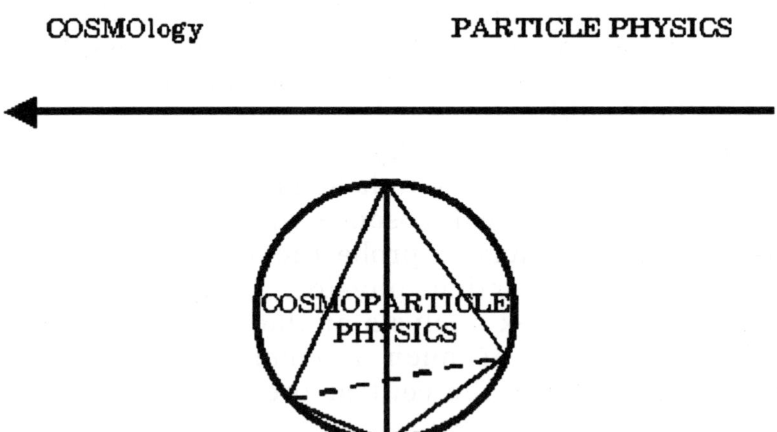

Fig.1. The "encircled pyramid' as the possible solution of the Urohboros puzzle.

The paradoxical form, that takes the body of cosmoparticle physics relative to traditional fields of science, may be illustrated by Esher-like "encircled pyramid", demonstrating the multidimensional way out from the

Urobohros puzzle. Urohboros's head-eating-tail-place corresponds to the foundations of the world system, that should be probed by multidimensional set of indirect macro- and microphysical tests.

The same picture can serve as an illustration for the astroparticle physics (the linear relation between cosmological and particle physics parts, corresponding to the head and tail parts of the snake) or for theories of everything (head-eating-tail point). Moreover, the junction of the pyramid and the snake body exhibit new aspects that the cosmoparticle physics brings to traditional fields of science.

Zero- and one-dimensional character of the most popular and widely recognised views on the links between cosmology and particle physics (the theories of everything and the astroparticle physics, respectively) can not provide weighted and self-consistent viewpoint on the whole subject.

The choice of underlying cosmological and particle models is dominantly the matter of taste and popularity. The criterion of *naturality* can hardly be useful. As A.Linde reasonably mentioned, Natural is what follows from the Laws of Nature, the ones we do not know and try to discover both in cosmology and particle physics.

In fact, the cosmoparticle physics studies the world as a whole in the fundamental relationship of its structure at smallest and largest scales. It treats *world systems*, in which the foundations of cosmology and particle physics are interrelated, so that the complete cosmological scenario is based on the unified particle theory, and the particle theory is cosmologically consistent. It is the first principle of cosmoparticle physics that the world system does exist.

The world system maintains the *correspondence* between the fundamental parameters, describing the phenomena of particle physics, astrophysics and cosmology and establishing quantitative links between microscopical and macroscopical effects.

The existence of the network of such links between properties of fundamental particles and astrophysical and cosmological parameters is the essence of the second principle of cosmoparticle physics.

Finally, the amount of parameters, involved in the world system, should be smaller than the amount of its signatures in particle physics, astrophysics and cosmology, thus providing the *completeness* of cosmoparticle tests, which is the third principle of cosmoparticle physics.

Cosmoparticle physics seems to reproduce the general feature of the fundamental physics in the largest and smallest physical scales. It represents the new level in fundamental relationship between microscopic and macroscopic descriptions, say, between thermodynamics, atomic theory, hydrodynamics and kinetics, or between the fundamental macroscopic and microscopic quantities, for instance, between the Avogadro number and the mass of proton.

However, losing the possibility of direct experiments at cosmological scales and in the superhigh-energy physics, that underlie the modern cosmology, the cosmoparticle physics has to evolve a set of nontrivial indirect probes.

Cosmoarcheology, the search for the footprints of new physical phenomena in the astrophysical data may be viewed as already existing branch of the proper cosmoparticle physics (Sakharov, 1989; Khlopov, 1989; 1996b,c). All its components are mixed up in the nontrivial manner and result in a set of astrophysical probes, proving the existence and investigating the possible properties of hypothetical particles, fields, objects and phenomena, predicted as the cosmological consequences of the particle theory.

The cosmoarcheology treats the Universe as a unique natural accelerator laboratory, so that the astrophysical data play here the role of specific experimental sample in *Gedanken Experiments*.

As in any experiment, in order to achieve the meaningful result, one should have clear understanding of the experimental device used, as well as the methods of data sampling and analysis.

The problem is that in the Universe both the sources and detectors particles are out of control.

Astrophysical processes can not be directly reproduced in laboratories, but however complicated the combination of effects is, theoretical astrophysics uses, as a rule, in its analysis the natural laws, proven in experiment.

The trouble is that in theoretical treatment of the inflationary Universe with the baryosynthesis and the nonbaryonic dark matter, the basic physical laws are not known. It makes the self-consistent formulation of cosmoarcheological approach to be, in general, model-dependent. One should account for the relationship between the hypothetical particle or field, probed by the astrophysical data, and the physics, underlying inflation, baryosynthesis and nonbaryonic dark matter. Since the latter is model-dependent, one should consider cosmological consequences of the considered hypothesis referred to the picture of cosmological evolution, based on the chosen particle model, underlying these necessary elements of the modern cosmology.

It means, that the cosmological trace of the hypothetical particle or field may be multi-step, following the nontrivial cosmological path, dictated by the model.

On the other hand, provided that the inflational baryon asymmetrical cosmology with the nonbaryonic dark matter is really the proper basis of the Universe, the real picture of its evolution *should* be much more complicated, than the original Gamow's Big Bang Universe scenario. One should expect it to be generally more sophisticated, than the simple addition of inflation, baryosynthesis and nonbaryonic dark matter to this scenario.

The reason is, that any physically motivated theoretical framework, giving rise to the necessary elements of cosmology, is generally much more extensive, supplementing these elements by a number of additional cosmologically relevant details which may be called the hidden parameters of the modern cosmology.

Testing these details, the cosmoarcheology extends the might of observational cosmology, connecting the true theory of the Universe to observations.

One may take apart the cosmological necessities and turn to the cosmological impact of the true particle theory.

Esthetical arguments favour the unification of fundamental forces. However, the higher is the level of unification the more extensive is the corrsponding particle model and the larger is the hidden sector it implies. It brings into consideration more and more exotic cosmological implications, deserving serious attention in view of their physical motivations.

There are more practical reasons to go beyond the standard model of particle interactions, recovering its internal theoretical inconsistencies. It also leads to specific cosmological effects, induced by cosmological consequences of corresponding extensions.

In the both cases, one uses cosmoarcheology as an astrophysical test. In the unified models one has the fundamental possibility of self-consistent treatment of the cosmological evolution and effects of new particles and fields. Separate consideration of some particular consequences of particle model, used generally in practical extensions, looses the grounds of self-consistency.

The superstring models, claimed to be the theories of everything, can not serve at the present level of their development as an example of a world system in the true cosmoparticle sense. Owing to the existing variety of their realisations, it is practically impossible to analyse the more or less complete set of physical, astrophysical and cosmological predictions.

It makes interesting the alternative phenomenological approach to the world systems. The main properties of elementary particles and the cosmologically relevant parameters, corresponding to the physical mechanisms of inflation, baryosynthesis and dark matter, may be considered within the unique quantitatively definite theoretical framework.

Such an approach may be illustrated by the recently proposed model of horizontal unification, to be discussed in the present book (see Sakharov, Khlopov (1994b, 1996) and Refs. therein).

It was shown, that the extension of the standard $SU(2)\text{x}U(1)\text{x}SU(3)_c$ model of electroweak and strong (QCD) interactions of elementary particles to the gauge symmetry $SU(3)_H$ of quark and lepton families provides not only reasonable theoretical description of the established existence of three families of quarks and leptons

$$\begin{pmatrix} \nu_e \\ e \\ u \\ d \end{pmatrix};\ \begin{pmatrix} \nu_\mu \\ \mu \\ c \\ s \end{pmatrix};\ \begin{pmatrix} \nu_\tau \\ \tau \\ t \\ b \end{pmatrix} \tag{2}$$

but in its realisation turns out to be the theoretical framework, incorporating in an unique scheme the physical grounds for the inflation, the baryosynthesis and the dark matter existence.

This model, offering the alternative (horizontal) way to unification, is in no case alternative to the more popular GUT or supersymmetric extensions of the standard model.

The internal problems of the minimal horizontal unification imply its further supersymmetric and GUT extensions, which are expected to give better consistency of its astrophysical and cosmological predictions with the observations.

But even in the present form the model reflects the main cosmoparticle principles. On the base of local gauge model with the spontaneous symmetry breaking it provides the phenomenology of world system, putting together practically all the main known particle properties and the main necessary cosmological parameters, related to the hidden sector of particle theory. Finally, the amount of free parameters of the model turns out to be much less, than the amount of its signatures in particle processes, astrophysics and cosmology, thus providing its definite test and exhibiting its completeness.

So, this model illustrates the might of cosmoparticle approach. Its fundamental scale of horizontal symmetry breaking is *a priori* unknown and corresponds to the hidden sector of the particle theory, but the complex analysis of the set of its physical, astrophysical and cosmological predictions makes it possible to fix the value of this scale in two rather narrow windows (around 10^6 and around 10^{10} GeV).

The second solution, corresponding to higher energy scale, seems to reproduce all the main features of the widely assumed as standard cosmological scenario with inflation, baryosynthesis and cold (axionic) dark matter.

The practical realisation of this scenario, which in no case reflects complete physical basis, shows, that even the most simple reduced cosmological scenario does contain some additional elements. They are: the postinflational dust-like stage, at which the primordial black hole (PBH) formation is possible, the successive PBH evaporation at RD stage after primordial nucleosynthesis, the formation of primordial percolation structure of archioles, etc.

This example favours the conclusion, that in *no* cases new cosmological elements, based on the hypothetical effects of particle physics, are reduced to inflation, baryosynthesis and dark matter *only*. It also resembles the system of nontrivial crossdisciplinary links that should be used in order to probe the true world system using the methods of cosmoparticle physics.

CHAPTER 2

THE HIDDEN SECTOR OF PARTICLE THEORY

1. Particle Theory - the Standard Model

Dramatic history of the search for the most fundamental constituents of matter has resulted in the modern quark-lepton picture of elementary particles and in the so-called standard model of their electroweak and strong interactions.

This picture followed from the revolutionary step, made in the early thirties of our century by the mutual efforts of the quantum theory and relativity and resulted in the radical change of the very idea on "elementary particle", "field" and "charge".

Since ancient world until the beginning of the XX century the fundamental notion of "elementarity" was essentially the same.

In accordance with the traditions of atomism, elementary particle was assumed to be an eternal non-destructible and non-creatable entity, bearing the minimal possible quantity of the corresponding quality.

Conservation of fundamental quantities, i.e., conservation of electric charge, was in this picture the trivial reflection of the eternal entity of the elementary charged particle, electron.

Elementary particles changed their names — from molecules to atoms and then to elementary constituents of atom — electrons and nuclei. The principle was the same – all observed changes in material objects were ascribed to the result of recombination of elementary particles in the same manner, as in the LEGO game.

Such an approach turned out to be very fruitful in chemistry, where the whole variety of chemical compositions was reduced to various combination of about 100 different chemical elements. However, the same approach lead to serious troubles in the analysis of the structure of atomic nuclei and nuclear transformations.

The discovery of positron and the detection of production and annihilation of electron-positron pairs, the problem of energy-momentum conservation in nuclear beta decay, the "nitrogen catastrophe", followed from the assumption of the presence of electrons inside the nucleus, seriously contradicted the idea of eternal electron at the nuclear scale.

Electrons, emitted in nuclear beta decay, being eternal, should have been located inside the nucleus before the decay, i.e. localised in the space region with the size of nucleus

$$r_N \leq 10^{-13} \text{ cm} . \tag{3}$$

But the uncertainty principle of quantum mechanics

$$\Delta p \cdot \Delta r \geq \hbar \tag{4}$$

predicts the uncertainty in electron momentum,

$$p_e \sim \Delta p \geq \frac{\hbar}{\Delta r_N} \sim 40 \text{ MeV}/c \tag{5}$$

which exceeds the momentum of electrons, emitted in the beta decay, by the factor of 5–10.

The uncertainty in electron energy, being typically of the order of

$$E_e \approx p_e c \geq 40 \text{ MeV} \tag{6}$$

exceeds by the same factor the typical binding energy per nucleon in the nucleus, which is about

$$E_b \sim 7 \text{ MeV} . \tag{7}$$

On the other hand, if the binding energy of "nuclear electrons" is taken equal to the energy of nucleons, the size of the region, in which the electrons are localised, turns out to exceed the size of nucleus by the order of the magnitude.

Moreover, the account for spin of electrons inside nucleus lead to the problem of nuclear spin–statistics correspondence, that lead to the "catastrophe" in the case of ^{14}N nuclei. Nuclear charge is

$$Q = Z|e| ,$$

where e is the charge of electron. In the case of nitrogen, the nuclear charge is equal to

$$Z = 7 ,$$

and it was explained by the compensation of the charge of 14 protons by 7 electrons inside the nucleus. However, in this case the total spin of odd (21) number of half-integer spin constituents (protons and electrons) should be half-integer in the units of

$$\hbar = \frac{h}{2\pi} ,$$

whereas the statistical properties of ^{14}N nuclei correspond to Bose-Einstein statistics of particles with integer spin.

Finally, the spectrum of electrons in beta decay turned out to be continuous and not linear, as expected for a single particle, emitted in the nuclear transition.

The hypothesis of neutrino, to be emitted in pair with electron in beta decay, offered the solution both for the "nitrogen catastrophe" and for the energy-momentum conservation in beta decay. The additional 7 half-integer spin neutrinos inside the nucleus made the total amount of constituents to be even, that resulted in the integer spin of nitrogen nucleus. On the other side, neutrinos were emitted together with electrons and carried away some energy and momentum, thus making beta spectrum continuous.

But this solution caused only more troubles, because for light neutrinos inside nucleus the same momentum uncertainty should have taken place as for electrons.

The discovery of neutron could not help, since the problem of localisation of light particles (neutrinos and electrons) inside nucleus reappeared as the problem of localisation of electron and neutrino in neutron before its beta decay.

In the beginning of the XX century, the discovery of electromagnetic field quantum — photon, which can be created in emission and annihilated in absorption processes, offered the first example of elementary particle, which can be created and annihilated.

The similar concept was not easily accepted in the case of constituents of matter.

To save the idea of eternal electron, the conception of "Dirac sea" was put forward.

"Creation" of electron-positron pair was interpreted in this picture as the transition of electron from the sea state to the state of free electron, accompanied by the hole in the sea, which was associated with positron.

The process of electron-positron "annihilation" was viewed in this picture as the occupation of the "hole" state in the sea by a free electron.

The fermion-hole formalism found applications in the theory of nuclear structure and in the theory of semiconductors, but could not stand the development of relativistic quantum field theory, based on the alternative idea of particle creation and annihilation.

In this alternative picture, the notion of charge and its relationship with conservation laws and symmetries takes on non-trivial and fundamental significance.

Charge conservation has lost the triviality of a property, inherent to eternal elementary charged particle.

In the world with elementary constituents, which can be created and annihilated, the laws of particle transformations and the particle properties, conserved in these transformations, reflect the fundamental symmetry of the microworld.

In the development of quantum field theory the particle symmetry and corresponding generalised charges have put the basement to the set of theoretical ideas referred to as the standard model of particles and their strong, weak and electromagnetic interactions.

Following the main aim of this book, we do not intend to give the substitute for the well-known textbooks.

Instead of giving systematic detailed presentation of the standard model, we concentrate attention on those aspects of the particle theory, which are important for extension of the standard model and relevant for cosmology.

Note that the ideas of quantum field theory, which appeared in the early thirties of our century, follow the idea on nonstationarity of the Universe, which was put forward and found its first observational proof in the twenties. This correspondence seems not to be occasional.

Beginning from the antique philosophy the stability of elementary particles found psychological grounds in the stability of world system.

Therefore the change in the world order – the expansion of the Universe – gave psychological background for the radical change in the description of elementary particles.

1.1. Quantum Electrodynamics

To trace the logical steps, leading to the ideas of the modern particle theory, let us consider the change of the notion of particle, field and charge in the transition from classical to quantum field theory of electromagnetism.

To the end of the XIX century the classical physical grounds for electricity, magnetism and light were unified in the Maxwell's electrodynamics.

The physical nature of interaction of electric charges, magnetic action of electric currents and optical phenomena was found in the existence of electromagnetic field, acting on electric charges and currents and originated in them.

Mathematically, the Maxwell equations relate spatial and time derivatives of the electromagnetic potential to electric charge and current densities and their spatial and time derivatives.

In classical electrostatics, electric charge density, $\rho(t,\mathbf{x})$, is the product of electric charge of elementary charged particle, electron e, and their number density, $n(t,\mathbf{x})$, given by

$$\rho(t, \mathbf{x}) = en(t, \mathbf{x}) \ . \tag{8}$$

The Lagrangian density $L_{el}(t,\mathbf{x})$ of the interaction of electric charge with the scalar electric potential $\varphi\ (t,\mathbf{x})$ has the form

$$L_{el} = en(t, \mathbf{x})\varphi(t, \mathbf{x}) \ , \tag{9}$$

where the charge density and electric potential are, in general, functions of spatial variables $\mathbf{x}$ and time t.

Taking into account the observed absence of magnetic charges, electrodynamics ascribes magnetic effects to the motion of charges, so that the corresponding Lagrangian density, describing the interaction of the (density of) electric current

$$\mathbf{j} = en(t, \mathbf{x})\mathbf{v} \tag{10}$$

with the vector potential $\mathbf{A}(t,\mathbf{x})$, is given by

$$L_{mag} = en(t, \mathbf{x})\mathbf{v}\mathbf{A}(t, \mathbf{x}) \; . \tag{11}$$

Relativistic theory is based on the relativity of space and time variables, treating them as space-time events

$$x = (t, \mathbf{x}) \; , \tag{12}$$

as well as on the relativity of electricity and magnetism, therefore the classical Lagrangian density of electromagnetic interactions $L_{em}(x)$ is built as the interaction of the electromagnetic four-current

$$j_\mu = (\rho c, \mathbf{j}) \tag{13}$$

with the electromagnetic four-potential

$$A_\mu(x) = (\varphi(x), \mathbf{A}(x)) \tag{14}$$

in the form

$$L_{em} = ej_\mu A^\mu \; . \tag{15}$$

In the non-relativistic quantum mechanics, the number density of electrons has the meaning of the density of probability to find an electron in the fixed time moment t in the given place $\mathbf{x}$, and is related to the wave function of electron $\Psi(t,\mathbf{x})$ as

$$n(t, \mathbf{x}) = \Psi^* \Psi = |\Psi|^2 \ . \tag{16}$$

With this modification, the Lagrangian density for electrostatic interaction takes on the form

$$L_{el}(t, \mathbf{x}) = e\phi(t, \mathbf{x})\Psi^*(t, \mathbf{x})\Psi(t, \mathbf{x}) \ . \tag{17}$$

In the quantum field theory, instead of the wave function $\Psi(t,\mathbf{x})$, defining the probability to find an eternal electron in the considered region, creation and annihilation of electrons and positrons is considered.

Operators $\Psi(x)$ and their conjugates $\Psi^+(x)$ are superpositions of creation and annihilation operators of electron and positron states.

Instead of the classical electromagnetic field potential $A_\mu(x)$, the operator of the electromagnetic field $\mathbf{A}_\mu(x)$ is introduced as the superposition of creation and annihilation operators of electromagnetic quanta. The quantum field operator of electromagnetic interaction is given by

$$\mathbf{L}_{\mathbf{em}} = e\mathbf{j}_\mu \mathbf{A}^\mu, \tag{18}$$

where the quantum field operator of the electromagnetic current has the form

$$\mathbf{j}_\mu = \Psi^+ \gamma_0 \gamma_\mu \Psi \ . \tag{19}$$

Here γ_μ with

$$\mu = 0,1,2,3$$

are the Dirac gamma matrices.

Thus, all processes of electromagnetic interaction in quantum electrodynamics are reduced to the combination of elementary acts of emission and absorption of electromagnetic quanta by charged particles.

The difference between quantum-mechanical and quantum field description of electromagnetic interaction, between their descriptions of emission and absorption of electromagnetic quanta lies in the way they treat the charged particle, emitting or absorbing such quanta.

Quantum mechanical description of photon emission or absorption by an atomic electron uses the transition current describing the change of the wave function of eternal electron in the initial and final state.

As a result of emission or absorption, the electron transfers from one atomic state to another, with the respective change in the density of probability to find it in the considered place and time.

The quantum field theory considers the same process as annihilation of electron in the initial state and creation of electron in the final state with respective creation (annihilation) of emitted (absorbed) electromagnetic quantum.

The evident difference in these approaches arises in the description of electron-positron pair creation or annihilation.

Quantum mechanical approach should have considered the electron transition to or from Dirac sea with the interpretation of the "hole" in the sea as a positron.

Quantum field theory considers electron and positron states on equal footing. It treats the creation of a pair as the creation of electron and positron states from physical vacuum and the annihilation of a pair as the annihilation of corresponding electron and positron states.

The fundamental reason for the possibility to consider electron in terms of its creation and annihilation is the identity of elementary particles of the same type.

Following quantum field theoretical approach quantum electrodynamics describes all quantum processes of interaction of charged particles and electromagnetic quanta on the base of one elementary act, treated from different viewpoints. The same diagram (Fig.2), considered from different sides, describes emission and absorption of photon by electron (from left to right) or positron (from right to left), creation of electron-positron pair by electromagnetic quantum (from downside upwards) and one-photon electron-positron annihilation (from upside down). The arrow directing from initial to final state corresponds to the electron. The opposite direction denotes the positron.

The elementary act of interaction in quantum electrodynamics represents the succession of processes of annihilation and creation of charged particles and electromagnetic quanta.

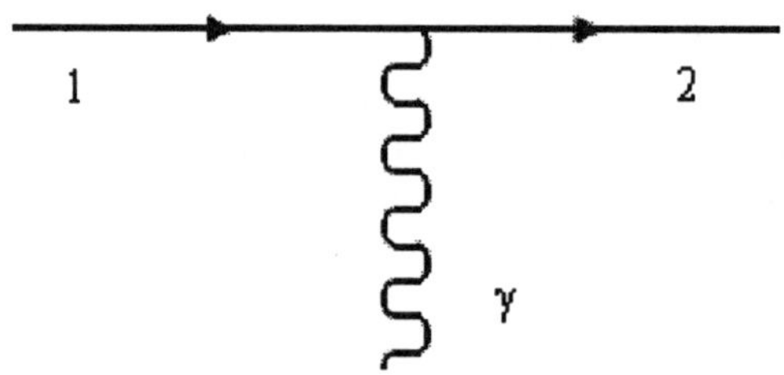

Fig.2. The elementary act of quantum electrodynamics

For instance, an elementary act of photon emission from electron (Fig.2, viewed from left to right) is the process, in which the electron 1 in the initial state is annihilated, and the electromagnetic quantum and the final state electron 2 are created.

Using the elementary diagram (Fig.2), one can reduce any process of electromagnetic interaction to the appropriate combination of elementary acts, viewed from different sides.

So, the electromagnetic scattering of two electrons occurs in this picture as a result of two elementary acts of interaction (Fig.3):
1) annihilation of initial electron 1 and creation of final electron 3, associated with creation of electromagnetic quantum γ, and 2) annihilation of electromagnetic quan-

tum in transition, in which initial electron 2 is annihilated and final electron 4 is created.

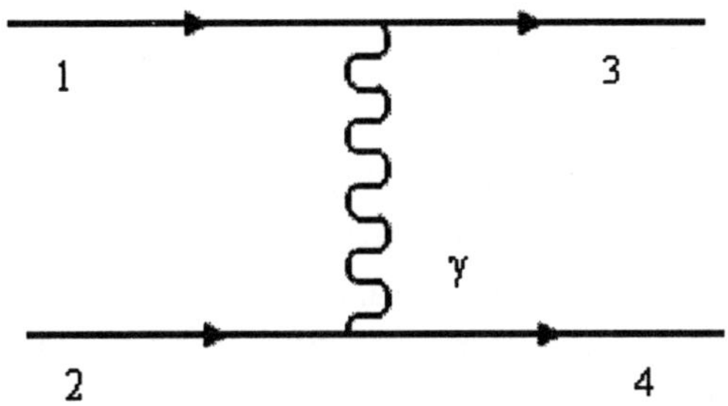

Fig.3. Electron scattering in quantum electrodynamics.

The possibility of creation and annihilation of electron states changes the meaning of electron properties.

The electron charge and its electromagnetic field are no more eternal properties of eternal particles.

Charge conservation and the electromagnetic field as the necessary attribute of electric charge loose their grounds in the eternity of elementary charged particles.

Both the electric charge and its electromagnetic field should find their nature in the laws, governing creation and annihilation of electron states, in the laws of symmetry of the quantum field theory.

1.2. Gauge Symmetry of Quantum Electrodynamics

In quantum mechanics, the charged particle is described by the complex wave function.

The probability to find an electron is given by the square of the modulus of the wave function. The phase of the wave function is unobservable. One can assume it to be arbitrary and make arbitrary change of the phase at different points in space and time without any observable effects.

Thus quantum mechanics of electron should possess invariance relative to phase transformations of the electron wave function given by

$$\Psi \to \Psi \exp\left(ie\chi(x)\right), \tag{20}$$

where e is the electric charge. Such transformations with the constant χ correspond to transformations of global $U(1)$ symmetry. Local phase transformations with $\chi(t,\mathbf{x})$ depending on space-time co-ordinates correspond to local $U(1)$ symmetry.

In the wave equation of quantum theory, local phase transformations add arbitrary terms to all terms with space and time derivatives:

$$\partial_\mu \Psi(x) \to \partial_\mu \Psi(x) + ie\partial_\mu \chi(x) \ . \tag{21}$$

These terms are proportional to derivatives of $\chi(x)$ and have no physical meaning.

To preserve invariance relative to local phase transformations of electron wave function, local gauge transformations of the electromagnetic potential A are involved. The local phase transformation, given by (20),

$$\Psi \to \Psi \exp\left(ie\chi(x)\right)$$

is accompanied by the simultaneous local gauge transformation of the electromagnetic potential, given by

$$A_\mu \to A_\mu + e\partial_\mu \chi(x) \ . \tag{22}$$

Provided that the derivatives of the wave function have the form of so called "long derivatives"

$$\partial_\mu \Psi(x) \to \partial_\mu \Psi(x) + ieA_\mu(x) \ , \tag{23}$$

the invariance relative to the phase transformation may be easily proven.

Gauge transformation compensates the unphysical terms arising from derivatives of the wave function.

On the other hand, phase rotations compensate the effects of gauge transformations, making the theory gauge invariant.

The condition of gauge invariance with the use of long derivative offers an aesthetically appealing way to introduce the electromagnetic interaction in the theory of free charged particles.

Introducing long derivative, one specifies the form of the electromagnetic interaction of particles with different spins.

For instance, in the quantum theory of electron, it specifies the form of the electromagnetic interaction with electric charge and magnetic moment of electron.

Moreover, the idea of local gauge invariance opens the way to relate the fundamental forces of Nature with the symmetry of fundamental particles of matter.

The logical chain here is transparent.

From the Noether theorem, the exact symmetry of the Lagrangian corresponds to the existence of the conserved charge.

Assuming that the symmetry is local, we have to introduce the gauge field, interacting with the charge, to preserve the symmetry of the Lagrangian.

Thus, in the framework of the Lagrange field theory, we can consider the symmetry between particles, relate some generalised charge to this symmetry, and introduce the gauge field, mediating the interactions, induced by these charges.

In the framework of the quantum field theory, one should consider the symmetry between the field operators of particle annihilation in the initial state and particle creation in the final state.

The transition from the initial to the final state formally corresponds to the symmetry transformation, induced by the corresponding generator of the symmetry group.

The set of charges, corresponding to the considered symmetry, can be related to the set of group generators, which are the sources of quanta of respective gauge field.

Hence one can expect that the true symmetry between different types of particles can provide mathematical description of their interactions, leading to the true picture of processes, studied by the high energy physics at accelerators.

1.3. Symmetry of Fundamental Particles

Starting from the late thirties of the XX century, the development of experimental particle physics lead to discovery of large amount of new particles. The way to their modern systematics was not straight.

In addition to electrons and nucleons, being constituents of all surrounding matter us, as well as electron neutrinos — the inevitable participants of beta processes, large amount of various particles was experimentally discovered.

Almost all these particles turn out to be unstable. With respect to fundamental interactions they represent two groups: leptons and hadrons. These groups also differ with respect to the internal particle structure.

One refers to leptons the electron e , and the electron neutrino ν_e , and two other pairs of a charged lepton and its neutrino: the muon μ^-, and the muon neutrino ν_μ, the tau lepton τ^- , and the tau neutrino ν_τ . One also refers to leptons their antiparticles: the positron, μ^+ , τ^+ and the electron, muon and tau antineutrinos. It was experimentally proved that neutrino is not identical to antineutrino and that electron, muon and tau neutrino are different particles.

In the entire observed particle processes, the lepton number L_l is conserved for each type of leptons l: the number of leptons minus the number of antileptons is constant.

The creation of electron in beta decay is accompanied by the creation of the antilepton — the electron antineutrino.

The same holds true in all observed reactions with muons and tau leptons.

Leptons are created and annihilated either in pair with corresponding antileptons, or such process is accompanied by the annihilation or the creation of lepton of the same type.

For instance, the muon is produced in pair with the antimuon or with the muon antineutrino.

The single muon appears as a result of reaction induced by the initial muon or the muon neutrino.

Charged leptons differ strongly in their mass. The mass of the electron is equal to

$$m_e = 0.511 \text{ MeV} , \tag{24}$$

whereas the masses of the muon and the tau lepton are equal to

$$m_\mu = 105.7 \text{ MeV} , \tag{25}$$

$$m_\tau = 1777 \text{ MeV} . \tag{26}$$

Owing to their large mass, the muon and the tau lepton are unstable.

There are only experimental upper limits on the mass of neutrino. In the Tables of Particle Properties (Barnett et al., 1996) these limits are given as

$$m_{\nu_e} < 7 \text{ eV} , \tag{27}$$

$$m_{\nu_\mu} < 0.27 \text{ MeV} , \tag{28}$$

$$m_{\nu_\tau} < 31 \text{ MeV} . \tag{29}$$

As we shall see, the question on the neutrino mass is of great significance for cosmology.

Charged leptons participate in weak and electromagnetic interactions.

Neutrinos participate only in weak interactions.

Leptons are considered as fundamental particles, having no internal structure.

On the contrary, hadrons, being much more numerous, are considered as composite particles. The structure of hadrons is explained on the base of the quark model.

All hadrons participate in strong interaction.

There are two groups of hadrons: baryons and mesons.

Baryons are fermions. They have half-integer spin. One refers to baryons the proton, the neutron, hyperons, baryon resonances and their antiparticles, called antibaryons.

In all observed particle reactions, the baryon number B is conserved. It means that in any process the number of baryons minus the number of antibaryons does not change. Another side of the baryon conservation is that the proton, being the lightest baryon, is stable.

Tables of particle properties (Barnett et al., 1996) give the lower limit on the proton lifetime

$$\tau_p > 1.6 \cdot 10^{25} \text{ years}, \tag{30}$$

which comes from the geochemical analysis and is independent on the possible mode of proton decay, and

$$\tau_p > 10^{31} \div 5 \cdot 10^{32} \text{ years}, \tag{31}$$

which comes from negative results of searches for particular modes of proton decay.

Such lifetimes, exceeding the age of the Universe by 15-20 orders of the magnitude, define the probability of proton decay.

For instance, the strongest restriction in Eq.(31) that the proton lifetime exceeds 5 10^{32} years, corresponds to the constraint, that in the average there are no proton decays during a year in 500 tons of the matter.

We shall see that the hypothetical baryon non-conserving processes are of great importance for the explanation of the origin of the matter in the Universe.

Mesons have an integer spin and are bosons. Pions, K-mesons, D-mesons and B-mesons are the examples of relatively long-living mesons.

The most numerous group of hadrons contains the so-called resonances. The lifetime of a resonance with the mass m is of the order of

$$\tau \sim \frac{\hbar}{mc^2} \qquad (32)$$

what is the typical time scale for the reactions due to strong interaction. The most part of the Tables of Particle Properties (Barnett et al., 1996) is occupied by baryon and meson resonances.

All hadrons are classified by their internal quantum numbers: spin, isospin, strangeness, charm, parity etc.

The set of these numbers determines the flavour of a particle.

With the use of symmetry between flavours, the classification of hadrons is possible.

The flavour symmetry provides the arrangement of hadrons in multiplets, i.e. the identification of groups of different particles as corresponding representations of the considered group of the flavour symmetry.

For instance, the proton and the neutron are considered as a doublet relative to isotopic spin (or, isospin)

$SU(2)$ symmetry, and pions π^-, π^+ and π^0 form an isotopic triplet.

Strong interaction, and, in particular, nuclear interaction is invariant relative to the isospin symmetry.

Based on the isospin invariance, one can consider different particles within one isospin multiplet as different quantum (isospin) states of one particle.

So, one can treat the proton and the neutron as "isospin-up" and "isospin-down" states of one particle — the nucleon N.

Nucleons and pions together with strange particles — hyperons, K-mesons, their antiparticles and resonances — were classified on the base of flavour $SU(3)$ symmetry.

The fundamental representations of this group (triplet and antitriplet) found no realisation among the known particles and were identified with the hypothetical u, d and s quarks and their antiparticles — antiquarks.

On the basis of the quark model it is possible to classify all hadrons as bound states of quarks and antiquarks.

The baryons are considered as the bound states of three constituent quarks, the antibaryons — as consisting of three antiquarks, and mesons — as the bound states of quark and antiquark.

For instance, the proton and the neutron are composed of u and d quarks as

$$p = (uud) \, , \tag{33}$$

$$n = (udd) \, , \tag{34}$$

the positively charged pion is

$$\pi^+ = (u\bar{d}) \, , \tag{35}$$

and the antiproton is the bound state of three antiquarks:

$$\overline{p} = \left(\overline{u}\overline{u}\overline{d}\right) . \tag{36}$$

In such an approach, the symmetry is used only for classification of particles, and the physical meaning of quarks is not clear.

It may be easily seen that in the units of elementary charge, defined as the absolute value of the charge of electron, quarks should have fractional electric charge, equal to

$$Q_u = +\frac{2}{3} \; ; \; Q_d = -\frac{1}{3} \; ; \; Q_s = -\frac{1}{3} \tag{37}$$

for quarks and

$$Q_{\overline{u}} = -\frac{2}{3} \; ; \; Q_{\overline{d}} = +\frac{1}{3} \; ; \; Q_{\overline{s}} = +\frac{1}{3} \tag{38}$$

for antiquarks.

Quarks are also assumed to have fractional baryon number equal to

$$B_q = +\frac{1}{3} \tag{39}$$

for quarks and

$$B_{\overline{q}} = -\frac{1}{3} \tag{40}$$

for antiquarks.

The results of experimental searches for fractionally charged particles are negative, giving the upper limit on the free-quark abundance in the Universe relative to hydrogen equal to

$$\frac{n_q}{n_H} < 10^{-21} \ . \tag{41}$$

On the other hand, calculations by Zeldovich et al. (1965) of the primordial abundance of free quarks in the Big Bang Universe, predict such abundance to be by 10 orders of the magnitude higher.

This contradiction leads to the conclusion, that there should be no free quarks in the Universe.

One should accept the idea, that quarks are always confined within hadrons and never appear as free particles.

Quarks should have additional degree of freedom, called colour. With the use of colour one resolves the problem of the quark statistics.

In the quark model, Δ^{++} resonance, having the spin 3/2 and the electric charge +2, should be the bound system of three identical u-quarks, being in the same spin-up state. If these quarks differ by colour, all three u-quarks can be in the same ground state with

$$l = 0 \ .$$

The colour plays the basic role in the gauge theory of strong interaction, called quantum chromodynamics (QCD).

The very name of this theory indicates the analogy with quantum electrodynamics. The colour is crucial in this analogy, playing the role of generalised charge in the dynamics of strong interaction.

The gauge colour symmetry is based on the colour $SU(3)_c$ group.

Colour transitions, determined by generators of $SU(3)_c$ group, are the sources of fields of the colour interaction.

These fields glue quarks in hadrons and the quanta of these fields are called gluons g .

The symmetry between the flavours of quarks and leptons is used to build the gauge theory of weak interactions.

The roles of weak charges play the transitions of the weak isospin.

By the analogy with the electrodynamics, weak charges are the sources of the field of weak interaction.

The quanta of this field, *W*- and *Z*- bosons, have large mass

$$m_W \cong 80 \text{ GeV} , \tag{42}$$

$$m_Z \cong 91 \text{ GeV} , \tag{43}$$

and their exchange is effective at short distances only.

It explains the short range of weak interactions.

So, gauge symmetries between fundamental particles of matter (leptons and quarks) lie in the base of the modern standard model of electromagnetic, weak and strong interactions.

1.4. Standard Model of Electroweak and Strong Interaction

The standard model is based on the group of gauge $SU(3)_c\text{x}SU(2)_L\text{x}U(1)$ symmetry, postulated for leptons and quarks.

The model assumes that the quark of each flavour has three colour states, i.e., is identified with the triplet representation of the $SU(3)_c$ group of symmetry. For the Lagrangian of the free quark field, the invariance relative to the gauge $SU(3)_c$ symmetry group implies the existence of gauge fields, mediating the colour interaction and called gluons.

Since the group $SU(n)$ has

$$N = n^2 - 1 \tag{44}$$

generators, there are 8 generators of the colour gauge group, which are the sources of 8 gauge fields of the colour interaction. Hence the gauge invariance of QCD implies the existence of 8 gluons.

The structure of the Lagrangian for the quark-gluon interaction has the form similar to electrodynamics:

$$L_{qg} = g_s \bar{q}_i \gamma_\mu q_j A^\mu_{ij} \ , \tag{45}$$

where

$$i, j = 1,2,3$$

are colour indices, and the gauge coupling constant g_s defines the dimensionless QCD constant

$$\alpha_s = \frac{g_s^2}{4\pi\hbar c} \ , \tag{46}$$

which plays in QCD the role, similar to the role of the fine structure constant

$$\alpha_{em} = \frac{e^2}{4\pi\hbar c} \tag{47}$$

in the quantum electrodynamics.

These dimensionless constants define the probability of elementary acts of quark QCD and QED interactions, respectively.

Owing to radiative effects, these constants are running. Their values depend on the distance between the corresponding charges, i.e., on the energy scale according to the quantum theory.

In quantum electrodynamics, the effect of renormalisation due to radiative corrections leads to the so called "zero charge" problem.

For any finite charge taken at very small distance, one obtains the value of this charge, measured at the large distance, equal to zero.

In quantum chromodynamics, the non-abelean character of the colour symmetry plays important role in radiative effects.

Gluon states transform as the octet of the colour group. It means, that they also bear colour charge and are the sources of gluon field.

In contrast to the tensor of the electromagnetic field defined as

$$F_{\mu\nu} = \partial_\mu A_\nu - \partial_\nu A_\mu \ , \tag{48}$$

the tensor of the gluon field is given by

$$G^a_{\mu\nu} = \partial_\mu A^a_\nu - \partial_\nu A^a_\mu + g_s f^a_{bc} A^b_\mu A^c_\nu \ , \tag{49}$$

where

$$a, b, c = 1, \ldots, 8$$

are the $SU(3)_c$ octet indices, $f_{bc}{}^a$ is the $SU(3)_c$ structure constant, antisymmetric under the permutation of the indices b,c.

So in contrast to the Lagrangian of the electromagnetic field, given by

$$L_{emf} = -\frac{1}{4} F_{\mu\nu} F^{\mu\nu} \ , \tag{50}$$

the Lagrangian of the gluon field, given by

$$L_{gf} = -\frac{1}{4} G^{a}_{\mu\nu} G^{a\mu\nu} , \tag{51}$$

contains the terms, which are trilinear and quadrilinear products of the gluon field operators, defining the direct three- and four- gluon couplings.

This specific features of gluon interactions lead to the nontrivial property of the running QCD constant.

It vanishes at small distances and turns into infinity at the scale of confinement.

One may parametrise the dependence of the QCD constant on the energy scale μ as follows:

$$\alpha_s(\mu) = \frac{4\pi}{\beta_0 \ln\left(\frac{\mu^2}{\Lambda^2_{QCD}}\right)} , \tag{52}$$

where the function β_0 depends on the number and the type of colour states contributing to the radiative effects, and Λ_{QCD} is the scale of confinement.

At large energy scales, corresponding to small distances, the QCD constant tends to zero, and the asymptotic freedom of colour interactions takes place.

Qualitative confirmation of the asymptotic freedom comes from the parton picture of deep inelastic lepton and photon scattering on nucleons.

At large transverse momenta, i.e., at small distances, leptons and photons scatter on free pointlike components of the nucleon – partons. The quark-parton model interprets partons in the terms of the momentum distribution of quarks inside nucleon.

The QCD confinement means, that no colour object, quark or gluon, can be in a free state.

Being created as free particles at small scales, quarks and gluons should be "dressed" at large distances and appear as jets of hadrons.

Thus, hadronic jets, induced by quarks, antiquarks and gluons, and observed in the high energy particle collisions, e.g., in electron-positron annihilation into hadrons, may be considered as the experimental proof for the existence of quarks and gluons.

The QCD scale Λ_{QCD} is interpreted in terms of the energy density of the gluon field inside hadrons, defining, in particular, the mass of constituent *u*-, *d*- and partially *s*- quarks.

In the standard model, the bare masses of these quarks are given by the theory of electroweak interaction.

According to QCD, at high temperature

$$T > \Lambda_{QCD} \tag{53}$$

the phase transition in the hadronic matter should take place to the phase of the quark-gluon plasma. It provides physical grounds for the relativistic equation of state in the so called "hadron era" of the early Universe.

Electromagnetic and weak interactions are unified in the standard model on the basis of the $SU(2)_L \mathrm{x} U(1)$ group of the gauge symmetry. They are considered as the unified electroweak interaction.

The left-handed states of quarks and leptons are put into doublet representations of the weak isospin group $SU(2)_L$. Their right-handed states are singlets relative to $SU(2)_L$ transformations.

The non-equivalence of left- and right- handed states reflects the parity nonconservation in weak interactions.

Three generators of $SU(2)_L$ group correspond to three W gauge bosons. Charged W^- and W^+ bosons mediate the interaction of charged weak currents. Neutral W-boson and gauge boson B of the $U(1)$ gauge group are combined in the photon and the neutral Z-boson, mediating the interaction of neutral weak currents.

The gauge interaction of weak currents, mediated by W- and Z- bosons, is reduced to the effective four-fermion interaction at the transfer moment much smaller than the mass of these bosons.

Thus, the charged current interaction, mediated by W^- and W^+ bosons, is reduced at low energies to the effective four-fermion Fermi interaction with the Fermi constant G_F , expressed via the gauge constant g_W and the mass of W boson m_W as

$$\frac{G_F}{\sqrt{2}} = \frac{g_W^2}{8m_W^2} \quad . \tag{54}$$

The underlying $SU(2)_L xU(1)$ symmetry of electroweak interaction is broken down to the $U(1)$ symmetry of the electromagnetic interaction by the Higgs mechanism of the spontaneous symmetry breaking.

This mechanism assumes that the auxiliary scalar fields φ exist besides quarks, leptons and gauge bosons. In the minimal model of electroweak interaction, the Higgs doublet of scalar fields

$$\varphi = \begin{pmatrix} \varphi^+ \\ \varphi^0 \end{pmatrix} \tag{55}$$

together with the doublet of their antiparticles

$$\overline{\varphi} = \begin{pmatrix} \overline{\varphi}^0 \\ \varphi^- \end{pmatrix} \tag{56}$$

are introduced.

The Lagrangian of these fields formally preserves the $SU(2)_L xU(1)$ symmetry, but the potential of self-interactions of a linear combination of neutral fields φ has the form making energetically more favourable the existence of the vacuum state with the nonzero vacuum expectation value υ:

$$V(\varphi) = \frac{1}{4}\lambda\left(|\varphi|^2 - \upsilon^2\right)^2 . \tag{57}$$

Thus, the spontaneous symmetry breaking takes place. The symmetry of Lagrangian is broken by the asymmetric ground (vacuum) state.

Relative to the true asymmetric vacuum, only one linear combination of neutral fields φ has the meaning of the physical scalar field. The other combination of neutral fields φ together with the charged fields φ acquire the meaning of the third, longitudinal components of the massive W- and Z- bosons.

The Yukawa couplings of the fields φ to fermions, which have the form

$$L_{Yuk} = h_f \varphi \bar{f} f \tag{58}$$

before spontaneous symmetry breaking, induce after spontaneous symmetry breaking the fermion mass terms, given by

$$L_m = m_f \bar{f} f , \tag{59}$$

where the masses of fermions are determined by the vacuum expectation value υ of the Higgs field

$$m_f = h_f \upsilon , \tag{60}$$

and the index f denotes

$$f = e, \mu, \tau, u, d, s, b, c, t .$$

At high temperature

$$T > T_{cr} \sim v \ , \tag{61}$$

due to the effect of thermal fluctuations, the minimum of the Higgs potential moves to the zero vacuum expectation value

$$\langle \varphi \rangle = 0 \ , \tag{62}$$

and the underlying $SU(2)_L xU(1)$ symmetry of electroweak interaction is restored. Thus, the Higgs mechanism of the spontaneous symmetry breaking implies phase transitions at high temperatures in the Big Bang Universe.

Phase transitions in the early Universe are the important cosmological consequence of the spontaneous symmetry breaking, involved in the particle theory.

In the standard model, the masses of fermions and W and Z are determined by the nonzero vacuum expectation value of the Higgs field. The restoration of the symmetry of the electroweak interaction at high temperature, leading to the zero vacuum expectation value, should result in the vanishing masses of fermions and gauge bosons. It means that according to the standard model not only physical conditions, but also the fundamental physical parameters (masses of elementary particles) should change in the early stages of the Big Bang Universe.

Most of the calculations in the standard model are based on the perturbation theory. For small dimensionless electromagnetic, weak and QCD coupling constants α, one may use in the calculations the perturbation series in the powers of these constants, provided that

$$\alpha < 1 \ . \tag{63}$$

However, the non-perturbative effects are important both in the QCD and in the theory of electroweak interactions.

Such effects lead in QCD to the θ-term, resulting in the non-vanishing and *a priori* not small effects of CP violation in the strong interaction.

Similar effects lead in the electroweak theory to the baryon and lepton number non-conserving transitions.

Such transitions, called sphalerons, are exponentially suppressed at low temperature but at the temperature of the order of the temperature of the electroweak phase transition and higher their probability is of order of 1. This phenomenon has important consequence for the theory of generation of the baryon excess in the initially baryon-symmetric Universe.

2. Esthetical and Practical Grounds for the Extensions of the Standard Model

Though the standard model of electroweak and strong (QCD) interactions of elementary particles does not meet any direct experimental contradiction, there are several reasons to consider it as incomplete and to go to its various extensions.

Esthetical arguments appeal to the unified theories of fundamental forces of Nature, in which the symmetry of standard model should be extended to the higher underlying symmetry, and the set of known particles should be completed to the full symmetry by a large variety of particles and fields, forming a "hidden sector" of the corresponding model.

The higher is the level of unification, the wider is the "hidden sector" it needs to complete the underlying symmetry.

Attractive as these reasons may be, the extensions of the standard model, following the needs to remove its internal inconsistencies, seem even more important from the pragmatic viewpoint.

Thus, one needs supersymmetry to remove quadratic divergence of radiative corrections to the mass of the

Higgs boson, as well as the axion solution for the CP violation problem in QCD.

Such extensions of the standard model are usually considered separately, and, being related to different parts of particle theory, are generally treated independently.

Note, that it is just this type of physical motivation, that is widely used as arguments for hot, cold, mixed, unstable etc. dark matter candidates.

Cosmologically relevant consequences of both aesthetically and pragmatically motivated extensions of the standard model are generally related to stable or sufficiently metastable particles or objects predicted in them.

Since (meta)stability is based in particle theory on some (approximate) conservation law, reflecting respective fundamental symmetry and/or the mechanism of symmetry breaking, the most fundamental new laws of Nature, predicted by extensions of standard model, lead to cosmological consequences, accessible to astrophysical probes.

Indeed, new symmetries, extending the symmetry of standard model, imply new exactly or approximately conserved charges, and the lightest particle possessing corresponding charge should be either stable or metastable.

The new charges may be related to local or global, continuous or discrete symmetry.

They may be topological, i.e. induced by the topology of the respective symmetry group.

In most cases the mass of hypothetical particles and objects reflects the scale, at which the assumed symmetry is broken.

Let us consider in more details the physical grounds and cosmologically relevant implications of the extensions of the standard model.

2.1. New Physics from the Grand Unification

The similarity in the description of the electromagnetic, weak and strong interactions in the standard model appeals for their unification.

In the grand unified (GUT) models, the gauge symmetry $SU(3)_c$x$SU(2)_L$x$U(1)$ of the standard model is embedded into the unifying group $G_{\rm GUT}$ of the grand unified gauge symmetry

$$G_{\rm GUT} \supset SU(3)_c \otimes SU(2)_L \otimes U(1) \tag{64}$$

Quarks and leptons in this construction are placed in the same representation of the unifying group $G_{\rm GUT}$, and the generators of the group related to lepto-quark transitions imply the existence of gauge bosons, mediating baryon and lepton number non conserving processes.

These processes play very important role in the mechanisms of generation of baryon excess in the initially baryon symmetric Universe.

The simplest unifying group, which can completely embed the symmetry of the standard model is the $SU(5)$ group.

The number of its generators is 24, so even in the simplest realisation of grand unification one should assume the existence of 12 new gauge bosons in addition to 12 bosons mediating strong (8 gluons) , weak (3 – W and Z bosons) and electromagnetic (photon).

In the $SU(5)$ gauge model these twelve new bosons mediate the processes violating baryon and lepton number conservation.

The mechanisms of GUT baryosynthesis may find experimental test in the searches for rare baryon non-conserving processes, such as proton decay.

In particular, it is predicted in the $SU(5)$ gauge model, the proton decay on the channel

$$p \to \pi^0 e^+ \tag{65}$$

with the lifetime

$$\tau_p \leq 10^{30} \div 10^{31} \text{ years} \tag{66}$$

that is excluded by the results of the searches for proton decay.

Historically, strong argument favouring the simplest scheme of grand unification followed from the analysis of the energy dependence of the running dimensionless constants α of weak, electromagnetic and strong interactions.

The early estimates of the weak coupling constant lead to the conclusion, that all three curves, describing the energy dependencies of these constants, intersect in one point corresponding to the energy near

$$E_{\mathrm{GUT}} \sim 10^{15} \text{ GeV} \tag{67}$$

provided that the particle content of the standard model does not change up to this energy.

However, more precise measurements of W- and Z-boson couplings to fermions in the direct studies of W and Z decays, resulted in the opposite conclusion. To reach the intersection of all three curves in one point, one should involve an additional set of particles in the theory, such as the supersymmetric particles.

In all GUT models, unifying electromagnetism with other interactions within a compact group of symmetry, magnetic monopole solutions appear.

These solutions follow from the topology of the GUT symmetry breaking. The GUT magnetic monopoles are stable topological defects, having the Dirac magnetic charge

$$g = \frac{\hbar c}{2e} \tag{68}$$

and the mass of the order of

$$m \sim \frac{\Lambda}{e} , \tag{69}$$

where Λ is the scale, at which electromagnetic $U(1)$ symmetry separates from the other interactions.

Taking the scale Λ equal to the scale of grand unification, being of the order of

$$\Lambda \sim E_{\rm GUT} \sim 10^{15} \text{ GeV} , \tag{70}$$

one obtains that the mass of the GUT magnetic monopole should be of the order of

$$m \sim 10^{16} \text{ GeV} \tag{71}$$

that makes the production of such monopoles impossible at accelerators and in collisions of cosmic rays with ultrahigh energy.

On the other hand, in the early Universe magnetic monopoles should have been produced as the result of the GUT phase transition. Their calculated primordial abundance turned out to be very high (Zeldovich, Khlopov, 1978; Khlopov, 1979; Preskill, 1979), and that lead to the serious problem of magnetic monopole overproduction. The solution of this problem implied the development of both cosmology and particle physics.

2.2. New Physics in the Modern Trends to the True Unification

In view of the evident insufficiency of the simplest realisation of the idea of grand unification on the basis of the minimal $SU(5)$ model, more sophisticated extensive GUT models, as well as more general grounds of unification should be considered.

Some GUT models imply nontrivial topology of the GUT symmetry breaking. It leads to the existence of topological defects.

In the case of spontaneously broken discrete symmetry, the two-dimensional topological defects — domain walls – should exist at the boundary of the degenerated vacua, labelled by the underlying discrete symmetry.

The energy density per unit surface of the domain wall is determined by the scale Λ of the discrete symmetry breaking, being proportional to

$$\rho_\sigma \propto \Lambda^3 . \tag{72}$$

If the continuous $U(1)$ symmetry is spontaneously broken, the one-dimensional defect (cosmic string) should exist as a result of symmetry breaking.

The corresponding energy density per unit length is proportional to

$$\rho_\mu \propto \Lambda^2 , \tag{73}$$

where Λ is the scale of the symmetry breaking.

The combination of continuous and discrete symmetry breaking leads to more sophisticated types of topological defects, such as wall surrounded by strings etc. The wall-surrounded-by-strings structure appears in the theory of invisible axion, to be discussed later.

One expects that the realistic grand unification naturally accommodates the theoretical ideas, leading to the extension of the standard model on different physical grounds.

The idea of supersymmetry is one of the most popular ideas in the grand unification. The symmetry between bosons and fermions is not only aesthetically appealing, but plays also an important practical role, removing the quadratic divergence in the radiative corrections to the Higgs boson mass, as well as providing physically natural grounds for the hierarchy of different symmetry-breaking scales.

For instance, with the use of supersymmetry the hierarchy in the electroweak and GUT symmetry-breaking scales, which differ by 13 orders of the magnitude, can be made stable relative to radiative effects.

The renormalization effects in the supersymmetric GUT model may induce the effect of the spontaneous electroweak symmetry breaking at low energies.

Since there is no exact symmetry between the known bosons and fermions, one assumes in supersymmetric models the existence of a set of new particles — supersymmetric partners for each known boson and fermion.

The supersymmetric partner has the same charge as the ordinary particle, but differs by spin, so that the supersymmetric partners of the known fermions are bosons, and the supersymmetric partners of the known bosons are fermions.

In supersymmetric models, one puts into correspondence to quarks and leptons scalar particles with zero spin, squarks (or quarkino) and sleptons, having the same charges as the ordinary quarks and leptons.

Supersymmetric partners of the gauge and Higgs bosons are fermions with spin 1/2, called gaugino and higgsino. One specifies photino, gluino, wino and zino as spin 1/2 supersymmetric partners of photon, gluon, W and Z bosons.

To explain the observed absence of supersymmetric particles in the Nature, one should assume that the supersymmetry is broken, so that supersymmetric particles are much heavier, than their ordinary partners.

In supersymmetric models a new quantum number – R-parity is considered, discriminating the ordinary and supersymmetric particles.

Exact R-symmetry guarantees the absolute stability of the lightest supersymmetric particle (LSSP). Its mass is generally related to the scale of supersymmetry breaking. In a wide class of supersymmetric models, the LSSP represents the linear combination of zino, photino and higgsino, and is called neutralino.

In some supersymmetric models, R-parity is not strictly conserved, so that the LSSP is metastable and decays on the ordinary particles. Sufficiently long-living metastable LSSP may also have important cosmological relevance.

In local supersymmetric models, the scale of supersymmetry breaking defines the mass of gravitino, the supersymmetric partner of graviton. Local supersymmetry makes it possible to unify weak, electromagnetic and strong interactions with gravity on the base of supergravity.

The models of supergravity are specified by the number of sets of supersymmetric partners and of different types of gravitino N . The $N = 1$ supergravity realises the simplest case of a single gravitino. The maximal number of gravitino is considered in the $N = 8$ supergravity.

In supergravity, the gravitino has the semigravitational coupling to other particles, inversely proportional to the Planck mass m_{Pl}. Due to extreme weakness of its interaction, such particle can not be experimentally studied even in the case when its mass does not exceed the energy threshold reached by accelerators. Cosmology turns out to be the unique source of information on the possible properties of such gravitino.

In the models of unification, more extensive than the minimal $SU(5)$ model, one can find natural place for the physics of neutrino mass. Taking apart the whole set of physical and cosmological phenomena related with the physics of neutrino mass, to be considered later, let us mention here only one interesting cosmologically relevant effect: the see-saw mechanism of neutrino mass generation.

The see-saw mechanism implies heavy right-handed neutrino state with the Majorana mass M_R determined by the scale of lepton number nonconservation. The mixing of this state with the state of the ordinary left-handed neutrino owing to the Dirac neutrino mass results in the generation of the Majorana mass of the ordinary left-handed neutrino

$$m_{\nu} = \frac{m_D^2}{M_R} , \tag{74}$$

where m_D is the Dirac mass of neutrino, typically related to the mass of the corresponding charged lepton.

The lifetime of the heavy right-handed neutrino is determined by its mixing with the ordinary left-handed neutrino ($\sim m_D/M_R$) and turns out to be inversely proportional to the mass of light left-handed neutrino. Thus, the see-saw mechanism of neutrino mass leads to the prediction of superheavy metastable neutrino.

The models of true unification should account for the solution of the problem of strong CP violation in QCD. Such solution generally involves Peccei–Quinn symmetry (see Sec. 3.3). The spontaneous breaking of this symmetry results in the existence of (pseudo)Goldstone boson, called axion, with the mass

$$m_{\alpha} = \frac{m_{\pi} f_{\pi}}{F} , \tag{75}$$

where m_{π} and f_{π} are the mass of pion and the pion constant, and F is the scale of the Peccei–Quinn symmetry breaking. Axion couplings to fermions are inversely proportional to F, and the axion lifetime relative to the decay on 2γ is of the order of

$$\tau(\alpha \to \gamma\gamma) = \frac{64\pi F^2}{m_{\alpha}^3} \propto F^5 . \tag{76}$$

The true unification should also take into account the problem of the equivalence of right- and left- handed co-ordinate systems. Starting from the paper by Lee and Yang (1956) the solution of this problem implies the existence of mirror partners of ordinary particles.

The mirror particles should not have ordinary gauge interactions and their own mirror interactions should be symmetric with the interactions of their ordinary partners. The mirror particles, having the same mass spectrum and the same internal mirror couplings, as their ordinary partners, are coupled to the ordinary matter only by gravitation. They represent very interesting form of the dark matter to be considered in details in Chapter 10.

The inclusion of the mirror partners together with the ordinary particles within the unique GUT symmetry leads after the GUT symmetry breaking and the separation of the ordinary and mirror sectors to the existence of the Alice strings, which are the cosmic strings, changing the relative mirrority of objects moving around them along the closed paths.

Note that the topological condition for the string solution as a result of the non-abelian symmetry breaking implies the existence of a strict discrete symmetry between the subgroups, remaining unbroken after the symmetry breaking. The symmetry between the ordinary and mirror particles is virtually the only known physically reasonable discrete symmetry, which may be exact in the particle theory. It makes the Alice strings to be the cosmic string candidate, well motivated from the viewpoint of the fundamental physics.

The hopes to build the final theory of everything are related with the development of superstring models. One finds in these models practically all the main trends to the unification: gauge symmetry, multidimensional space-time, supergravity and mirror symmetry, combined on the basis of the string theory. Strings are here the most fundamental elementary objects reduced to the effective quantum field theory in the four-dimensional space-time.

In such an approach, the fundamental constants of the quantum field theory are claimed to be finite. It may provide the quantitatively definite grounds for all the fundamental physics, but the problem is that such models contain very extensive hidden sector, to be probed by the means of cosmoparticle physics.

So, in the widely discussed heterotic string model, the initial E_8xE'_8 gauge symmetry is postulated in the 10 space-time-dimensional string model, assuming exact symmetry between the ordinary (E_8) and mirror (E'_8) worlds. The initial mirror symmetry is broken due to the combined action of compactification and gauge symmetry breaking, so that the shadow matter appears. The initially mirror partners lose the discrete symmetry with the ordinary particles.

In the 4-dimensional effective field model, to which the heterotic string model is reduced after compactification, the gauge symmetry of the ordinary matter comes from the E_8 symmetry, which is broken down to E_6 symmetry in order to compensate the effect of the compactification. The ordinary matter is accompanied in this case by the enormously large world of shadow particles and their interactions, corresponding to the (broken?) E'_8 gauge group.

Thus, even in the simplest realisation of the superstring models, one faces the problem of test for E'_8 symmetry model of the shadow world. To understand the complexity of the problem, remind that there are 248 fundamental fermions and 248 gauge bosons in the E_8 group.

The mechanism of the gauge symmetry breaking in compactification onto Calaby-Yao manifolds or orbifolds, used in superstring models, implies homotopically stable solutions with the mass

$$m \sim \frac{r_c}{\alpha'} , \tag{77}$$

where r_c is the radius of compactification and α' is the string tension. These objects are sterile relative to gauge interactions and may act on the ordinary matter by gravitation only.

So, even brief discussion of the modern trends to the Unification of the fundamental natural forces offers many examples of the particle zoo, one should face on the way to unification. The trouble is that the true unification should account for all these hypothetical objects, but all of them are related to the new phenomena, direct experimental search for which is either very hard or impossible in principle.

The aim of the cosmoparticle physics is to elaborate the cross-disciplinary system of studies available to probe these phenomena. In many cases, the cosmological effects are important or even unique sources of information on their possible existence.

3. Neutrino Mass and Invisible Axion

The general trends in the grand unification and practical extensions of the standard model will be complemented here by more detailed analysis of the new physics, related to the two widely discussed phenomena – the mass of neutrino and the invisible axion solution for the problem of the strong CP violation.

One may find from this discussion that these two phenomena seem to represent very different parts of the modern particle physics.

From the physical viewpoint, the Hot Dark Matter scenario, finding physical grounds in massive neutrino, and the Cold Dark Matter scenario substantially based on the model of invisible axion, can not be considered as alternatives and from the physical grounds should be treated together.

Based on the cosmoparticle physics, we'll see that the physics of neutrino mass and of the invisible axion can be put into the framework of the horizontal unification, which provides the basis for proper choice of the model of the cosmological dark matter in the physically self consistent way.

3.1. Physics of the Mass of Neutrino

The search for the mass of neutrino started immediately after the hypothesis on neutrino was put forward by W.Pauli (1930).

The only type of neutrino processes known in 30-th was nuclear beta decay. Neutrino was not detected directly, and the effect of neutrino mass was searched for indirectly by the precise measurement of the beta spectrum near the maximum of electron energy.

In the beta decay of a nucleus with the mass number A and the charge Z into the nucleus with the same mass number and the charge $(Z+1)$

$$(A,Z) \to (A,Z+1) + e^- + \overline{\nu}_e \, , \tag{78}$$

the maximal total energy of the electron is

$$E_{\max} = \left(M(A,Z) - M(A,Z+1) - m_\nu\right)c^2 \, , \tag{79}$$

so that formally the exact measurement of the maximal energy of electrons together with the precise data on the masses of the initial and final nuclei provides the determination of the mass of neutrino.

However, in practice, the most precise data on the mass difference of nuclei comes from the measurements of beta spectra. Therefore, in order to determine the mass of neutrino more sophisticated methods should be used.

If the neutrino mass is nonzero, the maximal energies of electrons correspond to the minimal energies of neutrino, and the massive neutrino become nonrelativistic in this energy range. Near the maximal electron energy, the form of the beta spectrum should change because one should use instead of the relativistic relationship between the neutrino energy and momentum

$$E_\nu \approx p_\nu c \tag{80}$$

the nonrelativistic relationship

$$E_\nu = \sqrt{(p_\nu c)^2 + m_\nu^2 c^4} \approx m_\nu c^2 + \frac{p_\nu^2}{2m_\nu} \ . \tag{81}$$

The first experimental claims on the existence of the nonzero mass of neutrino based on the change of the form of beta spectrum near the maximal electron energy appeared in the late 30-th (Alikhanian et al, 1938; Alikhanian and Nikitin, 1938). The puzzle was that the estimated value of neutrino mass, being of the order of 1 MeV, was not constant and varied from one decaying nuclei to another.

This puzzle was soon explained by Zavelsky (1939), who found that the theoretical form of beta spectrum, used by Alikhanian et al. (1938); Alichanian and Nikitin (1938) near the maximal electron energy was inapplicable for the whole spectrum. The correct description of the whole spectrum implied no change of its form near the maximal electron energy, thus giving only the upper limit on the neutrino mass.

It was the first but not the last episode in the dramatic story of the neutrino mass searches and "discoveries".

The last edition of the Tables of Particle Properties, putting aside the positive statements on the nonzero mass of neutrino, presents only upper limits on the masses of all the three types of neutrino.

The nonzero mass of neutrino implies the neutrino mass term in the Lagrangian of the form

$$L = m_\nu \overline{\Psi}_\nu \Psi_\nu \ , \tag{82}$$

linking the states of different helicity, i.e. left- and right-handed states.

Simply speaking, words the nonzero mass of neutrino means that its velocity may be very close to but never reaches the speed of light, so that the reference frame always exists, relative to which neutrino of the same polarisation moves in the opposite direction, i.e. has the opposite helicity.

In particular, the left-handed neutrino state is linked by the mass term to some right-handed "neutrino" state, which should also exist in the Nature.

In all known particle processes only left-handed neutrino and right-handed antineutrino participate. The neutrino mass term links the left-handed neutrino to either some new unknown state of right-handed neutrino or to the known state of right-handed antineutrino. The both possibilities lead us to the physics beyond the standard model.

In the first case, the new state of right-handed neutrino should be introduced into the particle theory.

Provided that such state exists, the neutrino mass term looks like the mass terms of all other fundamental fermions. This situation corresponds to the Dirac mass term.

In the second case, the neutrino mass term implies the transitions, in which the lepton number is not conserved, such as

$$\nu_L \to \overline{\nu}_R \ . \tag{83}$$

So the neutrino mass induces in the Lagrangian the lepton number changing term with

$$\Delta L = 2 \ . \tag{84}$$

In this case, neutrino and antineutrino may be associated with the different helicity states of one particle, the truly neutral Majorana neutrino, and the neutrino mass term is called the Majorana mass term.

It follows from the experimental upper limits, that if the neutrino mass is nonzero, it should be by few orders of magnitude smaller than the mass of the corresponding charged lepton.

The theoretical idea, explaining the possible large difference in the masses of neutral and charged leptons, is called the "see-saw" mechanism (Gell-Mann et al, 1979).

This idea is based on the fact that in contrast to all other fundamental fermions (charged leptons and quarks), neutrino is electrically neutral, so that the electric charge conservation does not forbid the possibility of the neutrino Majorana mass terms.

The see-saw mechanism assumes that the Dirac mass of neutrino m_D is of the same order of magnitude as the mass of the respective charged lepton, but the right-handed neutrino state involved in the Dirac mass term has the Majorana mass M, which is much larger than m_D (see Fig. 4).

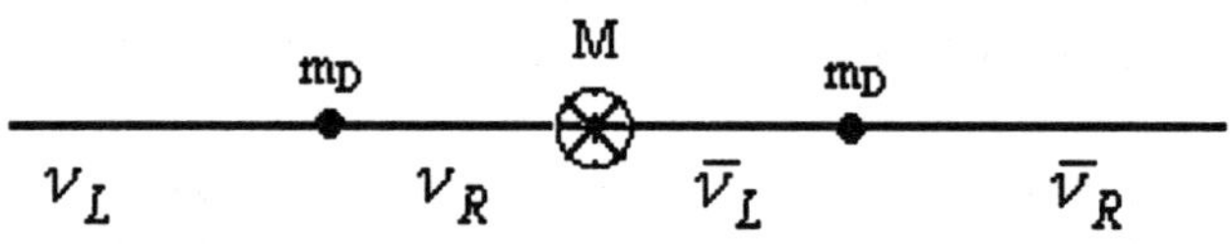

Fig.4. The succession of transitions leading to the see-saw mechansim of neutrino mass generation.

The Majorana mass term of the ordinary left-handed neutrino is generated due to the following succession of transitions

$$\nu_L \to \nu_R \to \bar{\nu}_L \to \bar{\nu}_R \tag{85}$$

and is equal to

$$m_\nu = \frac{m_D^2}{M} \; . \tag{86}$$

As a result, the two physically different neutrino mass states appear: very heavy right-handed neutrino with the Majorana mass

$$m_R \approx M \tag{87}$$

and the ordinary left-handed neutrino linked to the right-handed ordinary antineutrino by the Majorana mass

$$m_L = m_\nu = \frac{m_D}{M} m_D, \tag{88}$$

which is by the factor of $\sim (m_D/M)$ smaller than the mass m_D of the corresponding charged lepton.

Thus, the see-saw mechanism provides the mass of ordinary (left-handed) neutrino by few orders of magnitude smaller than the mass of the corresponding charged lepton, and the mass of the right-handed neutrino higher than the energy threshold, accessible in the experiments at the modern accelerators.

Formally, the right-handed neutrino can decay due to its small (of the order of $\sim (m_D/M)$) mixing with the state of the light left-handed neutrino. The probability of decay of right-handed neutrino into the channel

$$\bar{\nu}_L \to \nu_L Z \tag{89}$$

is estimated to be of the order of

$$W(\bar{\nu}_L \to \nu_L Z) \sim \frac{\alpha}{\pi}\left(\frac{m_D}{M}\right)^2 M = \frac{\alpha m_\nu}{\pi}, \tag{90}$$

where α is the dimensionless weak constant. For the small light neutrino mass the heavy right-handed neutrino is a long-living particle.

Moreover, the smaller is the mass of the ordinary light neutrino, the larger is the mass and the lifetime of its heavy right-handed partner. This property makes the superheavy metastable neutrino a very interesting object from the cosmological viewpoint.

Note, however, that in the case when the Dirac neutrino mass term is induced in the Lagrangian by the Higgs mechanism, that is the case in the standard model, neutrino couples with the Higgs boson, and this can induce rapid decay of heavy neutrino on the light neutrino and Higgs boson.

Though in its simplest form the see-saw mechanism results in the small Majorana mass for the ordinary left-handed neutrino, in more extensive versions this mechanism can provide also small Dirac masses for the light neutrino. It may be achieved by the multiple application of the see-saw mechanism and the redefinition of the right-handed antineutrino state as the states of right-handed light neutrino. So, in the general case, the neutrino mass Lagrangian may contain both Dirac and Majorana mass terms.

Note, that the existing experimental constraints do not put, in fact, too strong upper limits on the possible strength of the light right-handed neutrino interactions. The right-handed weak interaction, in which right-handed neutrino participate, can be mediated by the intermediate boson, being only 5-7 times heavier than W and Z bosons.

The Dirac mass term implies the existence of the neutrino magnetic moment. In the absence of right-handed currents the neutrino magnetic moment is of the order of

$$\mu_\nu \sim G_F m_\nu, \tag{91}$$

where G_F is the Fermi constant.

The existence of right-handed currents and mixing of bosons, mediating their interaction, with W and Z bosons may result in the neutrino magnetic moment, proportional to the mass of the charged lepton

$$\mu_\nu \propto G_F m_l \ . \tag{92}$$

3.2. *Neutrino Instability*

In the standard model, the states of electron, muon or tau neutrino are defined on the basis of the lepton number conservation for weak interaction processes. So, the electron neutrino is defined as the state which is created together with positron creation or electron annihilation.

Since the see-saw mechanism is based on lepton number nonconservation, one can expect that the proper mass states of neutrino do not coincide with the states with the definite lepton number.

For instance, in this case the "electron neutrino" created in beta decay is the superposition of several neutrino states with definite mass.

In general, the neutrino states with the definite lepton number ν_l , where l denotes e, μ or τ , are given by the superposition of the states ν_i with the masses m_i

$$\left|\nu_l\right\rangle = \sum_i a_{li}\left|\nu_i\right\rangle , \tag{93}$$

where a_{li} are the mixing coefficients.

Provided that the neutrino masses are different, the neutrino states, created with the given energy in the given superposition, propagate with different velocities, so that the superposition changes at some distance from the source.

It means, that one can observe the effect of neutrino oscillations: the muon or the tau neutrino appear at some distance from the pure source of the electron neutrino.

Neutrino oscillations are described in terms of the oscillation length. For neutrino with the energy E it is equal to

$$L = \frac{E}{\delta m_\nu^2} \, , \tag{94}$$

where δm^2 is the difference of the neutrino mass squared.

The second parameter is the amplitude of oscillations, determined by the neutrino mixing parameters. It defines the maximal admixture of the other neutrino type appearing in the neutrino flux from the source of neutrino with the definite lepton number.

The amplitude of neutrino oscillations also defines the maximal deficit of neutrino of the given type due to the transitions into unobservable neutrino states.

In principle, there are two possibilities to search for neutrino oscillations.

One may look for the other types of neutrino at some distance from the pure source of the given type of neutrino, or one may monitor the flux of the given type of neutrino with the distance and search for the deficit of these neutrino, induced by oscillations.

Neutrino propagation in the matter may cause the resonance enhancement of neutrino oscillations (Mikheev. Smirnov, 1981; Wolfenstein, 1980): at the resonant density, the amplitude of oscillations grows to the order of 1 even for very small vacuum mixing of neutrino. This effect is discussed as possible explanation for the observed deficit of the solar neutrino. Resonance conditions may also take place for neutrino oscillations in the longitudinal magnetic field (Akhmedov, Khlopov, 1988).

The other form of possible neutrino instability are the decays of heavier neutrino into lighter ones. It is easy to find that in the standard model the massive neutrino decays, for instance,

$$\nu_H \to \nu_L \nu_L \bar{\nu}_L \tag{95}$$

or

$$\nu_H \to \nu_L \gamma \,, \tag{96}$$

go through the creation of the virtual charged lepton and W-boson (see Fig.5 and Fig.6) and are strongly suppressed.

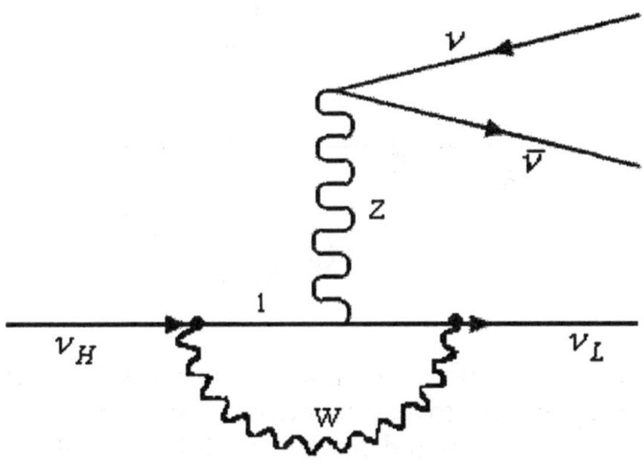

Fig.5. In the standard model the neutrino decay on three neutrino can go via creation of virtual pair of charged lepton and W-boson, which emit Z-boson decaying on neutrino pair.

The reason is that these processes are respectively the second-order weak and weak-electromagnetic, and their probabilities are of the order of

$$W\left(\nu_H \to \nu_L \nu_L \bar{\nu}_L\right) \propto G_F^4 m_H^9 \tag{97}$$

and

$$W(\nu_H \to \nu_L \gamma) \propto \alpha_{em} G_F^2 m_H^5, \tag{98}$$

where α_{em} is the electromagnetic fine-structure constant, G_F is the Fermi constant and m_H is the mass of the heavier neutrino ν_H.

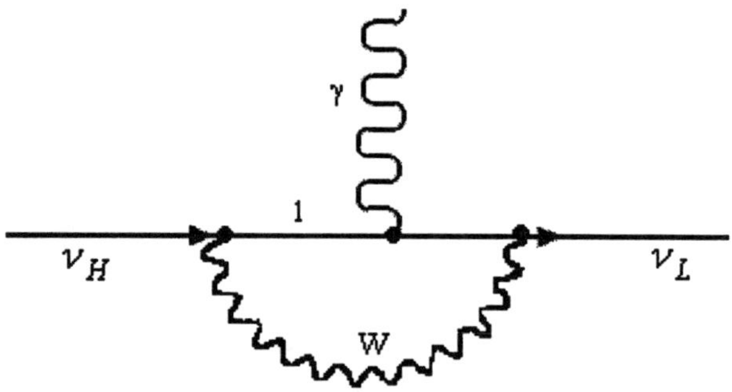

Fig.6. In the standard model, the neutrino decay on neutrino and photon can go via creation of virtual pair of W-boson and charged lepton, which emit photon

Moreover, in the orthogonal neutrino mass matrix, the mixing factors a_{il} are normalised by the condition

$$\sum_l a_{il}^+ a_{lj} = \delta_{ij} , \tag{99}$$

that leads to the vanishing amplitude for the non-diagonal transition

$$\nu_H \to \nu_L, \tag{100}$$

corresponding to

$$H \equiv i \neq j \equiv L. \tag{101}$$

The non-vanishing amplitude of the transition (100) can appear due to the account for the difference in the mass m_l of charged leptons in the intermediate state, that induces the additional suppression factor in the probabilities for the neutrino decays (95) and (96), which are given by

$$W(\nu_H \to \nu_L \nu_L \overline{\nu}_L) \propto G_F^4 m_H^9 \left(\sum_l a_{Ll}^+ a_{lH} \frac{m_l^2}{m_W^2} \right)^2 \tag{102}$$

and

$$W(\nu_H \to \nu_L \gamma) \propto \alpha_{em} G_F^2 m_H^5 \left(\sum_l a_{Ll}^+ a_{lH} \frac{m_l^2}{m_W^2} \right)^2 . \tag{103}$$

On the other hand, one finds, that the family symmetry breaking plays the important role in the mechanism of neutrino instability. Its effect is not reduced to the neutrino mass difference, because the charged lepton mass difference is also crucial. The nontrivial role of family symmetry breaking seem to arise in all other mechanisms of neutrino instability.

One may expect such mechanisms to come from the physics of neutrino mass generation.

Indeed, lepton number nonconservation, assumed in the see-saw mechanism, may be realised by the spontaneous symmetry breaking, what implies the existence of the Goldstone boson called majoron.

Neutrino coupling with the majoron M can induce the majoron decay modes of heavier neutrino, given by

$$\nu_H \to \nu_L M . \tag{104}$$

The so-called triplet majoron model (Gelmini, Roncandelli, 1981) assumes very low scale of majoron interaction, leading to strong interneutrino interactions, mediated by the majoron exchange. This model is practically excluded by the measurements of the width of Z-boson, in which the majoron effects should have contributed at the unacceptably high level.

The model of singlet majoron (Chikashige et al, 1980) is free of this problem, since the singlet majoron is decoupled from the processes with the known particles.

In the simplest singlet majoron model, the diagonalisation of the neutrino mass matrix leads simultaneously to the diagonalisation of neutrino-majoron couplings. So, the nondiagonal transitions given by Eq. (104) are nonvanishing in the next order of the small neutrino-majoron coupling constant

$$h = \frac{m_H}{F} \tag{105}$$

only, where m_H is the mass of neutrino ν_H , and F is the energy scale of the lepton number global symmetry breaking, which should exceed the electroweak scale, i.e.

$$F > 10^2 \text{ GeV}. \tag{106}$$

The probability of the decay (104) is then of the order of

$$W(\nu_H \to \nu_L M) \sim \frac{m_H^5}{F^4} \tag{107}$$

making this decay for the reasonable values of the scale F no more probable, than the photon decay (96).

The probability for the majoron decay (104) grows significantly if different signs of the lepton number are ascribed to different types of neutrino (Valle, 1983). In this case the probability of the majoron decay of the neutrino ν_H is given by

$$W(\nu_H \to \nu_L M) = \frac{h^2}{32\pi} m_H = \frac{m_H^3}{32\pi F^2}. \tag{108}$$

The differentiation in lepton number assignment, being crucial in the enhancement of the majoron mode of neutrino decay, is successfully realised in the models of spontaneously broken symmetry of quark and lepton families.

Spontaneous breaking of global family symmetry results in the prediction of Goldstone boson(s) called familon(s). In the case of global $SU(3)_H$ family symmetry breaking, the octet of massless familons is predicted. In the model of singlet familon (Anselm, Uraltzev, 1983), the family symmetry breaking results in the prediction of a single familon state.

In the familon model, the massive neutrino can decay due to neutrino coupling with familons f to the familon channel

$$\nu_H \to \nu_L f\,, \tag{109}$$

which has the probability, given by the same expression as Eq.(108) with the scale F being the scale of the family symmetry breaking.

The familon model finds experimental test in the searches for familon modes of decay of quarks and charged leptons, such as

$$\mu \to ef \tag{110}$$

$$\tau \to \mu f \tag{111}$$

or

$$s \rightarrow df \tag{112}$$

predicted in this model together with the neutrino decay, given by Eq. (109).

The physics of neutrino mass related with the see-saw mechanism of neutrino mass generation can be checked in laboratory searches for the lepton number L nonconserving processes with

$$\Delta L = 2 \ , \tag{113}$$

such as the double neutrinoless beta decay.

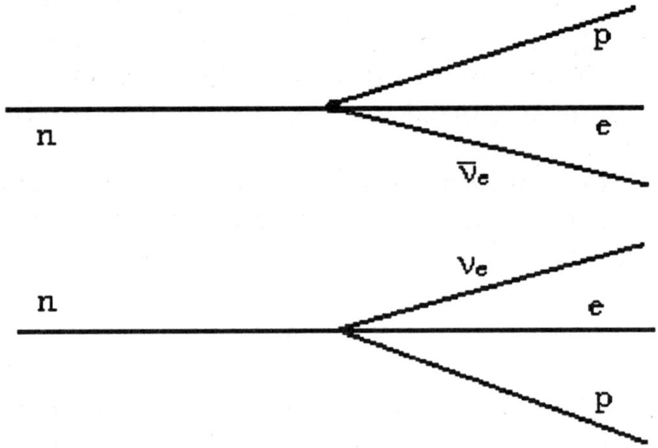

Fig.7. Double beta decay is the result of simultaneous decay of two neutrons in a beta-stable nucleus.

There exist nuclei (A,Z) that are stable relative to beta decays to neighbouring nuclei (A,Z+1) and (A,Z–1) but unstable relative to the double beta decay, when the two protons or the two neutrons (Fig.7) in the nucleus decay simultaneously, so that the processes take place

$$(A,Z) \rightarrow (A,Z-2) + 2e^{+} + 2\nu_e \ , \tag{114}$$

$$(A,Z) \rightarrow (A,Z+2) + 2e^{-} + 2\overline{\nu}_e \ . \tag{115}$$

The lepton number nonconservation, which follows the selection rule given by Eq. (113), leads to the double neutrinoless beta decays

$$(A,Z) \rightarrow (A,Z-2) + 2e^{+} \ , \tag{116}$$

$$(A,Z) \rightarrow (A,Z+2) + 2e^{-} \ . \tag{117}$$

In the case when neutrinoless double beta decay is induced by the Majorana mass of the electron neutrino, the amplitude of this process, represented by the diagram on Fig.8, is proportional to this mass.

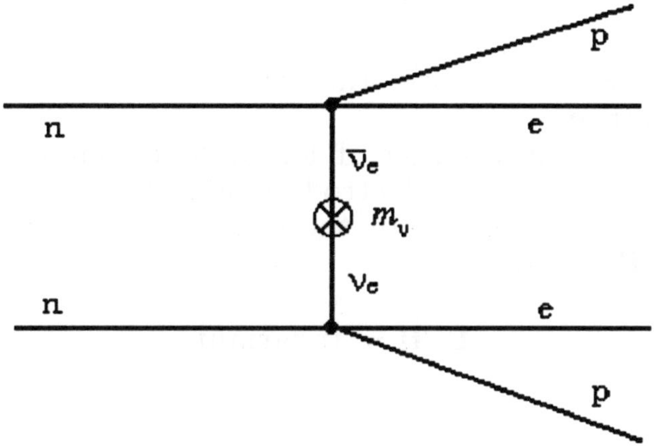

Fig.8. Owing to Majorana mass antineutrino, emitted in decay of one neutron, is captured as neutrino by the second neutron.

Hence, the upper limits on the probability for double neutrinoless beta decay impose constraints on the Majorana mass of electron neutrino. The model-independent analysis of the experimental data, made with all the precautions, gives the upper limit of about few eV.

3.3. The Axion Solution for the Problem of Strong CP violation in QCD

Instanton effects in the quantum chromodynamics induce complicated vacuum structure of QCD, resulting in the so-called θ term in the Lagrangian of the theory, given by

$$\Delta L = \frac{g^2}{16\pi}\theta_{QCD} G_{\mu\nu}\tilde{G}_{\mu\nu}, \tag{118}$$

where θ_{QCD} is the arbitrary and, in general, not small constant, g is the gauge constant of QCD, $G_{\mu\nu}$ is the tensor of the gluon field, and its dual tensor is defined as

$$\tilde{G}_{\mu\nu} = \varepsilon_{\mu\nu\lambda\rho} G^{\lambda\rho}. \tag{119}$$

One can easily see, that the θ term violates P- and CP- invariance.

Indeed, the product of the field tensor and its dual tensor corresponds in electrodynamics to the well known invariant scalar product of the electric and magnetic field strengths **E** and **H**:

$$\mathbf{E}\cdot\mathbf{H} = \text{ invariant} \tag{120}$$

for which

$$CP(\mathbf{E}\cdot\mathbf{H}) = -1 \ . \tag{121}$$

The interference of the similar CP-odd combination of gluon field strengths with the CP-even terms in the Lagrangian of QCD should induce effects of CP violation.

In the strong interaction processes, such effects are not observed. Moreover, from the negative result of the searches for the electric dipole moment of neutron, putting the upper limit

$$d_n < 10^{-25} e \cdot \mathrm{cm} \ , \tag{122}$$

it follows the upper bound on the value of the constant θ, given by

$$\theta < 10^{-9} \ . \tag{123}$$

The main difficulty in the indicated problem of the strong CP-violation comes from the fact, that even with the initial bare constant

$$\theta_{\mathrm{QCD}} = 0 \tag{124}$$

the existence of the effective θ term can not be excluded.

Such an effective term arises as the result of the electroweak symmetry breaking due to the existence of complex elements in the quark mass matrix in the standard model after unitary transformations to the physical basis, in which the mass matrix is diagonal.

Thus the effective value of θ in the Lagrangian, given by Eq. (118), turns out to be the sum of two terms of different physical nature

$$\theta = \theta_{\mathrm{QCD}} + \theta_{\mathrm{QFD}} \ , \tag{125}$$

where

$$\theta_{\mathrm{QFD}} = \det \hat{m} \ . \tag{126}$$

Here m denotes the full mass matrix, including all coloured fermions, which are present in the theory. It should take into account both the known quarks and all possible hypothetical coloured fermions.

The different nature of two terms in Eq. (125), each being, in general, not small, can not *a priori* provide their cancellation down to the level of

$$\theta < 10^{-9} \ .$$

In fact, this lies in the essence of one of the naturality problems in the particle theory.

The theoretical solution of this problem is related to the mechanism of the natural suppression of θ.

In order to provide such mechanism Peccei and Quinn (1977) assumed the additional chiral global $U(1)_{PQ}$ symmetry with nonvanishing colour anomaly.

If this symmetry is not broken, what corresponds to the existence of at least one massless quark, c.f., u-quark (Wilczek, 1977; Weinberg, 1978), all θ vacua are equivalent to the vacuum with

$$\theta = 0 \ . \tag{127}$$

In other words, the θ-term may be excluded from the Lagrangian by the chiral phase transformation of the u-quark field.

The possibility of massless u-quark theoretically is not excluded (Kaplan, Manohar, 1986), but it does not find serious fundamental grounds in the standard model and meets serious difficulties in the confrontation with the results of the current algebra and PCAC hypothesis.

However, the problem of the θ-term finds natural solution even in the case of the broken $U(1)_{PQ}$ symmetry. The constant θ acquires in this case the dynamical meaning of the amplitude of the pseudo-Goldstone field, related to the broken $U(1)_{PQ}$ symmetry, which, in turn, is manifestly broken by the colour anomaly. Hence, in the vacuum the condition

$$\langle \theta \rangle_{vac} = 0 \tag{128}$$

is automatically satisfied, thus providing the exact mutual compensation of the terms in Eq. (125). Such pseudo-Goldstone field is called axion.

The idea of the axion solution for the problem of the strong CP violation can be explained by the following simple argument. If the dynamical solution is found with the use of axion, it means that we should consider the Lagrangian (118) as the Lagrangian of the axion-gluon interaction. The nonvanishing colour anomaly means here that the constant of this interaction is nonzero. Zero vacuum expectation value for the axion field provides then the effective cancellation of the θ-term in the QCD vacuum.

The most important parameter, defining the properties of the axion, is the energy scale F_a at which the symmetry $U(1)_{PQ}$ is broken.

This scale appears in the Lagrangian of the axion-gluon interaction

$$L(\alpha gg) = \frac{g^2}{16\pi} \frac{\alpha}{F_a} G_{\mu\nu} \tilde{G}_{\mu\nu} , \tag{129}$$

where α is the axion field.

Lagrangian of axion interactions with the other gauge field can be expressed in the same form with the suitable change of the gauge coupling constants. So, the axion coupling C with the gauge bosons is generally defined by the scale F_a as

$$C \propto F_a^{-1} . \tag{130}$$

In particular, in the case of small, but nonzero axion mass, the axion-photon interaction leads to the axion decay

$$\alpha \rightarrow \gamma\gamma \, . \tag{131}$$

The lifetime of the axion relative to this decay is given in the complete analogy with the similar decay of neutral pion as

$$\tau(\alpha \rightarrow \gamma\gamma) = \frac{64\pi}{m_\alpha^3 F_a^2} \, . \tag{132}$$

Since neutral pion has the similar interaction with gluons, owing to the non-vanishing coupling to the colour anomaly, one can express the axion mass in terms of the scale F_a and the respective parameters of pion, so that the mass of axion is defined by this scale as

$$m_\alpha = A_a \frac{f_\pi}{F_a} m_\pi \, , \tag{133}$$

where the constant A_a depends on the choice of the specific axion model.

The scale F_a also determines the strength h of the axion interaction with fermions

$$h \propto F_a^{-1} \, . \tag{134}$$

The simplest variant of the axion model is the model of Weinberg-Wilczek axion (Wilczek, 1977; Weinberg, 1978) , in which the scale F_a coincides with the scale v of the electroweak symmetry breaking. This model was excluded by the combination of experimental and astrophysical constraints (Vysotsky et al, 1978; see review and Refs in Kim, 1987; Cheng, 1988). The analysis of these constraints lead to very high estimation of the lower limit for the axion scale

$$F_a \gg v \, . \tag{135}$$

Thus, the scale F_a should be related to some new high-energy scale in the particle theory. At this scale axion interactions turn out to be elusive, thus making the axion invisible.

3.4 The Models of Invisible Axion

In all models of the invisible axion, this particle appears as the Goldstone boson, related to the phase of a complex $SU(2)$x$U(1)$ singlet Higgs field.

The axion coupling to gauge bosons is generated in these models after $U(1)_{PQ}$ symmetry breaking due to mechanisms, which provide the existence of nonvanishing colour $U(1)_{PQ}$–$SU(3)_c$–$SU(3)_c$-anomaly.

In the most general case, the Lagrangian of the axion interaction with fermions (quarks and leptons) and photons has the form

$$L = g_{\alpha\beta} \alpha \bar{f}_\alpha \left(\sin\theta_{\alpha\beta} + i\gamma_5 \cos\theta_{\alpha\beta}\right) f_\beta + C_{\alpha\gamma\gamma} a F_{\mu\nu} \tilde{F}^{\mu\nu} \, , \tag{136}$$

where

$$\alpha, \beta = 1, \; 2, \; 3$$

are the indices of families of fermions f and the constants $\theta_{\alpha\beta}$,

$$g_{\alpha\beta} \propto F_a^{-1} \tag{137}$$

and

$$C_{\alpha\gamma\gamma} \propto F_a^{-1} \tag{138}$$

depend on the choice of the axion model.

Owing to the general theorem (Anselm, Uraltzev, 1983), the diagonal coupling of axion to fermions in Eq. (136) can be only pseudo-scalar, that gives

$$\theta_{\alpha\alpha} = 0 \tag{139}$$

and, neglecting the higher-order effects, the axion-mediated long-range interaction is absent.

The non-diagonal axion couplings, which in general are not excluded, can be both scalar and pseudoscalar or their combination, corresponding to

$$\theta_{\alpha\beta} \neq 0 \ , \tag{140}$$

depending on the choice of the model.

There are three main types of models of invisible axion.

1.Dine-Fishler-Srednitsky-Zhitnicki (DFSZ) axion (Dine et al, 1985; Zhitnicki, 1980). The model assumes only the known types of fermions (quarks and leptons) and extends the standard model only in the Higgs sector of the theory. The number of Higgs doublets φ is increased, as compared to the standard model, as well as a new singlet Higgs field σ is added.

In the DFSZ, the Lagrangian of Yukawa interaction of Higgs bosons with fermions has the form

$$L_{Yuk} = g^{(f)}_{\alpha\beta}\bar{f}_{L\alpha} f_{R\beta}\varphi_f + h.c. \ , \tag{141}$$

where $f_{L(R)}$ are the left-handed (right-handed) components of quark and lepton fields,

$$f = u, d, e \ ,$$

the indices of generations in the case of three families of quarks and leptons are

$$\alpha, \beta = 1,\ 2,\ 3$$

and φ_f are the Higgs doublets.

For the doublet

$$\varphi_u = \begin{pmatrix} \varphi_u^+ \\ \varphi_u^0 \end{pmatrix} \tag{142}$$

the vacuum expectation value of the neutral component is

$$\left\langle \varphi_u^0 \right\rangle = \frac{v_u}{\sqrt{2}} \tag{143}$$

and for the doublets

$$\varphi_d = \varphi_e = \begin{pmatrix} \varphi_d^+ \\ \varphi_d^0 \end{pmatrix} \tag{144}$$

the corresponding vacuum expectation value is

$$\left\langle \varphi_d^0 \right\rangle = v_d \sqrt{2}. \tag{145}$$

It is assumed that these vacuum expectation values are related by the condition

$$\left(v_u^2 + v_d^2\right)^{1/2} = v = \left(\sqrt{2} G_F\right)^{-1/2} . \tag{146}$$

In the most general case, the doublets φ_e and φ_d can be different. In particular, the choice

$$\varphi_e = \overline{\varphi}_u \tag{147}$$

is possible. However, the choice of these doublets, given here, is determined by the inclusion of the model into the framework of GUT models.

The Lagrangian (141) is invariant relative to global chiral phase transformations of the symmetry group

$$
\begin{aligned}
f_{L\alpha} &\to \exp(i\omega) f_{L\alpha} \\
f_{R\alpha} &\to \exp(-i\omega) f_{R\alpha} \\
\varphi_f &\to \exp(2i\omega) \varphi_f \\
\sigma &\to \exp(-2i\omega) \sigma
\end{aligned}
\tag{148}
$$

that is provided by the presence of the term in the Higgs potential, given by

$$L_H = \lambda \varphi_u \varphi_d \sigma^2 + h.c. \tag{149}$$

Assuming that

$$\upsilon << \langle \sigma \rangle = \frac{V}{\sqrt{2}} , \tag{150}$$

the axion field is mainly defined by the phase of the Higgs field σ. The admixture of the phases of fields φ_u and φ_d in the state of the axion field is suppressed as

$$\langle Arg\varphi_{u,d} | \alpha \rangle \sim \frac{\upsilon_{u,d}}{V} . \tag{151}$$

The nondiagonal couplings of such axion with fermions are absent, and the other parameters in the Eqs. (136) – (138) for the DFSZ axion are given by

$$A_a^{DFSZ} = 2N_g \frac{\sqrt{z}}{(1+z)}, \tag{152}$$

where

$$N_g = 3$$

is the number of quark-lepton families,

$$F_a = V\,,$$

$$g_f = \frac{m_f}{F_a}\left[X_f - \frac{(\delta_{fu} + \delta_{fd}z)N_g}{(1+z)}\right] =$$

$$= \frac{m_f m_\alpha}{m_\pi f_\pi A^{DFSZ}}\left[X_f - \frac{(\delta_{fu} + \delta_{fd}z)N_g}{(1+z)}\right],$$

$$C_{\alpha\gamma\gamma} = \frac{m_\alpha \alpha_{em}}{8\pi f_\pi m_\pi}\frac{1+z}{\sqrt{z}}\left(\frac{8}{3} - \frac{2}{3}\frac{(4+z)}{(1+z)}\right) =$$

$$= \frac{\alpha_{em} N_g}{4\pi F_a}\left(\frac{8}{3} - \frac{2}{3}\frac{(4+z)}{(1+z)}\right). \tag{153}$$

Here the mass ratio of u- and d-quarks, given by the current algebra estimation, is taken to be equal to

$$z = \frac{m_u}{m_d} \approx 0.56\,. \tag{154}$$

For all up-type quarks (u, c, t ...), the parameter X_{up} is equal to

$$X_{up} = \frac{2}{\left(x + \frac{1}{x}\right)x}, \qquad (155)$$

and for all charged leptons and down-type quarks $(d,s,b...)$, the corresponding parameters are

$$X_{lept} = X_{down} = \frac{2x}{\left(x + \frac{1}{x}\right)}, \qquad (156)$$

where the ratio of the vacuum expectation values is denoted by

$$x = \frac{v_u}{v_d}. \qquad (157)$$

In the parameters g_f and in the Kronecker symbols δ_{fu} and δ_{fd}, the fermion index f is running the values

$$f = u, d, e, s, c, \mu, b, \tau\ldots$$

The Kronecker symbols δ_{fu} and δ_{fd} in Eq. (153) mean that the corresponding terms are to be taken into account for u- and d-quarks only, and α_{em} is the electromagnetic constant.

2. The hadronic axion of Kim-Shifman-Vainshtein-Zakharov (KSVZ) (Kim, 1980; Shifman et al, 1980). Here the extension of the standard model takes place both in the Higgs and the fermion sectors of the theory: additional Higgs fields, as well as fermions are introduced. In addition to the ordinary quarks and leptons, the existence of new heavy quark colour triplet Q is assumed.

The Lagrangian of Yukawa interactions is given in this case by

$$L_{Yuk} = g\overline{Q}_L \sigma Q_R + h.c. \tag{158}$$

This Lagrangian is invariant relative to global chiral phase transformations

$$\begin{aligned} Q_L &\to \exp(i\omega) Q_L \ , \\ Q_R &\to \exp(-i\omega) Q_R \ , \\ \sigma &\to \exp(-2i\omega)\sigma \ . \end{aligned} \tag{159}$$

Such axion has no couplings with leptons in the tree approximation, and its coupling with light quarks occurs only due to its mixing with neutral pion.

The parameters of KSVZ axion in Eqs. (136)-(138) are given by

$$\begin{aligned} A_a^{KSVZ} &= \frac{\sqrt{z}}{(1+z)} \ , \\ g_{u,d} = \frac{m_u}{F_a(1+z)} &= \frac{m_u m_\alpha}{f_\pi m_\pi (1+z) A_a^{KSVZ}} \ , \end{aligned} \tag{160}$$

and

$$\begin{aligned} C_{\alpha\gamma\gamma} &= \frac{m_\alpha \alpha_{em}}{8\pi f_\pi m_\pi} \frac{1+z}{\sqrt{z}} \left(6Q_{em}^2 - \frac{2}{3}\frac{(4+z)}{(1+z)} \right) = \\ &= \frac{\alpha_{em}}{8\pi F_a} \left(6Q_{em}^2 - \frac{2}{3}\frac{(4+z)}{(1+z)} \right) \ , \end{aligned} \tag{161}$$

where Q_{em} is the electric charge of a heavy quark.

All Yukawa couplings of KSVZ axion with

$$f = e, s, c, \mu, \tau$$

are negligibly small.

3. The model of archion (Berezhiani, Khlopov, 1990a,b,c;1991; Berezhiani et al, 1992; Berezhiani, 1983; 1985) appears in the framework of the model of horizontal unification (MHU), which we consider in Chapter 11. The theory naturally embeds the global $U(1)_H$ symmetry. The spontaneous breaking of this symmetry results in the prediction of Goldstone boson of the invisible axion type.

The boson, that was called *archion* by Berezhiani and Khlopov (1990c), has both flavor diagonal and flavor nondiagonal couplings with fermions.

The global $U(1)_H$ symmetry in MHU can be identified with the symmetry of Peccei and Quinn, provided that there exist the triangle anomaly in the axial $U(1)_H$ current interaction with gluons.

In the simplest variant of the MHU gauge symmetry $SU(3)_c$x$SU(2)$x$U(1)$x$SU(3)_H$x$U(1)_H$ with the minimal set of heavy fermions, the anomaly is cancelled, and the archion remains almost massless similar to massless Goldstone boson, called arion (Anselm, Uraltzev, 1982). But in contrast to arion, coupled to photons, there is also no archion interaction with photons because of parallel cancellation of the corresponding triangle current–photon–photon anomaly.

In any realistic extension of the MHU model up to the GUT symmetry, c.f., up to the symmetry $SU(5)$x$SU(3)_H$, the cancellation does not take place due to additional heavy fermions, so that archion turns out to be similar to the hadronic axion with strongly suppressed coupling to leptons.

MHU symmetry leads to the following Yukawa couplings

$$L_{Yuk} = g_f \bar{f}_{L\alpha} F_{R\alpha} \varphi^0 + g_{nf} \bar{F}_{R\alpha} F_L^{\beta} \xi_{\alpha\beta}^{(n)} + \mu_f \bar{F}_L^{\alpha} f_R^{\alpha} + h.c., \quad (162)$$

where μ_f are the mass parameters of the same order of magnitude for

$$f = u, d, e \; .$$

In Eq. (162)

$$F = U, D, E \quad (163)$$

denote the additional heavy fermions and $\xi_{\alpha\beta}{}^{(n)}$ are the horizontal Higgs bosons, inducing the family symmetry breaking.

The Lagrangian (162) is invariant relative to global chiral $U(1)_H$ symmetry transformations given by

$$\begin{gathered} f_L \to \exp(i\omega) f_L \; , \quad f_R \to \exp(-i\omega) f_R \; , \\ F_L \to \exp(-i\omega) F_L \; , \quad F_R \to \exp(i\omega) F_R \; , \\ \xi^{(n)} \to \exp(2i\omega) \xi^{(n)} \; , \quad \varphi \to \varphi \; , \end{gathered} \quad (164)$$

Where

$$n = 0, \; 1, \; 2 \; .$$

In the archion model, the parameters defined in Eqs. (136)–(138) are given by

$$A_a = A_c \frac{\sqrt{z}}{(1+z)} ,$$

$$g_{33} = \frac{m_3^{(f)}}{V_1} , \quad g_{22} = \frac{m_2^{(f)}}{V_1}\left(S_{23}^{(f)}\right)^2 , \quad g_{11} = \frac{m_2^{(f)}}{V_1}\left(S_{13}^{(f)}\right)^2 ,$$

$$g_{23} = \frac{m_3^{(f)}}{2V_1} S_{23}^{(f)} , \quad g_{13} = \frac{m_3^{(f)}}{2V_1} S_{13}^{(f)} , \quad g_{12} = \frac{m_2^{(f)}}{2V_1} S_{13}^{(f)} S_{23}^{(f)} ,$$

$$C_{\alpha\gamma\gamma} = \frac{\alpha_{em} m_\alpha}{4\pi f_\pi m_\pi} \cdot \frac{z}{(1+z)} \left(\frac{A_{em}}{A_c} - \frac{2(4+z)}{3(1+z)} \right)^{-1} =$$
$$= \frac{\alpha_{em} A_c z^{2/3}}{4\pi F_a (1+z)^2} \cdot \left(\frac{A_{em}}{A_c} - \frac{2(4+z)}{3(1+z)} \right)^{-1} , \qquad (165)$$

where

$$S_{ij} = \sin \varphi_{ij} \quad \text{with} \quad i, j = 1, \; 2, \; 3$$

are the angle parameters of the unitary matrices V_f of rotation, diagonalizing the fermion mass matrices

$$V_{fL}^{+} m^f V_{fR} = \hat{m}_{diag}^f ,$$

and the phases are determined by the phase content of matrices V in the form

$$\varphi_{12} = \arg\left(\frac{V_{23} V_{13}}{V_{12}} \right) , \; \varphi_{13} = -\arg V_{13} , \; \varphi_{23} = -\arg(V_{12} V_{23}). \quad (166)$$

The scale

$$V_1 = F_a$$

is the scale of the $U(1)_H$ symmetry breaking and A_c and A_{em} are the color and electromagnetic anomalies, respectively.

In all three models of invisible axion, the Lagrangian of its interaction with nucleons has the form

$$L_{NN'} = i\overline{N}'\gamma_5\left(g^{(0)} + g^{(3)}\tau_3\right)N\alpha\ , \tag{167}$$

where

$$g^{(0)} = \frac{M}{2F_a}G_A^{(0)}\ ,\quad g^{(3)} = \frac{M}{2F_a}\cdot\frac{1-z}{1+z}G_A$$

for DFSZ axion, and

$$g^{(0)} = \frac{M}{2N_gF_a}\left(X_u + X_d - N_g\right)G_A^{(0)}\ ,$$

$$g^{(3)} = \frac{M}{2N_gF_a}\left(X_u - X_d - \frac{1-z}{1+z}\right)G_A$$

for hadronic axion and archion. Here M is the mass of nucleon and

$$G_A = G_A^{(0)} = 1.23$$

is the isotriplet axial nucleon form factor, determined in the experimental measurement of parameters of weak nucleon charged current.

CHAPTER 3

THE HIDDEN PARAMETERS OF THE MODERN COSMOLOGY

1. The Old Standard Big Bang Paradigm

1.1. The Expanding Universe.

Cosmology studies the Universe as a whole. Stars and interstellar medium, galaxies and intergalactic space, clusters of galaxies and galaxy superclusters, surrounding giant voids, in which galaxies are not observed, exhibit strongly inhomogeneous distribution of luminous matter in the Universe. However at scales, exceeding the largest scale of cosmological structure one finds the distribution of matter to be almost homogeneous and isotropic.

The simplest and reasonable conjecture is to consider homogeneous and isotropic cosmological model. Such model accounts for the presence of the matter in the modern Universe via the unique parameter — cosmological density ρ, which is the density of matter averaged over scales, exceeding the scale of the observed structure.

In early twenties of the XX century, the application of general relativity to the Universe as a whole found impossible the existence of stationary solutions of Einstein equations.

One could obtain stationary Universe only adding to the cosmological density the cosmological term, having no physical relation to the properties of the ordinary matter.

A.Friedman found nonstationary cosmological solutions of the equations of general relativity without cosmological term. Nonstationarity means that the Universe as a whole may contract or expand.

In the late twenties, astronomers found that several objects, treated before as faint nebulae in our Galaxy, are, in fact, distant galaxies, comparable with our Galaxy and containing billions of stars.

The discovery of other galaxies and the analysis of their luminosity lead to the discovery of systematic red shift in their spectra. Observed red shifts in spectra of distant galaxies, interpreted as the Doppler effect, correspond to "run away" of galaxies.

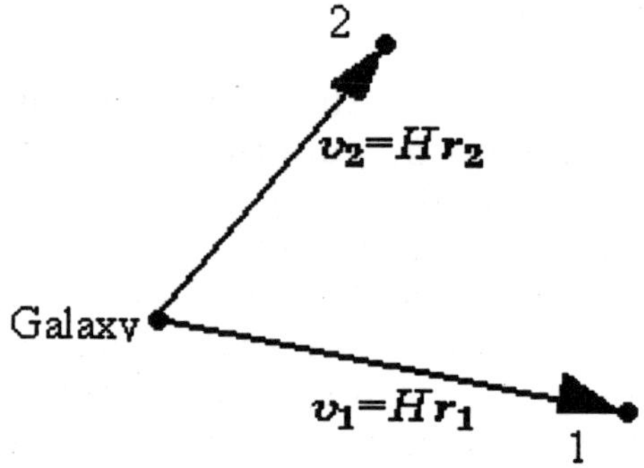

Fig.9. The "run away" of galaxies.

The "run away" velocity of a galaxy v was found to be proportional to its distance r according to the Hubble law

$$v = Hr \ , \tag{168}$$

where the coefficient H with the dimension of the inverse time is called the Hubble constant.

Though we observe the Universe, in which galaxies run away from us, it may be easily shown, that it does not mean the preferential position of our Galaxy in the world.

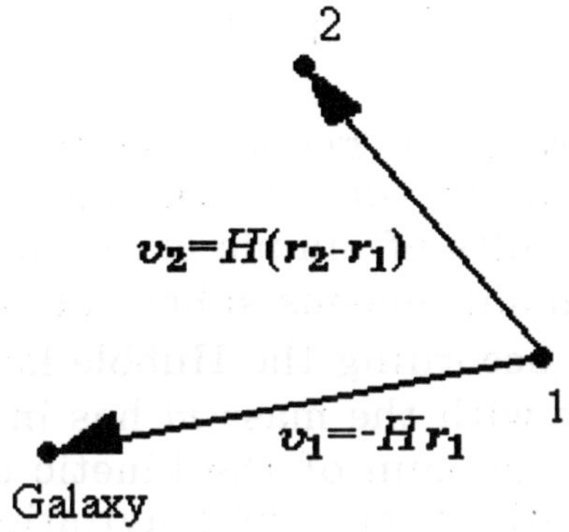

Fig.10. From the distant galaxy 1 the same picture of galactic "run away" is observed as from our Galaxy.

Simple consideration of vector algebra (see Fig.10) for the observer on the galaxy at the distance r, shows, that our Galaxy is observed by him as running away from his galaxy with the velocity, proportional to the distance, according to the Hubble law.

So, the phenomenon of the "run away" of galaxies is considered as the argument in favour of the model of expanding homogeneous and isotropic Universe. The Hubble constant H is the second important cosmological parameter of this model.

The Hubble constant is constant in space, not in time. Having the dimension of the inverse time, it defines the time scale of cosmological expansion.

The quantity, proportional to the inverse Hubble constant $1/H$, determines the time from the beginning of expansion t_U, called the age of the Universe. For the matter-dominated Universe, the age of the Universe is given by

$$t_U = \frac{2}{3} \cdot \left(\frac{1}{H}\right). \tag{169}$$

Followed from the general relativity, the model of expanding Universe in some cases finds exact correspondence with the results of simple classical consideration of an isotropic and homogeneous sphere of radius R and density ρ, expanding according the Hubble law.

A test body with the mass m has in this case the total energy, being the sum of the kinetic energy of expansion and the potential energy of gravitation:

$$E_{tot} = E_{kin} + E_{pot} = \frac{mv^2}{2} + \left(-\frac{GmM}{R}\right) \tag{170}$$

With the account for

$$v = \dot{R}$$

one finds, that the total energy is equal to

$$E_{tot} = \frac{4\pi GmR^2}{3}\left(\frac{3v^2}{8\pi GR^2} - \rho\right) = \frac{4\pi GmR^2}{3}\left(\frac{3H^2}{8\pi G} - \rho\right) \tag{171}$$

so that the sign of the total energy, determining the future regime of expansion, is defined by the relationship between the density ρ and the combination

$$\rho_c = \frac{3H^2}{8\pi G} \tag{172}$$

having the dimension of density and called the critical density.

If the cosmological density exceeds the critical density,

$$\rho > \rho_c \,, \tag{173}$$

Then the total energy is negative and the expansion should inevitably changed by contraction.

For the cosmological density, smaller than the critical one,

$$\rho < \rho_c \,, \tag{174}$$

expansion is going on forever.

In the first case, the model corresponds to the closed world, in the second — to the open Universe. The Universe with the density, exactly equal to the critical one, is called flat.

General relativity relates closed, flat and open cosmological models to the models with positive,

$$k = +1 \,, \tag{175}$$

zero,

$$k = 0 \,, \tag{176}$$

and negative,

$$k = -1 \,, \tag{177}$$

space-time curvature k, correspondingly.

In the nonstationary Universe, all distances change proportionally to the scale factor $a(t)$.

In the closed Universe, the scale factor has the meaning of the radius of the world.

In the flat and open cosmological models, the scale factor simply defines the time variation of the scales.

The comoving volume is proportional to $a(t)^3$, so that the density of matter, being inversely proportional to the comoving volume, decreases in the course of expansion as

$$\rho_b \propto a(t)^{-3}. \tag{178}$$

The time evolution of the scale factor follows from Einstein equations. In the case of dominance of matter in the Universe with

$$k = 0$$

equations of time evolution have the form:

$$\left(\frac{\dot{a}}{a}\right)^2 = \frac{8\pi G}{3}\rho\,, \tag{179}$$

$$\frac{\ddot{a}}{a} = -\frac{4\pi G}{3}\rho\,. \tag{180}$$

Solution of these equations gives the time dependence of the scale factor at the matter-dominant stage

$$a(t) \propto t^{2/3}\,. \tag{181}$$

One can easily find the correspondence between this exact form of equations, following from the general relativity, and the simple picture of expanding homogeneous and isotropic sphere, discussed above on the basis of Newtonian physics.

Taking into account that

$$\left(\frac{\dot{a}}{a}\right) = H \;, \tag{182}$$

one can deduce the first equation from the definition of the critical density, setting

$$\rho = \rho_{cr} \;. \tag{183}$$

The second equation follows from the Newtonian equation for the radial motion of the test body:

$$m\ddot{\mathbf{R}} = -\frac{GmM}{R^3}\mathbf{R} \;. \tag{184}$$

Astronomical measurements of both the cosmological density and the Hubble constant contain ambiguities, making at present impossible to give definite experimental proof for the choice of the cosmological model.

Indeed, the cosmological density in Einstein equations, in general, corresponds to any form of matter, present in the modern Universe.

Its estimation based on the luminosity of galaxies gives only the lower limit on the total density, which corresponds to the contribution of the visible matter, containing electrons and atomic nuclei. Atomic nuclei, i.e. baryons, dominate in this contribution.

Therefore, the estimations of luminous matter density give the lower limit for the baryon density in the Universe. One should treat it only as the lower limit, because the nonluminous baryon matter can exist, for which this estimation is not valid.

Dynamical effects offer more universal way to estimation of the cosmological density independent on its nature. The estimations based on gravitational lens effects can also be referred to such universal methods.

These estimations give the density, exceeding the density of luminous matter by more than one order of the magnitude. One should take into account the existence of the dark matter in the Universe. The elucidation of its physical nature is one of the most important problems of the cosmoparticle physics.

The value of the critical density is defined by the value of the Hubble constant.

The complications arising in experimental determination of H, are related to the proper account for systematic errors in the measurement of distances to galaxies and their velocities.

Nearby galaxies, distances to which can be estimated more accurately, can not represent the proper Hubble flow, since in the measured velocities dominate the effects of their motion in the local supercluster of galaxies.

"Run away" velocities of distant galaxies are more representative for the Hubble flow, but in measuring the distances to them one should use complicated procedure.

The change in the measured value of H by almost an order of magnitude, occured beginning from the early thirties up to the present time, is an illustration of the above mentioned complications.

Moreover, even now the values, measured by different groups, differ by the factor of 2.

The measured values are given with rather small statistical errors, so one should ascribe the existing difference in these values to systematical trends. The modern value of the Hubble constant lies in the interval

$$H = (50 \div 100)\frac{\text{km}}{\text{s} \cdot \text{Mpc}} . \tag{185}$$

Further we shall take for the definiteness the value

$$H = 50\frac{\text{km}}{\text{s} \cdot \text{Mpc}} . \tag{186}$$

The critical density for the chosen value of H is

$$\rho_{cr} = 5 \cdot 10^{-30} \frac{g}{cm^3} . \tag{187}$$

The dimensionless cosmological density Ω, widely used in cosmology, is

$$\Omega = \frac{\rho}{\rho_c} . \tag{188}$$

With the account for the existence of the dark nonluminous matter, one can define the density of the luminous matter to be of the order of

$$\Omega_{lum} = \frac{\rho_{lum}}{\rho_c} \approx 0.01 . \tag{189}$$

With the account for the dark nonbaryonic matter, one can define the relative contribution of baryons into the total density as

$$\Omega_b = \frac{\rho_b}{\rho_c} . \tag{190}$$

The comparison of the observed abundance of light elements with the predictions of the theory of Big Bang nucleosynthesis (see below and Chapter 5) is widely used for the estimation of the total baryon density, which can be contained in the modern Universe both in the luminous and nonluminous objects. Such estimations give

$$\Omega_b \approx 0.1 . \tag{191}$$

1.2. The Black Body Radiation Trace of the Big Bang

Tracing back the history of the expanding Universe, one inevitably comes to the conclusion, that the physical conditions in the Universe should have changed with time.

The modern picture of the Universe, full of galaxies and stars, should have appear as a result of the evolution of expanding dense nearly homogeneous and isotropic plasma.

Such plasma might have been in the state of cold degenerated matter, as it is now in white dwarfs and neutron stars.

The alternative idea of G.Gamov, that the primeval plasma was hot, found for more than three decades no observational confirmation, because it predicted the existence of relic thermal electromagnetic background radiation, which should be conserved in this model from the early hot stage of cosmological evolution.

The discovery in 1966 by Penzias and Wilson of the thermal electromagnetic radiation background favoured the model of hot expanding Universe.

The existence of the black body radiation background with the Planck intensity distribution $F_{em}(\nu)$, given by

$$F_{em}(\nu) = \frac{2h\nu^3}{c^2} \cdot \frac{1}{\exp\left(\frac{h\nu}{kT}\right) - 1}\,, \tag{192}$$

where the temperature

$$T = 2.7\ \mathrm{K}\,, \tag{193}$$

does not give any sizeable contribution to the modern cosmological density.

The electromagnetic background density, defined as

$$\rho_\gamma = \frac{\varepsilon_\gamma}{c^2} , \qquad (194)$$

where ε_γ is the black body background energy density, given by

$$\varepsilon_\gamma = \sigma T^4 \qquad (195)$$

where the Stephan-Boltzmann constant

$$\sigma = 7.57 \cdot 10^{-15} \frac{\text{erg}}{\text{cm}^3 \cdot \text{K}^4} , \qquad (196)$$

is in the modern Universe by about three orders of magnitude smaller, than the density of luminous matter, and plays no role in the dynamics of the modern cosmological expansion.

The observed remarkable isotropy of the thermal background, proven at the level of 10^{-5} , gives the strongest confirmation of the homogeneity and isotropy of the Universe.

The quantity

$$s = \frac{\sigma T^3}{k\rho} , \qquad (197)$$

where σ is the Stephan-Boltzmann constant and k is the Boltzmann constant, determines the specific entropy of the modern Universe, whereas the dimensionless ratio of photon and baryon number densities defines the entropy per baryon, equal to

$$S = \frac{n_\gamma}{n_b} = 1.3 \cdot 10^8 \Omega_b^{-1} . \qquad (198)$$

Here the number density of photons at the temperature T is equal to

$$n_\gamma \approx 20\left(\frac{T}{1K}\right)^3 \text{cm}^{-3} \tag{199}$$

that corresponds to

$$n_\gamma = 400 \text{ cm}^{-3} \tag{200}$$

at

$$T = 2.7 \text{ K},$$

and the number density of baryons is expressed in terms of the relative baryon density in the form

$$n_b = \frac{\rho_b}{m_N} = \frac{\rho_{cr}}{m_N}\Omega_b = 3 \cdot 10^{-6}\Omega_b \text{ cm}^{-3}, \tag{201}$$

where

$$m_N = 1.6 \cdot 10^{-24} \text{ g}$$

is the mass of nucleon, and the critical density is equal to

$$\rho_{cr} = 5 \cdot 10^{-30} \frac{\text{g}}{\text{cm}^3}$$

for the accepted above value of the Hubble constant

$$H = 50 \frac{\text{km}}{\text{s} \cdot \text{Mpc}}.$$

The quantity, inverse to S, is the ratio of baryon to photon number densities

$$r_b = S^{-1} = \frac{n_b}{n_\gamma} = 0.75 \cdot 10^{-8} \Omega_b \ . \tag{202}$$

It is considered as the third important cosmological parameter, being the measure of the baryon asymmetry of the Universe.

The high specific entropy of the modern Universe reflects the existence of the hot period at earlier stages of the cosmological evolution.

Moreover, in the early Universe the role of thermal background radiation turns out to be dominant in the dynamics of expansion.

The reason lies both in the high photon to baryon ratio and in the different dependence of the matter and radiation density on the scale factor.

The density of matter ρ_m is inversely proportional to the cube of the scale factor, whereas in the case of radiation density one should also take into account the red shift in the average energy of photons.

The frequency of a photon, emitted in the early Universe, experiences the red shift z, defined as

$$z = \frac{\nu_{em}}{\nu_{abs}} - 1, \tag{203}$$

where ν_{em} and ν_{abs} are the frequencies of emitted and absorbed photon, respectively.

Similarly the black body spectrum is red shifted in the course of expansion, so that the radiation temperature $T(t)$ at some earlier time t is related to the modern temperature T_0 as

$$T(t) = \left(1 + z(t)\right)T_0 = \frac{a(t = t_U)}{a(t)} T_0 \ , \tag{204}$$

where the red shift $z(t)$ is determined by the ratio of the scale factors, taken now at

$$t = t_U$$

and at the moment t.

With the account for the temperature decrease in the course of expansion, one obtains

$$\rho_\gamma \propto \varepsilon_\gamma(t) \propto T(t)^4 \propto a(t)^{-4}. \tag{205}$$

Going back in time (to large red shifts and respectively small scale factors) we come inevitably to the period of radiation dominance, when the energy density of radiation dominated in the total cosmological density and in the dynamics of cosmological expansion.

At radiation-dominant stage, the density of matter turns out to be negligible, as compared to the energy density of radiation, i.e.

$$\rho_\gamma >> \rho_m \ . \tag{206}$$

Thermodynamics of the black body radiation defines the radiation pressure

$$p = \tfrac{1}{3}\varepsilon_\gamma = \tfrac{1}{3}\rho_\gamma c^2 \ . \tag{207}$$

Hence, at radiation-dominant stage the equation of state of the Universe is

$$p = \tfrac{1}{3}\varepsilon \,, \tag{208}$$

and one should take into account the pressure in the equations of expansion.

At radiation-dominant stage the Einstein equations have the form:

$$\left(\frac{\dot{a}}{a}\right)^2 = \frac{8\pi G}{3}\rho \,, \tag{209}$$

$$\frac{\ddot{a}}{a} = -\frac{4\pi G}{3}\left(\rho + 3p/c^2\right) . \tag{210}$$

With the account for the relationship between pressure and density, given by the equation of state, one finds from these equations the time dependence of the scale factor at the radiation-dominant stage as

$$a(t) \propto t^{1/2} \,. \tag{211}$$

Note, that in the Einstein equations for the matter-dominant stage one neglects the pressure, that corresponds to the cosmological equation of state

$$p = 0 \,. \tag{212}$$

Of course, it does not mean, that the matter has no pressure. The dust-like equation of state of the Universe (212) means, that the nonzero pressure of nonrelativistic matter is of no cosmological meaning, since the pressure is negligible in the equations of expansion, as compared to the energy density of the matter.

Indeed, consider the ideal nonrelativistic gas of particles with the mass m and the number density n at the temperature

$$T << mc^2 \ . \tag{213}$$

Its equation of state has the form

$$p = \tfrac{3}{2} nT \ . \tag{214}$$

Taking into account that the number density n is related to the matter density as

$$n = \frac{\rho}{m} \ , \tag{215}$$

and the temperature T is related to the thermal velocity

$$v_T << c \tag{216}$$

as

$$T = \frac{mv_T^2}{2} \tag{217}$$

one obtains

$$p = \frac{3}{2} nT = \frac{3}{2} \frac{\rho}{m} \frac{mv_T^2}{2} = \frac{3}{4} \rho v_T^2 << \rho c^2 = \varepsilon \ . \tag{218}$$

1.3. Physical Cosmology of Old Big Bang Scenario

The simplest physically reasonable approach to treat the history of expanding hot Universe was related to the physically self-consistent picture of cosmological evolution of radiation and baryon matter, originated from the Big Bang.

In such picture, only the known types of elementary particles are considered in the modern Universe, and their early history is determined by thermodynamics applied to the expanding hot Universe.

On the basis of the known laws of atomic and nuclear physics, one may easily prove, that the baryon matter in the early Universe was in the state of plasma in equilibrium with radiation. In the early Universe, the temperature of the thermal electromagnetic background radiation was equal to the temperature of all particles in equilibrium.

The expanding Universe can be treated as the closed system with thermodynamic parameters varying due to expansion.

For the two-component mixture of matter and radiation, one can easily check that the expansion occurs adiabatically. In the early Universe the rate of expansion

$$\Gamma \sim t^{-1} , \tag{219}$$

where t is the cosmological time, much slower, than the rate of processes, establishing the equilibrium between matter and radiation.

It means, that on the radiation-dominant (RD) stage the condition

$$n(\sigma v)t \gg 1 \tag{220}$$

is valid, where n is the number density of electrons and nuclei, and the product (σv) of the cross section σ and the relative velocity v is the rate of their interaction with radiation. Hence, the thermal equilibrium is established on the RD stage.

One could follow the thermodynamic equilibrium conditions up to the earliest stages of the expansion, when the formal cosmological solution of the general relativity leads to singularity.

However, there exists the physical constraint on the classical relativistic cosmological solution at time interval

$$t \sim t_{Pl} \ . \tag{221}$$

The Planck time

$$t_{Pl} = \left(\frac{G\hbar}{c^5} \right)^{1/2} = 5 \cdot 10^{-44} \mathrm{s} \tag{222}$$

is the dimensional combination of the speed of light, Planck constant and gravitational constant, defining the time scale, at which the quantum effects should be involved into the space-time description.

At the Planck time the size of the cosmological horizon, i.e., the distance at which causal connection can be established by the light signal, is equal to the Planck length

$$l_{Pl} = \left(\frac{G\hbar}{c^3} \right)^{1/2} = 1.5 \cdot 10^{-33} \mathrm{s} \ . \tag{223}$$

The Planck time and length are the natural border for the interpolation of the classical picture of the expansion, given by the general relativity.

So one could trace the classical evolution of the expanding Universe during the period

$$t > t_{Pl} \tag{224}$$

and at the distances

$$l > l_{Pl} \tag{225}$$

only.

In the hot expanding Universe the Planck time

$$t \sim t_{Pl}$$

corresponds to the Planck temperature

$$T = T_{Pl} = m_{Pl}c^2 = 10^{19} \text{ GeV} . \tag{226}$$

Here m_{Pl} is the Planck mass, being the third important dimensional combination of the fundamental physical constants

$$m_{Pl} = \left(\frac{\hbar c}{G}\right)^{1/2} = 2 \cdot 10^{-5} \text{g} . \tag{227}$$

In the old Big Bang cosmology, the Planck scale implied the physically motivated cut in the "smooth" interpolation of the observed expansion to the beginning.

The period

$$t_{Pl} < t < 1 \text{ s} \tag{228}$$

is referred to as the very early Universe.

Physical processes in this period are determined by the high-energy physics. In the lack of the information on the particle properties at superhigh energies, the old Big Bang scenario simply postulated the relativistic particle dominance in the very early Universe and the general trend of the cosmological evolution given by

$$\rho = \frac{3}{32\pi G t^2} = 4.5 \cdot 10^{-5} \left(\frac{1\text{s}}{t}\right)^2 \frac{\text{g}}{\text{cm}^3} . \tag{229}$$

In the units

$$\hbar = c = 1 \tag{230}$$

the time dependence of the cosmological density (229) has the form

$$\rho = \frac{3}{32\pi} \cdot \frac{m_{Pl}^2}{t^2} \,. \tag{231}$$

The simplest assumption was that at very high temperatures

$$T >> m \tag{232}$$

all particles were relativistic and in the equilibrium with plasma and radiation.

Thermodynamics of the relativistic gas leads to the following relationship between the energy density of the Universe ε and its temperature T :

$$\varepsilon = \rho c^2 = \kappa\sigma T^4, \tag{233}$$

where the Stephan-Boltzmann constant is

$$\sigma = \frac{\pi^2 k^4}{15\hbar^3 c^3} = 7.57 \cdot 10^{-15} \frac{\text{erg}}{\text{cm}^3 \cdot \text{K}^4} \,, \tag{234}$$

k is the Boltzmann constant

$$k = 1.38 \cdot 10^{-16} \frac{\text{erg}}{\text{K}}$$

and κ is the effective number of species of relativistic particles with account for their statistical properties.

One finds that

$$(kT)^4 = \frac{1}{\kappa} \cdot \frac{45}{32\pi^3} \cdot \frac{\hbar^3 c^5}{Gt^2} \tag{235}$$

that results in the time dependence of the temperature

$$T = 1.5 \cdot 10^{10}\ \mathrm{K}\ \kappa^{-1/4}\left(\frac{1\ \mathrm{s}}{t}\right)^{1/2} . \tag{236}$$

For the temperature, expressed in MeV, one obtains

$$T = 1.3\ \mathrm{MeV}\ \kappa^{-1/4}\left(\frac{1\ \mathrm{s}}{t}\right)^{1/2} \tag{237}$$

and in the units

$$\hbar = c = k = 1$$

this relationship is given by

$$T = \frac{1}{\kappa^{1/4}} \cdot \left(\frac{45}{32\pi^3}\right)^{1/4}\left(\frac{m_{Pl}}{t}\right)^{1/2} . \tag{238}$$

The same relationship may be used to express the cosmological time in terms of the temperature

$$t = 2.25 \cdot 10^{20}\ \mathrm{s}\left(\frac{1\ \mathrm{K}}{T}\right)^{2} \kappa^{-1/2} \tag{239}$$

or in terms of the temperature expressed in MeV

$$t = 1.7\ \mathrm{s}\left(\frac{1\ \mathrm{MeV}}{T}\right)^{2} \kappa^{-1/2} . \tag{240}$$

In the units

$$\hbar = c = k = 1$$

one obtains

$$t = \frac{1}{\kappa^{1/2}} \cdot \left(\frac{45}{32\pi^3}\right)^{1/4} \cdot \frac{m_{Pl}}{T^2} \ . \tag{241}$$

The total number density of all relativistic species, taking for the average energy of particles

$$\langle E \rangle \sim 3kT \tag{242}$$

is given by

$$n_r = \frac{\varepsilon}{3kT} = 0.01 \cdot \kappa^{1/4} \cdot \left(\frac{c}{\hbar G}\right)^{3/4} t^{-3/2} =$$

$$= 5 \cdot 10^{31} \mathrm{cm}^{-3} \kappa^{1/4} \left(\frac{1\ \mathrm{s}}{t}\right)^{3/2} \ . \tag{243}$$

The thermodynamic equilibrium is established in any system, if the rate of processes, maintaining the equilibrium, exceeds the rate of the variation of the parameters of the system, such as its density, temperature etc.

In the expanding Universe the latter rate coincides with the rate of expansion.

If the time scale of the physical process exceeds the cosmological time scale, the system goes out of the equilibrium relative to the considered process.

The question on the going out of the equilibrium is one of the most important questions of the physical cosmology. So let us consider it in more details.

If the system contains several types of different particles, and there exist transitions between the particles of different type, the equilibrium distribution is established under the constant conditions. This distribution further does not change with time.

If the parameters of the system change more rapidly, than the most rapid process, leading to the creation of particles of given type or to their annihilation, the particles of this type go out of the equilibrium with other particles.

Consider the particular case of creation and annihilation of particle-antiparticle pairs.

The charged particle has the antiparticle. Particle and antiparticle can annihilate. Pairs of the considered particle and antiparticle can be created in the collisions of other particles. Non-pair creation or annihilation of the particles is also possible in the interactions with the other particles.

In particular, when the most rapid is the process of pair annihilation, the particles go out of the equilibrium when the characteristic time of annihilation τ exceeds the cosmological time scale t i.e. when

$$\tau > t \, . \tag{244}$$

For particles and antiparticles with the concentration

$$n_p = n_a = n \tag{245}$$

and relative velocity v the characteristic time of the process of their annihilation with the cross section σ is

$$\tau = (n\sigma v)^{-1}. \tag{246}$$

If

$$\tau > t \, ,$$

the further change of the particle concentration is determined only by the expansion of the Universe. In this case the freezing out of the considered particles takes place.

Decoupling of weakly interacting particles represents the specific case of the particle freezing out. Here τ has the meaning of the characteristic time of the particle interaction with the other particle, so that n is the concentration of other particles and (σv) is the rate of their interaction with the weakly interacting particles.

In the process of decoupling at

$$t \sim \tau \,, \tag{247}$$

the equilibrium distribution of weakly interacting particles does not change, and in the successive adiabatic expansion the gas of decoupled particles follows the equilibrium distribution until the particles become nonrelativistic at

$$T \sim m \,, \tag{248}$$

where m is the mass of these particles.

In the old Big Bang scenario, the physical motivation for the simple thermodynamically equilibrium picture of the very early Universe was based on the fact that among the known particles only nucleons, electrons, photons and neutrino are of the cosmological significance. The hundreds of all other particles, mentioned in Tables of particle properties, are unstable.

In particle physics, unstable particles (e.g. resonances) have the lifetime of the order of

$$\tau \sim \left(\frac{\hbar}{mc^2}\right) \tag{249}$$

and even the so-called "metastable" particles, being very long-living relative to the particle physics time scale

$$\tau >> \left(\frac{\hbar}{mc^2}\right), \tag{250}$$

are very short-living relative to the cosmological time scale for the particles with the mass m, that corresponds to the cosmological time, when $T \sim m$, and is given by

$$t_c \sim \left(\frac{m_{Pl}}{m}\right)\cdot\left(\frac{\hbar}{mc^2}\right). \tag{251}$$

In the cosmology it is reasonable to consider only stable known particles. The list of these particles is virtually exhausted by photons, electrons and positrons, nucleons and antinucleons, neutrino and antineutrino.

The cosmological effect of all other known particles is reduced to some slight quantitative modifications in the relativistic equation of state (208) of the very early Universe. Only exponentially large set of such particles could cause significant influence on the cosmological evolution, for instance, the existence of the maximal Hagedorn temperature, but the development of the quark model and quantum chromodynamics found no physical grounds for these possibilities.

To the end of the first second of expansion the net result of the very early Universe in the old Big Bang scenario was practically reduced to the conditions of the equilibrium for the gas of photons, electron-positron and neutrino-antineutrino pairs and to the small but fundamentally important admixture of nucleons with the concentration of the order of

$$n_b \sim 10^{-9} n_\gamma .$$

The period of the radiation-dominated stage

$$t > 1 \text{ s} \tag{252}$$

corresponds to the temperature

$$T < 1 \text{ MeV} \ , \tag{253}$$

at which natural conditions for the well studied in the laboratories nuclear and atomic processes should take place.

The evident difference in cosmological and laboratory conditions for these processes is the very low density of matter in the Big Bang Universe.

One finds from Eqs. (199), (202) and (243) that the baryon number density is

$$n_b = 3.8 \cdot 10^{23} \text{cm}^{-3} \Omega_b \left(\frac{1 \text{ s}}{t} \right)^{3/2} \tag{254}$$

that virtually corresponds to the conditions of laboratory vacuum

$$n_b < 3.8 \cdot 10^{16} \text{cm}^{-3} \tag{255}$$

at

$$t > 10^4 \text{s} \tag{256}$$

and

$$\Omega_b \sim 0.1 \ . \tag{257}$$

It explains the specifics of particle, nuclear and atomic processes in the Big Bang Universe.

During the period

$$t \sim 0.1 \div 1 \text{ s} \tag{258}$$

corresponding to the temperature

$$T \sim 3 \div 1 \text{ MeV} , \tag{259}$$

the time scale of reactions of weak interaction begin to exceed the cosmological time scale, so that neutrino decouple from other particles.

It results in the existence of the thermal neutrino background with the concentration determined by the equilibrium conditions in the period of decoupling.

At the temperature T, the relativistic gas of neutrino-antineutrino pairs has the number density

$$n_{\nu\bar{\nu}} = 2\int \frac{1}{\exp\left(\frac{pc}{kT}\right)+1} \cdot \frac{d^3 p}{(2\pi\hbar)^3} . \tag{260}$$

In the same conditions the number density of photons is determined by the Planck distribution and is given by

$$n_{\gamma} = 2\int \frac{1}{\exp\left(\frac{pc}{kT}\right)-1} \cdot \frac{d^3 p}{(2\pi\hbar)^3} . \tag{261}$$

One can find simple analytical relationship between the equilibrium number densities of neutrino-antineutrino pairs and photons, using the following mathematical trick (see Okun,1981).

The thermodynamic functions of relativistic gas of fermion-antifermion-pairs are given by the integrals over the Fermi-Dirac distribution.

The fermion number density is

$$n_f = g_f \int \frac{1}{\exp\left(\frac{pc}{kT}\right)+1} \cdot \frac{d^3 p}{(2\pi\hbar)^3} , \tag{262}$$

where g_f is the statistical weight and the fermion energy density for relativistic fermions

$$E \approx pc$$

is equal to

$$\varepsilon_f = g_f \int pc \cdot \frac{1}{\exp\left(\frac{pc}{kT}\right)+1} \cdot \frac{d^3 p}{(2\pi\hbar)^3} . \tag{263}$$

One can express n_f and ε_f in terms of the integrals over the Fermi-Dirac distribution function:

$$n_f = \frac{g_f}{2\pi^2} \cdot \left(\frac{kT}{\hbar c}\right)^3 \cdot I_{2f}\left(\frac{pc}{kT}\right) \tag{264}$$

and

$$\varepsilon_f = \frac{g_f}{2\pi^2} \cdot \left(\frac{kT}{\hbar c}\right)^3 \cdot (kT) \cdot I_{3f}\left(\frac{pc}{kT}\right) , \tag{265}$$

where the integrals I_{nf} are defined as

$$I_{nf}(x) = \int_0^\infty \frac{x^n}{\exp(x)+1} \cdot dx . \tag{266}$$

In the same manner, the corresponding quantities for relativistic Bose gas can be expressed in terms of the integrals over the Bose-Einstein distribution as

$$n_b = \frac{g_b}{2\pi^2} \cdot \left(\frac{kT}{\hbar c}\right)^3 \cdot I_{2b}\left(\frac{pc}{kT}\right) \tag{267}$$

and

$$\varepsilon_b = \frac{g_b}{2\pi^2} \cdot \left(\frac{kT}{\hbar c}\right)^3 \cdot (kT) \cdot I_{3b}\left(\frac{pc}{kT}\right), \tag{268}$$

where g_b is the statistical weight of the considered bosons and the integrals I_{nb} are given by

$$I_{nb}(x) = \int_0^\infty \frac{x^n}{\exp(x) - 1} \cdot dx . \tag{269}$$

The values of the integrals I_{nf} and I_{nb} are tabulated, but it turns out that their difference has very simple analytical form.

Indeed, consider the difference

$$I_{nb}(x) - I_{nf}(x) = \int_0^\infty \frac{x^n}{\exp(x) - 1} \cdot dx - \int_0^\infty \frac{x^n}{\exp(x) + 1} \cdot dx . \tag{270}$$

Since the integrals converge one has

$$I_{nb}(x) - I_{nf}(x) = \int_0^\infty \left(\frac{x^n}{\exp(x) - 1} - \frac{x^n}{\exp(x) + 1} \right) \cdot dx =$$

$$= \int_0^\infty \frac{2 \cdot x^n}{\exp(2x) - 1} \cdot dx = \frac{1}{2^n} \cdot \int_0^\infty \frac{(2x)^n}{\exp(2x) - 1} \cdot d(2x) =$$

$$= \frac{1}{2^n} \cdot I_{nb}(x) \tag{271}$$

and, finally, we obtain the analytical formula relating the integrals for relativistic fermions and bosons

$$I_{nf}(x) = \left(1 - \frac{1}{2^n}\right) \cdot I_{nb}(x) \ . \tag{272}$$

Taking into account this relationship one can relate the equilibrium number densities of photons and pairs of left-handed neutrinos and right-handed antineutrinos of one type

$$n_{\nu\bar{\nu}} = \left(1 - \frac{1}{2^2}\right) \cdot n_\gamma = \frac{3}{4} n_\gamma \ , \tag{273}$$

and their equilibrium energy densities

$$\varepsilon_{\nu\bar{\nu}} = \left(1 - \frac{1}{2^3}\right) \cdot \varepsilon_\gamma = \frac{7}{8} \varepsilon_\gamma \ . \tag{274}$$

In the similar relationship between photons and relativistic electron-positron pairs one should take into account that the number of degrees of freedom in the gas of photons (left- and right- handed states) is two times smaller, than this number for the gas of electron-positron pairs (left- and right handed electrons and positrons).

So the relativistic electron-positron pair equilibrium number density is given by

$$n_{e^-e^+} = 2 \cdot \left(1 - \frac{1}{2^2}\right) \cdot n_\gamma = \frac{3}{2} n_\gamma \tag{275}$$

and their equilibrium energy density is equal to

$$\varepsilon_{e^-e^+} = 2 \cdot \left(1 - \frac{1}{2^3}\right) \cdot \varepsilon_\gamma = \frac{7}{4}\varepsilon_\gamma \ . \tag{276}$$

During the period of neutrino decoupling, their number density is related to the number density of photons in the equilibrium.

After decoupling at

$$t > 10^2 \text{ s} \tag{277}$$

when in the process of expansion the temperature falls down to

$$T < 100 \text{ keV} \ , \tag{278}$$

the equilibrium electron-positron pairs annihilate, thus increasing the number density of photons without any significant influence on the neutrino density.

One can use the conservation of the specific entropy for calculation of the corresponding change in the neutrino-to-photon ratio.

At the temperature

$$T = T_1 \ , \tag{279}$$

just before neutrino decoupling, the equilibrium mixture of photons and electron- positron pairs has the specific entropy

$$S_\gamma + S_{e^-e^+} = \frac{4}{3} \cdot \frac{\sigma T_1^3 + \frac{7}{4}\sigma T_1^3}{\rho_1} \tag{280}$$

and the pairs of neutrino and antineutrino of the given type have the specific entropy

$$S_{\nu\bar{\nu}} = \frac{4}{3} \cdot \frac{7}{8} \frac{\sigma T_1^3}{\rho_1} \; . \tag{281}$$

At some lower temperature

$$T = T_0 \; , \tag{282}$$

after the neutrino decoupling and the electron-positron annihilation, the specific entropy of photons is equal to

$$S_\gamma = \frac{4}{3} \cdot \frac{\sigma T_{0\gamma}^3}{\rho_0} = S_\gamma + S_{e^-e^+} = \frac{4}{3} \cdot \frac{11}{4} \cdot \frac{\sigma T_1^3}{\rho_1} \tag{283}$$

and the neutrino entropy is

$$S_{\nu\bar{\nu}} = \frac{4}{3} \cdot \frac{7}{8} \frac{\sigma T_{0\nu}^3}{\rho_0} = \frac{4}{3} \cdot \frac{7}{8} \frac{\sigma T_1^3}{\rho_1} \; , \tag{284}$$

where we separately used the conservation of entropy in the photon-electron-positron mixture and in the decoupled gas of neutrino-antineutrino pairs in the course of the adiabatic expansion.

So, the neutrino-to-photon ratio decreases and one obtains finally the Big Bang prediction for the modern concentration of primordial gas of left-handed neutrino and right-handed antineutrino of one type to be equal to

$$n_{\nu\bar{\nu}} = \frac{3}{4} \cdot \frac{4}{11} \cdot n_\gamma = \frac{3}{11} n_\gamma \; , \tag{285}$$

where the factor 4/11 accounts for the photon number density increase due to electron-positron annihilation.

Another important consequence of neutrino decoupling when the weak reaction time scale exceeds the cosmological time scale is that the beta processes, in which protons transform into neutrons and neutrons transform into protons,

$$\begin{aligned}
&\nu_e n \rightarrow ep \ , \\
&\overline{\nu_e} p \rightarrow e^+ n \ , \\
&ep \rightarrow \nu_e n \ , \\
&e^+ n \rightarrow \overline{\nu_e} p \ ,
\end{aligned} \tag{286}$$

went out of the equilibrium.

At the temperature

$$kT < \Delta m_{np} c^2 \ , \tag{287}$$

where the mass difference of neutron and proton

$$\Delta m_{np} c^2 = m_n c^2 - m_p c^2 = 1.28 \text{ MeV} \ , \tag{288}$$

these reactions maintained the equilibrium ratio of neutron and proton concentrations, given by

$$\frac{n}{p} = \exp\left(-\frac{\Delta m_{np} c^2}{kT} \right) . \tag{289}$$

When beta processes went out of the equilibrium, this ratio was frozen out and should not change until

$$t \sim \tau_n \sim 10^3 \text{ s} \tag{290}$$

when neutrons decay.

However, most of neutrons do not decay, because at

$$t \sim 10^2 \text{ s} \tag{291}$$

corresponding to the temperature

$$T \sim T_D \sim 10^2 \text{ keV} \tag{292}$$

they react with protons and form deuterium in the process

$$n + p \to D + \gamma \, . \tag{293}$$

At higher temperature

$$T > T_D \, , \tag{294}$$

the inverse reaction of photodesintegration

$$\gamma + D \to n + p \tag{295}$$

destroys all produced deuterium, but at

$$T < T_D \tag{296}$$

photodesintegration is not effective, and the produced deuterium participates successive thermonuclear transformations

$$\begin{aligned} D + D \to T + p \quad &\Rightarrow \quad T + D \to {}^4He + n \, , \\ D + D \to {}^3He + n \quad &\Rightarrow \quad {}^3He + n \to T + p \, . \end{aligned} \tag{297}$$

As a result of these transformations, the so-called "primordial chemical composition" is formed. Its main components are hydrogen and helium-4, with weight fractions about 75% and 25%, respectively. There is small ($\sim 10^{-5}$ - 10^{-4}) fraction of deuterium and helium-3 and much more smaller amount of lithium-7.

The absence of stable nuclear states with the atomic number

$$A = 5$$

puts practically unachievable threshold for further nuclear transformations in Big Bang nucleosynthesis.

The primordial light element abundance play very important role in our further discussion.

At the temperature

$$T_{rec} \sim 4 \cdot 10^3 \text{ K} , \tag{298}$$

corresponding to the period of the recombination

$$t_{rec} \sim 10^{13} \text{ s} , \tag{299}$$

when electrons and protons form neutral atoms of hydrogen, photons cease to interact with the matter.

At the temperature

$$T_m \sim \frac{1}{3} \cdot \frac{n_b}{n_\gamma} \cdot m_p \sim \frac{1}{3} \cdot r_b \cdot m_p , \tag{300}$$

the density of the baryonic matter

$$r_b = m_p n_b \tag{301}$$

exceeds the density of radiation

$$\rho_\gamma \sim 3 T_\gamma n_\gamma . \tag{302}$$

The moment t_m when the matter domination takes place,

$$\rho_b > \rho_\gamma \tag{303}$$

is determined by the value

$$r_b = \frac{n_b}{n_\gamma} = 7.5 \cdot 10^{-9}\Omega_b \tag{304}$$

and is rather close to the period of recombination.

Thus, at the time

$$t > t_{RD} \sim t_m \tag{305}$$

the modern matter-dominated stage of the cosmological evolution begins.

The gravitational instability of neutral gas evolves on this stage. Small matter density fluctuations grow, forming the observed structure of inhomogeneities, to which belong galaxies, their clusters and superclusters, stellar clusters, stars, etc.

The growth of density fluctuations lasts considerable time. Long after the moment t_{RD} matter expands almost homogeneously, and the growth of inhomogeneities is reduced to the growth of density contrasts between different regions.

Only relatively not long ago, at

$$t_s \geq 10^{16} \text{ s}, \tag{306}$$

there were formed the first inhomogeneities, separated from the general cosmological expansion.

Their successive evolution resulted in galaxy formation. Thermonuclear reactions in stars, being formed at this stage, lead to heavy element formation. Star explosions enrich interstellar medium by heavy elements. So, as a result of stellar nucleosynthesis, the chemical elements, which are important for the existence of the bodies around us, as well as for our own existence, formed.

Based on the physical laws, well proven in laboratories, the old Big Bang scenario is physically self-consistent. Moreover, its main predictions for decades found qualitative confirmation in observations. The discovery of black body background radiation, measurements of the light element abundance seemed to confirm this picture.

However, the development of particle theory and the attempts to apply its prediction to cosmology initiated the critical analysis of the old Big Bang model.

The process started in the late 70-th with the optimistic hope to use GUT models for the explanation of the observed baryon asymmetry of the Universe. The mechanisms of baryosynthesis, based on the simple $SU(5)$ GUT model, were expected to find experimental proof in the searches for proton decay.

In the early 80-th, the experimental claims on the nonzero electron neutrino mass stimulated the development of the theory of cosmological structure formation. These trends seemed to find physical grounds in the measurements of the neutrino mass.

And though neither the rapid proton decay, nor the nonzero electron neutrino mass found experimental confirmation, the theoretical analysis of the cosmological consequences of particle theory has uncovered the internal inconsistency of the old Big Bang model.

The set of troubles, unrecoverable in the framework of the old cosmological paradigm, was clarified in this analysis, and this opened the way to the new paradigm — the inflationary models with baryosynthesis and nonbaryonic dark matter.

The physical grounds of these models lose the possibility of the direct experimental tests, and need special methods for their probe. Development of physically self-consistent cosmological models and complex test for their physical grounds is the main task of cosmoparticle physics.

2. Inflational Cosmology with Baryosynthesis and Dark Matter

It followed from the old Big Bang Scenario that the early cosmological evolution should have taken place at superhigh temperatures. This conjecture made the conditions in the very early Universe dependent on the laws of superhigh-energy physics.

Our modern knowledge of these laws is based on the predictions of particle theory. Thus, the development of particle theory implies natural influence on the picture of the very early Universe. In principle, all cosmologically significant phenomena, predicted by particle theory, can find their place in the history of the Universe.

Since the true theory of superhigh-energy physics is not developed yet, we can only make some reasonable guess on the phenomena that should exist in cosmology.

The development of particle theory and its application to the Big Bang model found at least three phenomena, which are widely accepted as being necessary in the Big Bang scenario both on esthetical and practical reasons.

It led to the change of the cosmological paradigm. The modern Big Bang model is now generally referred to as the inflational scenario with baryosynthesis and dark matter.

The development of this scenario offers the exciting possibility to explain the main cosmological parameters by physical mechanisms.

So, the choice of open, closed or flat cosmological model is related to the mechanism of inflation. In the simplest cases, this mechanism leads to the prediction of a flat Universe with

$$\Omega = 1 \ .$$

The observed baryon-to-photon ratio is considered as a result of baryosynthesis, defining the modern baryon density.

The difference between the modern total and baryon density is ascribed to the nonbaryonic dark matter. In the simplest case, the dark matter density is determined by the mass and the concentration of frozen out weakly interacting particles.

The attractive idea to determine cosmological parameters through the parameters of particles and fields is not the only advantage of the new paradigm. Inflation, baryosynthesis and dark matter recover internal inconsistencies of Big Bang cosmology.

However, the price is the unknown physical grounds and the wide variety of possible realisations on the base of different approaches. Our hope to remove this ambiguities is related to cosmoparticle physics.

Assuming that the inflational baryon-asymmetrical cosmology with the nonbaryonic dark matter is more realistic, than Gamow's original Big Bang scenario, one faces the problem of observational evidences, specifying the choice of the inflational model, the mechanism of baryosynthesis and with the proper form of nonbaryonic dark matter. We refer to this problem as to the problem of hidden parameters of the modern cosmology.

In cosmoarheology, the hidden parameters specify the properties of the Universe as the natural accelerator. We should invoke the details of the considered cosmological model to make our statements on the possible properties of hypothetical particles and fields self-consistent.

Of course, we may ignore the hidden parameters, but some reasonable measure of the uncertainty of cosmological statements should be introduced in this case.

Let us consider now the esthetical and practical reasons for the extension of the old Big Bang model, resulting in the modern cosmological paradigm. Taking into account the existence of numerous reviews and books (see, c.f., the books by A. Linde, 1990; Kolb, Turner, 1990) we can limit our discussion to brief review of these ideas, paying more attention to their physical meaning and nontrivial ways for their test.

2.1. Magnetic Monopoles in the Old Big Bang Model

The existence of isolated magnetic poles – magnetic monopoles – was discussed at almost every stage of the development of the fundamental theory of magnetism.

C.O.Coulomb developed his magnetostatics on the basis of exact symmetry between electric and magnetic forces. He suggested that the long-range Coulomb forces should exist between isolated magnetic charges.

This symmetry is violated in classical electrodynamics, and magnetic properties are ascribed to the motion of charged particles. But on the next step of development – in the framework of quantum mechanics – the symmetry between electricity and magnetism was considered again.

P.A.M.Dirac (1931,1948) pointed out that the existence of magnetic monopoles with the magnetic charge

$$g = \frac{\hbar c}{2e} \qquad (307)$$

can explain the quantization of electric charge.

Similar ideas were put forward by J. Schwinger and D.Mandelstam in the context of quantum field theory. In this case, the elementary magnetic charge was found to be two times larger than the Dirac's quantum-mechanical value (307).

In 1974, t'Hooft and Polyakov showed that magnetic monopole solutions can appear in the gauge models of electroweak interaction. However, such solution required the compact group of the unifying symmetry, in which the electromagnetic $U(1)$ symmetry is embedded. That is not the case in the standard model of electroweak interaction.

The common feature of all these approaches to the problem of magnetic monopoles was their exotic position in the corresponding theoretical framework. Magnetic monopoles were possible, appealing, but not inevitable elements of electromagnetic theory.

Experiments, resulting in increasingly restrictive upper limits on monopole concentration in the terrestrial and lunar matter and in cosmic rays, as well as more and more severe lower limits on the mass of monopole in accelerator research, seem to disfavour the idea on the existence of magnetic monopoles.

The status of magnetic monopoles has changed drastically in the framework of GUT models. In these models the $U(1)$ symmetry of electromagnetic interaction was embedded in the compact group of GUT symmetry, making the existence of magnetic monopoles with the Dirac magnetic charge (307) an inevitable topological consequence of the GUT symmetry breaking. The predicted mass of such monopoles is of order of

$$m \sim \frac{\Lambda}{e} , \tag{308}$$

where Λ is the scale of GUT symmetry breaking.

For

$$\Lambda \sim 10^{15} \text{ GeV} \tag{309}$$

the monopole mass is

$$m \sim 10^{16} \text{ GeV} \tag{310}$$

what explains the negative results of accelerator searches for monopoles.

Moreover, no possible artificial or natural source of energetic particles in the known Universe can provide the condition for production of particles with such a huge mass. Only the extrapolation to the earliest stages of the cosmological expansion seem to find natural conditions for magnetic monopole production.

According to the old Big Bang model any species of particles with the mass m should be present in the equilibrium at the temperature

$$T > m, \tag{311}$$

provided that the particle interaction is sufficiently strong to maintain the equilibrium. It means that the particle reaction rate (σv) is large enough to satisfy the condition (compare with the Eq. (220))

$$n(T)(\sigma v) > \Gamma, \tag{312}$$

where $n(T)$ is the particle number density at the temperature T and (compare with the Eqs. (219) and (241))

$$\Gamma \sim \frac{T^2}{m_{Pl}}$$

is the rate of the cosmological expansion.

When the temperature falls down to

$$T < m, \tag{313}$$

and the equilibrium condition (312) is valid, the particle number density in the equilibrium is

$$n(T) = \left(\frac{2}{\pi^3}\right)^{1/2} (mT)^{3/2} \exp\left(-\frac{m}{T}\right). \tag{314}$$

At the temperature

$$T \sim T_f \,, \tag{315}$$

when the rate of particle interaction is equal to the expansion rate

$$n(T_f)(\sigma v) = \Gamma(T_f) \,, \tag{316}$$

the particles go out of equilibrium. Their concentration deviates from the equilibrium concentration (314). As a result, the particles freeze out and their relative concentration does not change any more.

The idea of particle freezing out is based on the nonstationarity of the Universe.

Consider, for example, some stable charged particles with their antiparticles being in the equilibrium at high temperatures. Their equilibrium concentration is maintained by the processes of pair creation and annihilation.

When the temperature falls down in the course of expansion, the pair production is suppressed and the equilibrium concentration falls down exponentially following the law (314). The annihilation rate also decreases. When the rate of expansion exceeds the annihilation rate particles and antiparticles should freeze out.

On the basis of this picture, one may easily estimate the primordial concentration of any species of stable particle.

In the case of monopoles, the conservation of magnetic charge implies their absolute stability. On the same reason, the production of monopole should be accompanied by the production of antimonopole with the opposite magnetic charge.

For monopoles in the early Universe, one can apply the above arguments, assuming that at high temperatures the monopole-antimonopole pairs were in the equilibrium. With the decrease of the temperature, the rate of expansion should exceed the rate of monopole-antimonopole annihilation, that results in the frozen out concentration of primordial magnetic monopoles.

Assuming the monopoles to be point-like particles with the mass m and the magnetic charge g , Domogatsky and Zheleznykh (1969) calculated their Big Bang primordial abundance and put constraints on these parameters from the observational constraints on the concentration of monopoles in the Universe.

However, in the case of the GUT magnetic monopoles, the calculation of primordial monopole abundance should be seriously modified. The reason is that GUT monopoles are not the ordinary particles, i.e. they are not quanta of corresponding quantum fields, but appear as topological defects as a result of the GUT symmetry breaking. Let us consider on a simple example of a triplet Higgs field the production of monopoles in the course of GUT phase transition (Kibble, 1976, Khlopov, 1988).

Consider a triplet scalar field

$$\Phi = \{\Phi_1, \Phi_2, \Phi_3\} \tag{317}$$

with the Higgs potential, depending on

$$|\Phi|^2 = \Phi_1^2 + \Phi_2^2 + \Phi_3^2 \tag{318}$$

and achieving the minimum at zero temperature for the nonzero vacuum expectation value of the field

$$\left\langle |\Phi|^2 \right\rangle = \Lambda^2. \tag{319}$$

At high temperature, thermal fluctuations change the form of this potential, so that at

$$T > \Lambda \tag{320}$$

its minimum corresponds to the vacuum expectation value

$$\left\langle |\Phi|^2 \right\rangle = 0 \,, \tag{321}$$

and the symmetry is restored. In the successive expansion, the temperature sinks below the critical value

$$T \sim T_{cr} \sim \Lambda \,, \tag{322}$$

at which the transition takes place to the phase with the vacuum expectation value, given by Eq. (319) and corresponding to the broken symmetry.

But the asymmetric vacuum is degenerate, since any set of fields satisfying the condition

$$\left\langle |\Phi|^2 \right\rangle = \Phi_1^2 + \Phi_2^2 + \Phi_3^2 = \Lambda^2 \tag{323}$$

corresponds to the minimum of the Higgs potential. In the field space, the minimum is achieved on the sphere

$$\Phi_1^2 + \Phi_2^2 + \Phi_3^2 = \Lambda^2 \,. \tag{324}$$

The co-ordinates on this sphere may be defined by two variables, φ and θ, being polar and azimuth angles. Any value of φ and θ corresponds to the minimum of the Higgs potential.

In the course of the phase transition, the vacuum rearrangement takes place in the ordinary space. In each spatial point the ground state, being the physical vacuum, moves to the asymmetric phase, defined by Eqs. (319) and (323). However, the values of φ and θ are arbitrary and their values may be different in the different spatial points.

To minimise the effect of gradient terms in the Lagrangian of the Higgs field

$$L = \left(\partial_\mu \Phi\right)^2 - V\left(\left|\Phi\right|^2, T\right), \tag{325}$$

the values of $\varphi(\mathbf{x})$ and $\theta(\mathbf{x})$ should be slow varying over the scales smaller than the correlation length

$$r_c \sim \frac{1}{e\Lambda}. \tag{326}$$

But along the paths, much longer than r_c , large variations of $\varphi(\mathbf{x})$ and $\theta(\mathbf{x})$ are possible, since these angle variables are not correlated. If they change on 2π along the closed path, the contraction of such a path leads to a singular point with ambiguous value of φ and θ .

To provide the continuity of the triplet field Φ , one should put zero vacuum expectation value in this singular point and impose significant deviation of the Higgs potential from its minimal value within the distance of the order of r_c from this point. Thus significant energy density is concentrated around this singular point, resulting in the formation of a massive object — monopole.

The object looks like a hedgehog. The gradients of $\varphi(\mathbf{x})$ and $\theta(\mathbf{x})$ determine the strength of its magnetic field, corresponding to magnetic charge $+g$. It may be easily shown that the same argument leads to the formation of a neighbouring hedgehog with the opposite magnetic charge, $-g$. So the non-local creation of monopole–antimonopole pair takes place.

Note that the variation of $\varphi(\mathbf{x})$ and $\theta(\mathbf{x})$ corresponds to the spatial variation in the definition of electrically charged and neutral states. "Having combed the hedgehog", i.e. having chosen in the entire space the unique definition of the charged and neutral states, one obtains the Dirac (1931,1948) magnetic monopole solution with a singular string of magnetic flux, going from the monopole to infinity.

The GUT symmetry is restored inside the GUT monopole. In particular, the baryon charge is not conserved in the monopole interior. It turns out that in the singular field of magnetic monopole the proton decay

$$p \to e^+ e^- e^+ \tag{327}$$

is induced (Rubakov, 1981; Kallan, 1982) with the cross section determined by the size of the proton as

$$\sigma \sim 10^{-28} \text{ cm}^2 . \tag{328}$$

The specific properties of GUT monopoles resulted in the nontrivial picture of their cosmological evolution in the old Big Bang model.

Before the GUT phase transition massive magnetic monopoles simply do not exist, since the symmetry is restored. They appear in the Universe after the phase transition both as point–like singularities, described above, and as pairs due to the mechanism of pair production.

The latter can take place in the phase transition of the second order. The mass of monopoles is determined by the vacuum expectation value. At the temperature smaller but near the temperature of the phase transition

$$T \to T_{cr} \sim \Lambda$$

the vacuum expectation value, being the parameter of ordering, should follow the Curie–Weiss temperature dependence. It results in the same temperature dependence for the mass of magnetic monopoles given by

$$m(T) \propto \sqrt{T_{cr}^2 - T^2} \tag{329}$$

near the critical temperature (Zeldovich, Khlopov, 1978).

However, the estimated frozen out concentration of monopoles turns out to be independent on all these details as well as on the precise value of the initial monopole concentration. Provided that it is sufficiently high, say, as high as it may be estimated for the topological mechanism of the non-local monopole production (Preskill, 1979)

$$n_m \sim \frac{1}{r_c^3} \sim e^3\Lambda^3 \tag{330}$$

the successive diffusion of monopoles to antimonopoles and their annihilation washes out the information on the initial monopole concentration (Zeldovich, Khlopov, 1978).

Following Zeldovich and Khlopov (1978) let us consider the theory of monopole–antimonopole annihilation in the early Universe.

The annihilation cross section is determined by the Coulomb attraction of magnetic charges. At the temperature T the Coulomb attraction is significant at the distances

$$r \le r_0 \equiv \frac{g^2}{T}. \tag{331}$$

If the mean free path of monopoles relative to scattering in plasma exceeds r_0 ,

$$\lambda >> r_0 \ , \tag{332}$$

The free monopole–antimonopole annihilation can be considered. In the opposite case

$$\lambda << r_0 \ , \tag{333}$$

the annihilation should be calculated in the diffusion approximation.

In any case, the initial concentration of monopoles is by few orders of the magnitude smaller than the concentration of relativistic plasma particles, and the mean free path of monopoles is defined by their scattering on the charged particles and is given by

$$\lambda = \frac{1}{n_{ch}\sigma} . \tag{334}$$

The cross section of the monopole multiple scattering on 90° is given by

$$\sigma \sim \frac{(ge)^2}{Tm} \tag{335}$$

(Zeldovich, Khlopov, 1978; Preskill, 1979). We obtain that the condition (333), under which the diffusion approximation is valid, is satisfied at

$$t < t_1 \sim \frac{m_{Pl}}{\alpha^2 m^2} , \tag{336}$$

that corresponds to the temperature

$$T > \alpha m \sim e\Lambda \sim eT_{cr} , \tag{337}$$

where α is the fine structure constant and m is the mass of monopole.

The rate of the monopole–antimonopole annihilation can be found in the diffusion approximation by the consideration of the diffusion of particles with magnetic charge −g to the absorbing sphere with the radius

$$a \le r_0$$

and with the magnetic charge +g.

The spherically symmetric diffusion equation has the form

$$\frac{\partial n(r,t)}{\partial t} = D \cdot \frac{1}{r^2} \cdot \frac{\partial}{\partial r} r^2 \left[\frac{\partial n(r,t)}{\partial r} + \frac{g^2}{T \cdot r^2} \cdot n(r,t) \right], \tag{338}$$

where the diffusion coefficient D is given by

$$D = \frac{1}{3} \lambda v \, . \tag{339}$$

The solution of (338) for the stationary distribution

$$\frac{\partial n}{\partial t} = 0$$

of diffusing particles with the boundary conditions

$$n(\infty) = n_0 \equiv n_m$$

and

$$n(a) = 0$$

is given by

$$n(r) = \begin{cases} 0 \, , & r \le a \\ n_0 \dfrac{1 - \exp\left(\dfrac{r_0}{r} - \dfrac{r_0}{a} \right)}{1 - \exp\left(- \dfrac{r_0}{a} \right)}, & r > a \end{cases} \tag{340}$$

For the diffusion flux one obtains

$$q = 4\pi r^2 D \frac{\partial n(r)}{\partial r} \approx 4\pi D r_0 n_m \tag{341}$$

and the monopole annihilation rate in the diffusion approximation is given by the expression

$$\left(\frac{dn_m}{dt}\right)_{ann} = -n_m^2 4\pi D r_0 \tag{342}$$

(not by

$$\left(\frac{dn_m}{dt}\right)_{ann} = -n_m^2 \pi a^2 v \tag{343}$$

as it is the case for free monopoles).

With the account for (342), the equation for the relative concentration of monopoles

$$\nu = \frac{n_m}{n_r} ,$$

where n_r is the concentration of relativistic particles, has the form

$$\frac{d\nu}{dt} = -4\pi D r_0 n_r \nu^2 = -A\theta^{1/2} \nu^2 , \tag{344}$$

where

$$A = \frac{4\pi}{3} \cdot \frac{g}{e} m , \tag{345}$$

and the dimensionless temperature is defined as

$$\tau = \frac{t}{t_0} \tag{347}$$

and taking into account

$$\theta(t) = \theta(t_0)\tau^{-1/2} \equiv \theta_0\tau^{-1/2} , \tag{348}$$

one obtains the solution of the equation (344) in the form

$$\nu(t) = \frac{\nu(t_0)}{1 + \frac{4}{3} A\theta_0^{1/2} t_0 \left(\tau^{3/4} - 1 \right) \nu(t_0)} \tag{349}$$

for

$$t_0 < t < t_1 . \tag{350}$$

The diffusion approximation is not valid after

$$t > t_1 \approx \frac{g^8 e^4 m_{Pl}}{m^2} \tag{351}$$

when the rate of free monopole annihilation is given by the Eq. (343).

Following Zeldovich and Khlopov (1978), the quantity a, which determines the free monopole annihilation cross section, is the maximal impact parameter for which the monopole motion turns out to be finite due to bremsstrahlung. It was found that the free monopole annihilation practically does not change the result of the diffusion annihilation. The concentration of primordial magnetic monopoles is given by

$$\nu(t >> t_1) = \nu(t_1) \ .$$

Finally, provided that the initial monopole concentration satisfies the condition

$$\nu_0 >> \frac{3}{4A\theta_1^{1/2} t_1} \ , \tag{352}$$

the primordial magnetic monopole concentration is independent on the initial concentration and is given by

$$\nu_m = \frac{n_m}{n_r} = \frac{m}{g^5(eg)m_{Pl}} \approx 10^{-9}\left(\frac{m}{10^{16}\,\text{GeV}}\right) . \tag{353}$$

One easily finds from the Eq. (353) that for the mass of magnetic monopole

$$m = 10^{16} \ \text{GeV} \tag{354}$$

the primordial magnetic monopole concentration exceeds the baryon number density

$$\frac{n_m}{n_\gamma} > \frac{n_b}{n_\gamma} \tag{355}$$

so that the magnetic monopole density

$$\rho_m = m \cdot n_m \tag{356}$$

should be by 16 orders of magnitude larger than the density of the baryon matter. To solve the problem of the monopole overproduction, the inflational cosmological models were suggested, where the initial monopole concentration is strongly suppressed (see the next Section).

Note that in general the question on the formation and the evolution of the magnetic field structure in the Universe is of special interest in the presence of magnetic monopoles.

In particular, the mechanism of topological monopole formation in the course of phase transition proceeds so that first the monopole field is formed in space and only then the monopole is localised as the singularity in this field. It is clear that the local field, being formed in the phase transition, does not have any initial orientation. Thus, loops of magnetic field may be formed together with monopoles and antimonopoles (Khlopov, 1988). The self-consistent evolution of these primordial fields together with the monopole–antimonopole plasma has not yet been studied.

2.2. *Inflational Recovery of the Old Big Bang Model*

Inflation is the necessary element of cosmological picture. Inflational models explain, why the Universe expands. They provide solution for horizon, flatness, magnetic monopole, etc., problems (Guth,1981; Linde, 1984). The solution is based on superluminous expansion, which takes place in the case of the cosmological equation of state

$$p < -\frac{1}{3}\varepsilon\ . \tag{357}$$

Under the condition (357), the acceleration in the cosmological equation (compare with Eq. (210))

$$\frac{\ddot{a}}{a} = -\frac{4\pi G}{3c^2}(\varepsilon + 3p) \tag{358}$$

is positive and the formal solution for the law of expansion is given by

$$a(t) \propto \exp\left(\int H dt\right) , \tag{359}$$

where H is defined from the Eq. (209) as

$$\left(\frac{\dot{a}}{a}\right) = H = \sqrt{\frac{8\pi G\varepsilon}{3c^2}} . \tag{360}$$

Neither the matter, nor the radiation dominance can lead to the equation of state with the negative pressure. One needs some hypothetical phenomena to occur in the very early Universe, inducing unstable negative pressure stage of cosmological evolution.

The simplest possibility, mentioned by Sakharov (1965) and Gliner (1965), was first considered by Gliner (1970) and Gliner, Dymnikova (1975), who assumed the initial state of the cosmological expansion to be maximally symmetric in space-time. It corresponds to the De Sitter vacuum equation of state given by

$$p = -\varepsilon = -\Lambda = \text{const} . \tag{361}$$

The time dependence of the scale factor is then given by the exponential law

$$a(t) \propto \exp(Ht) \tag{362}$$

with the Hubble constant H defined as

$$H = \sqrt{\frac{8\pi G\Lambda}{3c^2}} . \tag{363}$$

Bugrii and Trushevsky (1976) found the possibility for the exponential stage of expansion in the hadron stage but the conditions they assumed were not supported by the hadron physics based on QCD.

The conditions close to the equation of state (361) were found by Starobinsky (1980) to appear in the early Universe due to R^2 effects induced by the true vacuum polarisation of the classical gravitational field (Gurovich, Starobinsky, 1979; see review in Zeldovich, 1980).

However, the general clear understanding that inflation is necessary both for cosmology and particle theory, came only after the work of Guth (1981). Inspired by the solution of the cosmological GUT magnetic monopole overproduction problem, this work has stipulated in the transparent form the list of internal inconsistencies of the old Big Bang model to be removed by inflation.

The old inflational scenario (Guth, 1981) assumed the GUT strong first-order phase transition, so that it resulted in the supercooling of the Universe in the phase with the restored GUT symmetry.

At low temperature the Higgs potential is given by the Eq.(57). The symmetric phase corresponds to the zero vacuum expectation value of the Higgs field. Then the Higgs potential is reduced to constant and its dominance in the energy density leads to the equation of state (361). Owing to the time dependence of the scale factor, given by Eq. (362), the correlation radius grows exponentially that results in the exponentially suppressed topological production of magnetic monopoles.

After successive reheating of the Universe, the temperature is much smaller than the temperature of GUT phase transition, so that the monopole–antimonopole concentration due to the pair production is also exponentially small.

Thus the inflational scenario results in the suppression of the initial concentration of magnetic monopoles, that removes the problem of GUT monopole overproduction.

Exponential growth of the scale factor at the inflational stage recovers several problems of the initial conditions inherent to the old Big Bang model.

The horizon problem is related to the remarkable homogeneity of the Universe within the modern cosmological horizon originated from different initially causally disconnected regions.

To illustrate the fantastic fine tuning of the initial conditions that is needed in the old Big Bang model, let us estimate the initial size of the region within the modern cosmological horizon l_h , which at present time t_U is of order of

$$l_h = ct_U \sim 10^{28} \text{ cm} . \tag{364}$$

In the old Big Bang model, the scale factor changes inversely proportional to the temperature. Therefore one finds, that, say, at the Planck time t_{Pl} , when the temperature was equal to the Planck temperature T_{Pl} , this region should have the size

$$l_h(t_{Pl}) = l_h(t_U) \cdot \frac{a(t_{Pl})}{a(t_U)} = l_h(t_U) \cdot \frac{T_0}{T_{Pl}} \approx 2.5 \cdot 10^{-4} \text{ cm}, \tag{365}$$

where we take the modern temperature of the thermal background radiation to be equal to

$$T_0 = 2.7 \text{ K} .$$

On the other hand, the size of causally connected region at the Planck time is equal to the Planck length (compare to Eq. (223))

$$l_{Pl} = ct_{Pl} = 1.5 \cdot 10^{-33} \text{ cm} . \tag{366}$$

Hence, the observed homogeneity within the modern cosmological horizon should be provided by the fine tuning of initial conditions given in the

$$N \sim \left(\frac{l(t_{Pl})}{l_{Pl}} \right)^3 \sim 3 \cdot 10^{87} \qquad (367)$$

different causally disconnected regions.

In the case of a non-flat Universe the horizon problem is strengthen by the **flatness problem**.

The existing ambiguity in the total cosmological density allows the existence both the model with positive and negative curvature, and with density within an order of the magnitude around the critical density.

Such uncertainty in the modern density corresponds to very fine tuning in the initial deviation from the flat Universe.

The deviation from the critical density in the non-flat Universe is inversely proportional to the square of the scale factor. It means that the order of modern uncertainty corresponds to the initial deviation from the critical density, which should be at the Planck time of the order of

$$\Omega(t_{Pl}) - 1 \leq \left(\frac{a(t_{Pl})}{a(t_U)} \right)^2 \cdot O(10) \leq 10^{-63} \ . \qquad (368)$$

Thus, the analysis of the global properties of the modern Universe (Guth, 1981) uncovered the extreme fine tuning in the initial conditions for the non-flat Universe. It should be close to the flat Universe with the accuracy, given by the Eq. (368) in each one of more than 10^{87} different causally disconnected region.

One can easily find that the exponential growth of the scale factor naturally removes both problems.

Provided that the e-folding, defined as the power in the exponent in Eq. (359) or (362), exceeds 72, the size of the region, causally connected at the Planck time, grows so that it embeds the size of the modern cosmological horizon. It explains the similarity of the initial conditions for the observed part of the modern Universe.

For the non-flat Universe, such e-folding provides the exponential growth of the scale factor making the contribution of the curvature effect in deviations from the flat Universe compatible with the existing uncertainty in the cosmological density.

It may be said (see, e.g., Zeldovich, Khlopov, 1984) that the cosmological expansion itself is explained by inflation. The acceleration at the inflational stage (its positive sign in the Eq. (358)) gives the initial momentum to the expansion, i.e. it is that very Big Bang the Big Bang Universe originated from.

However the appealing idea of inflation finds serious problems in cosmological and physical applications.

In the "old inflational model", the strong first-order phase transition, driving inflation, results in the "boiling" of the Universe. The dominance of the false vacuum finishes, when the bubbles of the true vacuum nucleate, expand and collide.

The energy release in bubble collisions leads to the reheating of the Universe. But the thermalisation of this energy needs the small size of bubbles, that corresponds to rapid bubble nucleation.

On the other side, the e-folding, being necessary for the explanation of the global properties of the modern Universe, assumes sufficiently slow bubble nucleation. The large size of bubbles in this case means that all energy released in the phase transition transforms into the kinetic energy of bubble walls. It leads to the large empty space inside the true vacuum bubble which can be hardly reheated after the bubble wall collisions. The Universe turns to be strongly inhomogeneous.

In the "new inflational scenario" (Linde, 1982; Albrecht, Steinhardt, 1982; for review, see Linde, 1984; 1990), the transition from the inflational stage proceeds due to slow rolling of the effective potential down to the true vacuum state. This scenario removes the problem of inhomogeneities induced by the bubble nucleation. But the fluctuations of the scalar field induce density fluctuations. To provide sufficiently low amplitude of these density fluctuations, the effective potential should be very flat at

$$\langle \Phi \rangle \to 0 \, .$$

It implies very slow rolling of the field from the symmetric phase down to the true vacuum of the asymmetric phase.

The same condition of slow rolling appears in the chaotic inflation scenario (Linde,1984). The idea of this scenario may be easily illustrated on the base of the classical scalar field theory (Zeldovich, Khlopov, 1984).

Consider the Lagrange theory of classical scalar field

$$L = \frac{1}{2}\left(\partial_\mu \varphi\right)^2 - V(\varphi) \, , \tag{369}$$

and assume for simplicity that the potential $V(\varphi)$ has the form

$$V(\varphi) = \frac{1}{2} m^2 \varphi^2 \tag{370}$$

and the considered free massive field is homogeneous

$$\varphi = \Phi \cos(mt) \, . \tag{371}$$

According to the Lagrange field theory, the energy–momentum tensor is given by

$$T^{\mu\nu} = \left(\partial^\mu \varphi\right)\left(\partial^\nu \varphi\right) - g^{\mu\nu} L \, . \tag{372}$$

In the case of the field, given by the Eq. (371), that leads to the energy density

$$\varepsilon = \frac{1}{2} m^2 \varphi^2 \tag{373}$$

and to the pressure

$$p = -\frac{1}{2} m^2 \varphi^2 \cos(2mt) \ . \tag{374}$$

One easily finds from Eqs. (373) and (374) that the equation of state of the scalar field at

$$t << \frac{1}{m} \tag{375}$$

is

$$p \approx -\varepsilon \ . \tag{376}$$

The equation of state (376) appears at the initial stage of cosmological evolution of any scalar field, provided that its derivative terms are negligible as compared with the potential. On the basis of the chaotic inflation scenario, a wide variety of possibilities seem to appear for the inflaton, i.e., the field driving the inflation.

To make the proper choice between these possibilities, or, at least, to make some reduction of their number, additional traces of inflational mechanism should be considered. Let us stipulate here (Khlopov, 1996) the possible tracers of inflational dynamics to be considered in more details below.

Fluctuations on inflational stage induce the spectrum of initial density fluctuations, giving rise to galaxy and large scale structure formation in respective scales. The amplitude of these fluctuations is constrained by the observed isotropy of the thermal electromagnetic background radiation.

It rules out all inflational models with high amplitude of predicted fluctuations, the most of GUT-induced phase transition scenarios, in particular.

In the simplest models with the equation of state on the inflational stage close to Eq. (376), the flat Harrison-Zeldovich form of the spectrum is predicted. Then the estimated amplitude of initial fluctuations at the modern scale of the cosmological large scale structure provides some information on the possible properties of inflaton, e.g. on the form and parameters of scalar field potential.

For more complicated inflational models, e.g., multicomponent inflation, the form of the predicted spectrum of fluctuations can differ from the simple flat one.

Phase transitions on the inflational stage lead to specific peaks or plateaux in the spectrum with the position and amplitude defined by the parameters of the model. One should also account for the phase transitions after the global inflational stage, in which the initial spectrum can be modified.

Both in R^2 and scalar-field-driven (e.g., chaotic) inflational scenarios long dust–like post–inflational stage appears, induced by coherent inflaton field oscillations.

For instance, in the case of massive homogeneous non-interacting scalar field, considered above, one easily finds that at the time t large as compared with the period of the field oscillations

$$t >> \frac{1}{m} \tag{377}$$

the pressure given by Eq. (374), averaged over the time intervals much larger than the period of oscillations, is

$$\langle p \rangle_{tm>>1} = \left\langle -\frac{1}{2} m^2 \phi^2 \cos(2mt) \right\rangle_{tm>>1} = 0 \ . \qquad (378)$$

The duration of such stages defines the maximal temperature of the Universe after reheating, when radiation dominance stage begins. It also defines the specific entropy of the Universe after reheating.

The initial density fluctuations grow on post–inflational dust–like stage. The growth follows the general law of development of gravitational instability on matter-dominated stage in the expanding Universe and is given by

$$\frac{\delta\rho}{\rho} \propto t^{2/3} \ . \qquad (379)$$

If the ratio of cosmological time scales, corresponding to the end t_1 and the beginning t_0 of the dust–like stage exceeds

$$\frac{t_1}{t_0} > \delta^{-3/2} \ , \qquad (380)$$

where δ is the amplitude of fluctuation, the corresponding inhomogeneities are formed. Evolution of such inhomogeneities can lead to primordial black hole (PBH) formation.

The spectrum of PBHs reflects the scales, at which inhomogeneities are formed, as well as the mechanism of PBH formation.

The minimal probability of the PBH formation corresponds to the direct formation of PBHs in contraction of a very small fraction of configurations, evolved from specifically isotropic and homogeneous fluctuations.

The account for PBH formation, as a result of evolution of the bulk of inhomogeneities, strongly increases the number of expected PBHs.

Peaks in the spectrum of density fluctuations, produced at inflational stage, can also induce the PBH formation even on radiation-dominant stage with the probability exponentially dependent on the amplitude of fluctuation.

It makes primordial black holes as important tracer of the inflational physics. We shall consider it in details in Chapter 4.

To conclude, one can find different realisations for the inflational stage.

The internal constraint on these realisations is put by the condition of "the graceful transition from inflation to Friedman expansion". In strong first-order phase transitions such transition should follow bubble collisions, in the slow rolling scenario the particle creation by coherent field oscillations should take place. One may consider such transition as the evaporation of the condensate of physical particles (Dymnikova, Krawczyk, 1996; Dymnikova, Khlopov, 1998).

The other important problem is the physical nature of inflaton. The development of inflational models makes more and more elusive the hidden sector of particle theory, responsible for inflation. The condition of physical self-consistency of inflational physics with the baryosynthesis and the physics of dark matter may be important. As the solution the physical self-consistency of the initial conditions is formulated (Dymnikova, Khlopov, 1998).

Note in conclusion that the different physical motivations for inflation are in fact not alternative and may be combined in the true cosmology of the very early Universe. However, one important property seems to be valid in all inflational models.

The scale, corresponding to the maximal e-folding given by inflation, puts the physical constraint on the homogeneity of the Universe.

The other side of the inflational mechanism of the homogeneity of the observed part of the Universe is the inhomogeneity of the Universe far beyond the modern cosmological horizon. In studies of the Universe as a whole the physical mechanism offered as the basis for the homogeneous and isotropic cosmological model inevitably leads to much more complicated cosmological picture.

2.3. Baryosynthesis.

The generally accepted motivation for the baryon-asymmetric Universe is the observed absence of antimatter at macroscopic scales up to the scales of clusters of galaxies. In the baryon-asymmetric Universe, the observed baryonic matter is originated from the initial baryon excess, surviving after the local nucleon-antinucleon annihilation, that was taking place at the first millisecond of cosmological evolution.

The baryon excess is assumed to be generated in the process of baryogenesis (Sakharov, 1967; Kuzmin, 1970), resulting in the baryon asymmetry of the initially baryon-symmetrical Universe.

It turned out, that under some conditions virtually all existing mechanisms of baryogenesis may lead to the inhmogeneous baryosynthesis and even to the generation of antibaryon excess in some places. So, inhomogeneities of the baryon excess distribution, and even domains of antimatter in the baryon-asymmetric Universe can provide a probe for the mechanism of baryogenesis.

In the original Sakharov's scenario (1967) of baryosynthesis, the baryon excess originated from the CP-violating effects in out-of-equilibrium baryon–nonconserving processes.

To illustrate the idea of the Sakharov's mechanism consider the baryon-symmetric Universe in which the out-of-equilibrium decays of some particles X and of equal amount of their antiparticles take place.

The baryon non-conservation in these decays means that the products of decay along different channels have different baryon number. Assume, for definiteness, that there are two different decay modes of the particle X, namely, the channel

$$X \to qq \; , \tag{381}$$

and the channel

$$X \to \overline{q}l \; , \tag{382}$$

where q and l denote quark (e.g. d quark) and charged lepton. The corresponding channels for the antiparticle are

$$\overline{X} \to \overline{q}\overline{q} \tag{383}$$

and

$$\overline{X} \to q\bar{l} \; . \tag{384}$$

Owing to CPT invariance, the total widths of particle and antiparticle are strictly equal. However, due to CP-violation the relative probabilities, called branching ratios or simply branchings, for particular decay modes of particle and antiparticle do not coincide.

Let the total decay probability be equal to 1. The relative probability of the decay (381) is

$$Br(X \to qq) = r \; , \tag{385}$$

and the branching ratio for the channel (382) is

$$Br(X \to \overline{q}l) = 1 - r \; . \tag{386}$$

For the antiparticle, the branching ratio for the channel (383) is

$$Br(\overline{X} \to \overline{q}\overline{q}) = \overline{r} \tag{387}$$

and for the channel (384)

$$Br(\overline{X} \to q\bar{l}) = 1 - \bar{r} \; . \qquad (388)$$

With the account for baryon charges of quarks and antiquarks

$$B_q = +\frac{1}{3} \; , \quad B_{\bar{q}} = -\frac{1}{3} \; , \qquad (389)$$

one finds that as a result of the out-of-equilibrium decays (381)-(384) in the baryon symmetric Universe with the equal initial concentrations of particles and antiparticles

$$n_X = n_{\overline{X}} \qquad (390)$$

the baryon excess is generated, equal to

$$n_b = (r - \bar{r}) \cdot n_X \; . \qquad (391)$$

The magnitude of the baryon excess is determined by the concentration of the decaying particles as well as by the difference of branching ratios of respective modes for particles and antiparticles. This difference is determined by the magnitude and the sign of the CP-violating phase ϕ.

If the magnitude and even the sign of $\phi(\mathbf{x})$ varies in space, in the regions, where the CP-violating phase $\phi(\mathbf{x})$ changes its sign, the same out-of-equilibrium baryon-non-conserving processes leading to the baryon asymmetry, result in spatial dependence of the baryon density $n_b(\mathbf{x})$ and even in the antibaryon excess

$$B(\mathbf{x}) < 0 \; . \qquad (392)$$

The spatial dependence of ϕ is predicted in the model of spontaneous CP violation or in models, where CP-violating phase is associated with the amplitude of the invisible axion field. The size and the amount of antimatter in domains, generated in this case, is related to the parameters of models of CP violation and/or of the invisible axion (see review in Chechetkin et al, 1982; Khlopov, Chechtkin, 1987; Khlopov, 1996) .

Supersymmetric (SUSY) extensions of GUT models offer another possible origin of the baryon asymmetry of the Universe. Recall (see Chapter 2) that the supersymmetry puts ordinary quarks into correspondence with their partners with zero spin. These scalar quarks are bosons and can form the Bose condensate. Affleck, Dine (1983) and Linde (1983) found that the superpotential is flat relative to the baryon charge, and the existence of scalar quark condensate is not forbidden.

Such a condensate, being formed with the positive baryon charge ($B > 0$), induces the baryon asymmetry, after the decay of scalar quarks on quarks and gluinos. However, the mechanism does not fix the value and the sign of the baryon charge of condensate, opening the possibilities for inhomogeneous baryon charge distribution as well as for antibaryon domains.

It was already mentioned in Chapter 2, that at the high temperature the standard model of electroweak (EW) interactions predicts baryon charge nonconservation. This effect puts severe constraints on the GUT and SUSY GUT mechanisms of baryosynthesis.

Electroweak non-conserving processes follow the selection rule

$$B + L = 0 \tag{393}$$

and wash out any initial baryon asymmetry, which follows the selection rule

$$B - L = 0 \ . \tag{394}$$

In particular, the electroweak baryon nonconservation (EW BNC) makes impossible to generate the observed baryon asymmetry by the processes, predicted in the $SU(5)$ GUT model, since they obey the selection rule (394).

On the other hand, the EW BNC opens the new approach to baryosynthesis, because one can use it for the generation of the observed baryon asymmetry.

It was found (Shaposhnikov, 1996) that one can not generate the baryon excess sufficient to explain this asymmetry in the framework of the minimal standard model. The extensions of the standard model are necessary to provide the mechanism for baryosynthesis based on EW BNC. It proves the general statement that the physical grounds for baryosynthesis are related to the hidden sector of particle theory.

Among the possible extensions of the standard model needed for the baryosynthesis based on EW BNC, an interesting possibility exists related with the physics of the mass of neutrino. The Majorana mass of neutrino is induced by the violation of lepton number with the selection rule

$$\Delta L = 2 \ . \tag{395}$$

CP violation in the out-of-equilibrium lepton-number-non-conserving processes implied by neutrino mass physics can lead to lepton excess, which is transformed into baryon excess owing to EW BNC. Such scenario is realised in the model of horizontal unification to be discussed later in the present book.

Based on the extensions of the standard model, the mechanisms of EW BNC baryosynthesis also imply the possibility of antimatter domains, e.g., due to the spontaneous CP violation (Comelli et al, 1994).

Thus antimatter domains may appear in baryon asymmetric Universe and may be related to practically al the mechanisms of baryosynthesis, as well as to mechanisms of CP violation and to possible mechanisms for primordial baryon charge inhomogeneity.

The size of domains depends on the details of the respective mechanisms. With the account for inflation the domains may be as large as the modern horizon, what is the case for the models of the "island Universe" (Dolgov et al, 1987) with very large scale inhomogeneity of baryon charge distribution.

The averaged effect of the domain structure is determined by the relative amount of antimatter

$$\Omega_a = \frac{\rho_a}{\rho_{cr}}, \tag{396}$$

where ρ_a is the cosmological density of antimatter averaged over the large scales, and

$$\rho_{cr} = \frac{3H^2}{8\pi G}$$

is the critical density, and by the mean size of domains l, which are the characteristic scale of their spatial distribution or (for small domains) the time scale t_{ann} of their annihilation with the surrounding matter.

One can consider antimatter domains as a profound tracer of the origin of the matter in the baryon asymmetric Universe.

2.4. Nonbaryonic dark matter.

The main arguments favouring nonbaryonic nature of dark matter in the Universe are Big Bang nucleosynthesis (BBN) in the inflational cosmology and the formation of large scale structure of the Universe in conditions of the observed isotropy of relic radiation.

Since the both lines will be discussed in details in following chapters, let us consider briefly the troubles of the old Big Bang model to be recovered by the hypothesis of the nonbaryonic dark matter.

The first line of arguments comes from the predictions of the theory of Big Bang nucleosynthesis. The reasonable fit of this theory to the observed abundance of light elements is obtained at the baryon density

$$\Omega_b \leq 0.1 \div 0.2 \ , \tag{397}$$

whereas the total density, predicted by the simplest variants of the inflational cosmology, corresponds to

$$\Omega_{\text{tot}} = 1 \ . \tag{398}$$

One can ascribe the difference between the total and the baryon density to the nonbaryonic dark matter.

The second type of arguments is that the formation of the large scale structure is compatible with the observed isotropy of thermal electromagnetic background only if some weakly interacting form of matter triggers the structure formation with the minor effect in the angular distribution of relic radiation.

There are several scenarios of structure formation by hot (HDM), cold (CDM), unstable (UDM), mixed hot+cold (H+CDM), hierarchical decaying (HDS), etc. dark matter.

These scenarios physically differ by the ways and succession in which the elements of structure are formed, as well as by the number of model parameters. But having in mind the general independence of motivations for each type of dark matter candidates, one finds from the particle physics viewpoint that the hot, cold, unstable, etc. dark matter are not alternatives and rather supplementary options to be taken together, accounting for the whole set of reasonable physical arguments.

Indeed, one considers the would be eV-(10eV)- neutrino mass as physical motivation for the hot dark matter scenario. But, as we have seen in Chapter 2, massive neutralinos, predicted in supersymmetric models, or invisible axions, following from the Peccei–Quinn solution of the strong CP violation problem in QCD, being cold dark matter candidates, are based on physical grounds, which are in no way alternative to the physics of neutrino mass.

So mixed hot+cold dark matter scenarios seem to be physically more reasonable, than simple one-parameter HDM or CDM models.

However, all these motivations do not correlate with the problem of quark–lepton families, of the existence of three types of neutrinos.

Physical mechanisms of family symmetry breaking lead to new interactions, causing the massive neutrino instability relative to the decay on lighter neutrinos and light Goldstone boson, familon or singlet Majoron. Neutrino instability, intimately related to family symmetry breaking, provides physical grounds for unstable dark matter (UDM) scenarios (Doroshkevich, Khlopov, 1984; Gelmini et al., 1984; Turner et al., 1984).

At the expense of the additional parameter, namely, the lifetime of unstable particles, the UDM models remove the contradiction between the data on the total density within the inhomogeneities, estimated as

$$\Omega_{\text{inhm}} < 1 , \tag{399}$$

and the prediction of the inflational cosmology, given by Eq. (398). In these models, the difference in Ω is ascribed to the homogeneous background of decay products of unstable dark matter particles.

As we shall discuss in Chapter 9 the UDM models also recover the disadvantages of HDM scenarios, related to quite rapid evolution of the structure after its formation.

Owing to neutrino instability the large scale structure formed at red shifts corresponding to observed distant objects, survives after the decay of the most part of the dark matter, having formed the structure.

The actual multicomponent content of dark matter may be extremely richer, if one takes into account the hypothesis of the mirror and shadow matter. As we mentioned in Chapter 2, this hypothesis recovers the equivalence of left- and right- handed co-ordinate systems in Kaluza-Klein and superstring models.

For instance, in the heterotic E_8xE_8' string model one have to account for the whole set of 248 matter fields and their 248 interactions, arising from the shadow E_8' sector of this model.

Even the above far from complete list of options poses the serious problem of the proper choice of the true combination of various dark matter candidates in physically motivated multicomponent dark matter scenarios.

Thus, since physical grounds for all nonbaryonic dark matter candidates lie outside the standard model and lose the proper experimentally proven basis, we should either to take into account all the possible ways to extend the standard model, treating all the candidates as independent, or to find a quantitatively definite way to estimate their relative contribution.

CHAPTER 4

COSMOARCHEOLOGY OF THE VERY EARLY UNIVERSE

Following the new cosmological paradigm, one considers the very early Universe as the period when inflation, baryosynthesis and freezing out of dark matter particles has taken place. The uncertainty in the description of these three phenomena leads to the ambiguity in the very early history of the cosmological evolution.

To specify the picture of very early Universe, one should specify the physical laws, underlying the processes in this period. It needs special methods of linking the observational data to the effects of superhigh-energy physics, which governs the physical conditions in the very early Universe. The whole complex of problems related to elaboration of such methods is the subject of the part of cosmoparticle physics called cosmoarcheology.

Cosmoarcheology considers the observational data on the Universe as the data sample of the *Gedanken Experiment*, undertaken to test the cosmological models and particle theories, on which these models are based.

As to the cosmoarcheology of the very early Universe, the complexity of cosmoarcheological analysis is multiplied by the model dependence of the cosmological evolution.

To calculate the abundance of hypothetical particles or to analyse the formation of hypothetical objects, one has to make assumptions on the physical conditions in this period. The physical processes, as well as the cosmological conditions follow from the hidden sector of particle theory and should be treated self-consistently to obtain the reliable estimation. It means that cosmological consequences of the particle theory should be considered in the framework of the cosmological model based on the same particle theory.

In the framework of the inflational model with baryosynthesis and nonbaryonic dark matter, the old Big Bang model calculations of the freezing out of hypothetical particles or of the formation of topological defects in cosmological phase transitions are model dependent and lose their reliability.

However, we found in the previous Chapter the tracers of the very early Universe, which are less model dependent and are shared by different mechanisms of inflation and baryosynthesis. They are primordial black holes and antimatter domains, which we discuss in this Chapter.

1. Primordial Black Holes as the Cosmoarcheological Tool

1.1 Primordial Black Holes

According to the theory of gravitation, any object with the mass M can form a black hole, if this mass is put within its gravitational radius, given by

$$r_g = \frac{2GM}{c^2} . \tag{400}$$

The existence of black holes, following from the general relativity, was considered historically in the framework of the Newton theory of gravitation.

Still in the early XIX century P.S. Laplace pointed out, that the parabolic velocity may reach the speed of light on the surface of supermassive stars, so that such objects can exist but can not be observed.

Laplace has deduced the above expression for the gravitational radius r_g assuming that the sum of the kinetic energy of the particle of the light with the mass m, given by

$$E_k = \frac{mc^2}{2} ,$$

and the potential energy of its gravitational attraction to the body with the mass M , given by

$$E_{pot} = -\frac{GMm}{r} ,$$

is equal to zero", i.e. he found the value of the gravitational radius from the equality

$$\frac{mc^2}{2} - \frac{GMm}{r_g} = 0 \tag{401}$$

similar to one generally used to calculate the parabolic velocity of material bodies.

Though the argument of Laplace was based on the corpuscular theory of light, the quantitative answer turned out to be true even in general relativity.

Theoretical astrophysics considers black holes as the final stage of natural evolution of stars with the mass, exceeding few solar masses. For smaller objects, the natural physical conditions for their collapse into black holes can hardly be arranged in the modern Universe.

However, as it was pointed out by Zeldovich and Novikov, black holes with any mass, exceeding the Planck mass, can, in principle, be formed in the early Universe, since any mass within the cosmological horizon can naturally form black hole, provided that expansion stops in the considered region. Such black holes are called primordial black holes (PBH).

However, this general possibility to form primordial black hole is highly improbable in the homogeneously expanding Universe, since it assumes metric fluctuation of the order of unity. In the case, when metric fluctuations are given by the Gaussian distribution with the dispersion

$$\left\langle \delta^2 \right\rangle << 1 , \tag{402}$$

the probability for a fluctuation of the order of unity is given by the exponentially small tail of this distribution with the high amplitude.

In the Universe with the equation of state

$$p = \gamma\varepsilon , \tag{403}$$

where the numerical factor γ lies in the interval

$$0 < \gamma < 1 , \tag{404}$$

the probability to form the black hole by the fluctuation within the cosmological horizon is given by

$$W_{PBH} \approx \exp\left(-\frac{\gamma^2}{2\left\langle \delta^2 \right\rangle} \right) . \tag{405}$$

So, the PBH spectrum exhibits exponentially strong sensitivity to the amplitude of fluctuations.

In the inflationary cosmology, the spectrum of fluctuations is constrained by the theory of the large-scale structure formation and by the data on the isotropy of the black body background radiation. It corresponds to the scales varying from the scale of the modern large scale structure to the scale of the modern cosmological horizon. The PBH spectrum reflects the primordial inhomogeneity at much smaller scales, thus offering the probe for the far ultraviolet part of the primordial fluctuation spectrum.

The case

$$\gamma << 1 \tag{406}$$

corresponds to the dust-like stage of cosmological evolution with the equation of state

$$p = 0\ , \tag{407}$$

and the estimation (405) is not valid, because using of Eq. (405) one formally finds the wrong answer that in the limit

$$\gamma \to 0$$

the probability tends to unity.

Nevertheless, the detailed analysis of the probability of the PBH formation at the dust-like stage shows its strong enhancement, as compared to the case of radiation-dominant stage. It makes the PBH spectrum the sensitive probe for the existence of early dust-like stages in the early Universe and for the physical mechanisms, based on the particle theory and leading to such stages.

Another source of strong inhomogeneity in the early Universe, which may also result in black hole formation, is related to the first-order phase transitions in the early Universe.

The bubble nucleation gives rise to strongly inhomogeneous structure of bubble walls expanding in the false vacuum, so that the latent energy, released in the transition from false to true vacuum, converts into the kinetic energy of these walls. Bubble collisions may concentrate the mass equivalent to this energy within the gravitational radius, so that black holes are formed. It makes the PBH spectrum to be a sensitive probe for the character of phase transitions in the early Universe.

The important property of black holes is the independence of their properties on the form of matter, from which they are formed. Therefore, the parameters of superweakly interacting forms of matter, such as the shadow matter, giving rise to the PBH formation, can also be studied with the use of the PBH spectrum.

1.2. PBH Signatures of Dust-Like Stages in the Early Universe

Regardless to their origin, dust-like stages in the early Universe cause the development of gravitational instability within the cosmological horizon. Similar to the formation of the modern cosmological structure, the growth of small initial density fluctuations should have been resulted in the formation of inhomogeneities.

Though the size of inhomogeneities in the early Universe can be as small as the size of atoms or even nuclei, the classical theory of gravitational instability is valid for the analysis of their evolution.

The mass distribution and the other properties of gravitationally bound objects, formed at the dust-like stage, depend both on the initial spectrum of fluctuations in the considered scale and on the matter content.

The main point is that any such object is unstable relative to gravitational instability and should inevitably collapse into black hole. However, the time scale for the loss of the gravitational stability strongly depends on the evolution of the object, being determined by the mechanisms of energy dissipation in this object.

One can specify two in principle different types of objects: “galaxies” and “stars”, which can be formed at the dust-like stage and whose evolution can result in the black hole formation.

As we know, in the modern Universe both stars and galaxies are formed by the baryonic matter. However, the baryonic matter in stars can lose energy by emission of radiation. It induces very rapid stellar evolution as compared to the cosmological time scale. The existing stellar candidates to black holes are assumed to be the final stage of such evolution for stars with the mass, exceeding several solar masses.

Stars in galaxies represent the collisionless gas. The time scale of star collisions is much larger than the age of the Universe, and the evolution of galaxies takes much more time, than the evolution of stars.

Active galactic nuclei are the second example of the modern candidates to black holes. It is assumed that the high concentration of stars in the central part of galaxies provides their collapse into black holes with the masses of the order of

$$M \sim 10^6 \div 10^8 M_\odot \ . \tag{408}$$

These analogies may be useful in the analysis of possible inhomogeneities and their evolution on the dust-like stages of the early Universe.

For definiteness and for the sake of simplicity let us consider only one type of nonrelativistic matter in the early Universe. In the case, when its density ρ_m exceeds the density of relativistic particles ρ_r ,

$$\rho_m > \rho_r \ , \tag{409}$$

one obtains the matter-dominant stage with the cosmological equation of state

$$p = 0 \ .$$

Starting from the beginning of the matter-dominant stage at

$$t = t_0 \ , \tag{410}$$

density fluctuations in the considered non-relativistic matter grow within the cosmological horizon as

$$\frac{\delta\rho}{\rho} \propto t^{2/3} . \qquad (411)$$

If the initial amplitude of density fluctuations is equal to

$$\frac{\delta\rho}{\rho}(t_0) = \delta , \qquad (412)$$

at the time

$$t \sim t_f = t_0 \delta^{-3/2} \qquad (413)$$

the amplitude density fluctuations grows up to

$$\frac{\delta\rho}{\rho} \sim 1 , \qquad (414)$$

and the fluctuations form inhomogeneities, separated from the general cosmological expansion. These inhomogeneities are the gravitationally bound systems of the considered non-relativistic matter.

The successive evolution of such gravitationally bound systems are determined by the properties of the non-relativistic matter forming the inhomogeneities.

Weakly interacting massive particles form gravitationally bound systems of collision-less gas, being in the respect to successive evolution similar to the galaxies, composed by the collision-less gas of stars.

Energy dissipation in the gravitationally bound system of weakly interacting massive particles is very slow (Zeldovich, Poduretz, 1965). It is basically determined by the long process of evaporation of particles, whose velocity exceeds the parabolic velocity of the system.

In the case of particle binary collisions, the crude estimate of the time scale of evolution of such a system gives

$$t_{ev} = \frac{N}{\ln N} t_{ff} \tag{415}$$

for the gravitationally bound system of N particles, where for the system with the density ρ the free fall time is

$$t_{ff} \sim (4\pi G\rho)^{-1/2} \ . \tag{416}$$

According to Gurzadian and Savvidi (1987), the evolution time scale can be strongly enhanced due to collective effects in the gas of collisionless particles, so that for large N this time scale is of the order of

$$t_{ev} \sim N^{2/3} t_{ff} \ . \tag{417}$$

One finds that even in this case in the period, when the objects are formed from small initial density perturbations, the formation of black holes as a result of the evolution of gravitationally bound systems of weakly interacting particles lasts more than the cosmological time.

The latter time, being of the order of

$$t_f \sim (G\rho)^{-1/2} \ , \tag{418}$$

where ρ is the cosmological density in the period, when density fluctuation in the considered scale grows up to

$$\frac{\delta\rho}{\rho} \sim 1 \ ,$$

is of the order of the free fall time in the gravitationally bound system formed from this fluctuation

$$t_f \sim t_{ff} \, . \tag{419}$$

Since

$$t_{\rm ev} >> t_f \sim t_{ff} \, , \tag{420}$$

the duration of the dust-like stage should be sufficiently long and it should finish at

$$t_e > t_{\rm ev} >> t_f \tag{421}$$

to provide the formation of black holes in the result of the evolution of gravitationally bound systems of weakly interacting particles.

Nonrelativistic matter, coupled to relativistic particles and radiation, forms gravitationally bound systems, whose evolution is defined by the energy dissipation due to radiation, as it is the case for the usual matter in ordinary stars.

With the use of this analogy, one can naturally deduce, that the evolution time scale for such systems is comparable with the cosmological time of their formation, or even smaller, and as a result of this evolution primordial black holes are copiously produced even for relatively short periods of the matter dominance in the very early Universe, provided that

$$t_e > t_f \geq t_{\rm ev} \, . \tag{422}$$

It turns out that the minimal estimation of the PBH formation at early dust-like stages can be done independent on the form of the nonrelativistic matter, dominating in the early Universe.

1.3. Direct PBH formation at dust-like stages

The idea of direct PBH formation at the dust-like stage, first put forward by Khlopov and Polnarev (1980) (see Polnarev, Khlopov, 1985; Chechetkin, Khlopov, Sapozhnikov, 1982; Khlopov, Chechetkin, 1987 for review), can be expressed as follows.

As we have mentioned above, the mass within the cosmological horizon is just the mass within its gravitational radius, provided that there is no expansion. The original idea of Zeldovich and Novikov (1966) (see also Novikov et al., 1979, and Polnarev, Khlopov, 1985 for review) was to stop the relativistic expansion. In the case of homogeneous and isotropic Universe, that corresponds to exponentially small probability.

The other possibility is to consider the density fluctuation at dust-like stage and to study the conditions under which the growth of this fluctuation results in the formation of sufficiently homogeneous and isotropic configuration that contracts within its gravitational radius.

The direct PBH production means that after the growth of the fluctuation to the order of unity and separation from the cosmological expansion the configuration contracts within its gravitational radius.

To the moment t_1 when the contraction starts the configuration may be characterised by the following quantities:

1) the mean density ρ_1 equal by the order of the magnitude to the mean cosmological density at the time t_1 ;

2) the size of the configuration r_1 ;

3) the deviation from the spherical form s defined as

$$s = \max\{|\gamma_1 - \gamma_2|, |\gamma_1 - \gamma_3|, |\gamma_2 - \gamma_3|\} , \qquad (423)$$

where γ_1, γ_2, γ_3 define the deformation along the three main orthogonal axes of the configuration;

4) the inhomogeneity u of the density distribution inside the configuration defined as

$$u \sim \frac{\delta\rho_1}{\rho_1} . \tag{424}$$

The black hole formation as the result of contraction corresponds to the mean density

$$\rho_{BH} \sim \frac{M}{r_g^3} \sim \frac{\rho_1}{x^3} , \tag{425}$$

where

$$x = \frac{r_g}{r_1} \tag{426}$$

and r_g is the gravitational radius of the considered configuration with the mass M.

On the other hand, the maximal density, which may be reached in the contraction of non-spherical configuration is given by

$$\rho_{\max} \sim \frac{\rho_1}{s^3} . \tag{427}$$

Eq. (427) follows from the estimation of the minimal size of the configuration, which may be achieved as the result of contraction, and is given by

$$r_{\min} \sim \max\{t_c \cdot \Delta v, \Delta r\} , \tag{428}$$

where t_c is the time scale of contraction,

$$\Delta v \leq s \cdot \frac{r_1}{t_c} \tag{429}$$

is the characteristic difference of initial velocities of contraction along the different axes and

$$\Delta r \leq s \cdot r_1 \tag{430}$$

is the characteristic initial "flatness" of the configuration.

One finds from Eqs. (425) and (427) that in order to form the black hole the configuration should be nearly spherical symmetric,

$$s \leq x \ll 1 \,. \tag{431}$$

For most configurations $s > x$, and the contraction results in the formation of structures with small gravitational potential. The specific features of these structures depend on the properties of particles that participate in their formation (see the preceding Section).

When the density approaches ρ_{BH} in the course of contraction at the time t_{BH}, the equation of state inside the configuration can be relativistic,

$$p = \tfrac{1}{3}\varepsilon \,.$$

To provide the sufficient condition for the PBH formation, the pressure gradients should not exceed the gravitational forces. That condition imposes the constraint on the inhomogeneity of the configuration at the moment t_{BH} :

$$\frac{\delta\rho_{BH}}{\rho_{BH}} < 1 \,. \tag{432}$$

If the massive particles represent the collisionless gas, the equation of state may not change when the density approaches ρ_{BH}. To provide the PBH formation in this case, it is sufficient to assume that the configuration contracts inside the gravitational radius before the time $t_{\rm caus}$, when caustics forms in the centre of the configuration. The corresponding constraint has the form

$$t_{\rm caus} > t_{BH} \ . \tag{433}$$

Indeed, if the condition (433) is not valid, collisionless particles may pass the caustics and leave the region inside the gravitational radius so that the PBH may not form.

If to the moment t_{BH} the configuration is fragmented onto weakly interacting "pieces" with the mass

$$\Delta M << M \ ,$$

the account for the velocity dispersion of these pieces results in the condition of the PBH formation, also given by Eq. (433).

The contraction of nearly spherical dust-like configuration is described by the Tolmen solution. The analysis (Khlopov, Polnarev, 1980; Polnarev, Khlopov, 1985) of the Tolmen solution shows that both conditions (432) and (433) are reduced to the same constraint on the inhomogeneity of the configuration at the moment t_{BH} :

$$u \leq x^{3/2} \ . \tag{434}$$

The conditions of high spherical symmetry and homogeneity of the configuration given by Eqs. (431) and (434) provide the sufficient conditions of the direct PBH formation.

Khlopov and Polnarev (1980,1981,1982, 1985) used the following arguments to estimate the probability for configuration to satisfy these conditions:

1) In the case of the normal law of distribution of configurations in the inhomogeneity (u) with the dispersion of the order of unity, the probability of configuration with the anomalously small inhomogeneity

$$u \leq x^{3/2} << 1$$

is determined by the phase space corresponding to such configurations so that this probability is given by

$$W_u \sim u \sim x^{3/2}. \tag{435}$$

2) Before the tensor of deformations is transformed to the principal axes, one should consider its six independent components as random quantities. Therefore one should take into account that the sphericity condition, given by Eq. (431) implies that five independent conditions should be simultaneously satisfied.

Namely, one should impose three conditions that the nondiagonal components are very small, and two conditions that the three diagonal elements are very close to each other.

Assuming that the probability for spherically symmetric configuration satisfying the condition (431) is also determined by the phase space corresponding to configurations with

$$s < x ,$$

one finds that this probability should be proportional to the fifth power of x :

$$W_s \sim x^5. \tag{436}$$

It follows from Eqs. (435) and (436) that the direct production of black holes takes place on the dust-like stage with the probability no less than

$$W_{BH} \geq W_s \cdot W_u \sim x^{13/2} . \quad (437)$$

Eq. (437) determines the minimal fraction of matter, which enters black holes of the mass M at the dust-like stage in the period of the PBH formation.

The spectrum of mass of primordial black holes formed by the direct mechanism at the early dust-like stage can be related to the spectrum of density fluctuations, which is formed in particular at the inflational stage.

It can be shown (Khlopov, Polnarev, 1980) that for the fluctuations which enter the cosmological horizon before the beginning of the dust-like stage as well as for the fluctuations which have not grown up to unity before the end of this stage the PBH formation by the direct mechanism is strongly suppressed. The direct mechanism is effective only in the interval of PBH masses M

$$M_0 < M < M_{\max} . \quad (438)$$

The minimal mass M_0 is the mass within the cosmological horizon at the beginning t_0 of the dust-like stage and is equal to

$$M_0 = \frac{4\pi}{3} \cdot \rho(t_0) \cdot t_0^3 \sim m_{Pl} \cdot \frac{t_0}{t_{Pl}} . \quad (439)$$

The maximal mass is indirectly defined by the condition that the fluctuation of the mass M, entering the horizon with the initial amplitude $\delta(M)$, grows up to the amplitude of the order of unity just in the end of the dust-like stage at the moment t_e . This condition has the form

$$t_e \sim t(M_{\max}) \cdot \left[\delta(M_{\max})\right]^{-3/2} . \tag{440}$$

In the interval given by Eq. (438), the probability of the direct PBH formation is no less than

$$W_{PBH}(M) \geq \left[\delta(M)\right]^{13/2} . \tag{441}$$

Note that the direct mechanism of BH formation is universal because it does not depend on the form of non-relativistic matter and on the period of its dominance in the Universe.

However, its formal application to the modern Universe leads to very low minimal probability of the formation of black holes with the mass of the order of the mass of galaxy superclusters.

On the other hand, this mechanisms provides the universal model-independent probe for the inhomogeneity on early dust-like stages. The sensitivity of this probe on the basis of astrophysical data is strongly enhanced by the analysis of the possible effects of the PBH evaporation.

1.4. PBH evaporation

The possibility of the black hole evaporation, first discovered by Hawking (1975), is the most important property of primordial black holes related to their possible astrophysical effects in the Universe after the first second of expansion.

The black hole with the mass M has the gravitational radius, given by Eq. (400) which we rewrite here in the units (230) as

$$r_g = 2GM = 2\frac{M}{m_{Pl}^2} . \tag{442}$$

In the strong gravitational field near the black hole particles can be created (Zeldovich, Starobinsky, 1976).

Therefore, the particle emission from the black hole is possible due to quantum effects (Hawking, 1975). The black hole emission is described as the emission from the surface of a black body with the temperature T_{PBH}

$$T_{PBH} = \frac{1}{4\pi r_g} = \frac{m_{Pl}^2}{8\pi M} \tag{443}$$

(Hawking, 1975; Page, 1981).

The luminosity of black hole is by the order of magnitude equal to

$$\frac{dE}{dt} \sim T^4 r_g^2 \sim r_g^{-2} \sim \frac{m_{Pl}^4}{M^2} \ . \tag{444}$$

The energy loss, determined by Eq. (444), means that the black hole loses its mass with the rate

$$\frac{dM}{dt} = -\frac{dE}{dt} \sim -\frac{m_{Pl}^4}{M^2} \ . \tag{445}$$

One can easily find from Eq. (445) that after the time

$$t_{\rm evp} \sim \left(\frac{M}{m_{Pl}}\right)^3 t_{Pl} \tag{446}$$

the black hole with the mass M loses all its mass. It means that the black hole evaporates.

With the account for all the numerical factors, the time scale of the black hole evaporation is equal to

$$t_{evp} = 10^{-27}\,\mathrm{s}\left(\frac{M}{1\ \mathrm{g}}\right)^{3} \tag{447}$$

(Hawking, 1975; Novikov et al., 1979; Hawking, 1976; Page, 1976).

According to the theory of stellar evolution, black holes should be formed in the result of evolution of stars with the mass M exceeding at least two solar mass

$$M > 2M_{\odot} \ .$$

The time scale of evaporation of such black holes is of the order of

$$t_{evp} \sim 10^{66} \ \text{years} \ ,$$

and their evaporation can be neglected. But the masses of primordial black holes may be much smaller than the stellar masses. Virtually any value is possible for the PBH masses down to the Planck mass m_{Pl} (Markov, 1965; 1966) or even smaller than m_{Pl} (Zeldovich, 1981).

Primordial black holes with the mass, smaller than

$$M < M_{evp} \sim 10^{15} \ \mathrm{g} \ , \tag{448}$$

have the evaporation time smaller, than the age of the Universe. Being formed in the very early Universe, such black holes should have been completely evaporated to the present time.

But the effects of their evaporation may lead to observational consequences, thus providing a definite test of their existence in the past. Detailed discussion of PBHs and restrictions on their possible concentration in the early Universe may be found in papers of Carr (1975), Hawking (1974, 1975), Zeldovich, Starobinsky (1976), Page, Hawking (1975), Novikov et al. (1979), in the reviews of Carr (1981), Polnarev, Khlopov (1985), Chechetkin et al. (1982), Khlopov, Polnarev (1982), Khlopov, Naselsky, Polnarev (1985), Khlopov, Chechetkin (1987) and in Refs. therein.

The fluxes of particles, produced in PBH evaporation, may cause the same influence on the physical processes in the Universe as the products of decay of metastable particles. However, virtually all species with the mass

$$m < T_{PBH} ,$$

predicted by the particle theory should be produced in the PBH evaporation.

Irrespective to the values of their coupling constants, the production rate of any species of particles is determined by thermodynamic equilibrium at the temperature T_{PBH}.

Due to gravitational mechanism of particle production, the evaporating black hole is the unique universal source of all types of particles, existing in Nature, with the mass, not exceeding T_{PBH}.

2. Formation of Black Holes in the First-Order Phase Transitions.

In the first-order phase transitions, the bubble walls collisions can lead to the PBH formation (Hawking, Moss, Stewart, 1982; Moss, 1994), provided that the kinetic energy of colliding bubbles is concentrated within its gravitational radius. However, such concentration needs very special conditions for the collision, say, simultaneous collision of several walls, that strongly suppresses the probability for the PBH formation.

Recently, a new mechanism of PBH production in the collision of only two bubbles was suggested (Khlopov et al., 1998; Konoplich et al., 1998)), thus making PBH spectrum to be a sensitive indicator for the cosmological first-order phase transitions.

The simplest example of a field theory, leading to the cosmological first-order phase transitions with bubble creation, is a scalar field theory with two nondegenerate vacuum states. The state with the lower energy is the true vacuum and the higher-energy state corresponds to the false vacuum.

Being stable on the classical level, the false vacuum state decays due to quantum effects, that results in the nucleation of the true-vacuum bubbles and their subsequent expansion within the region of the false vacuum. The potential energy of the false vacuum is converted into the kinetic energy of bubble walls thus making them highly relativistic in a short time.

The bubble expands until it collides with another one. As it was shown by Hawking, Moss, Stewart (1982) and Moss (1994), black holes may be created directly in the simultaneous collision of several bubbles. In papers by Khlopov et al. (1998) and Konoplich et al. (1998) the mechanism was found of BH formation in the collisions of only two bubbles with a probability of the order of unity. It induces the huge production of BH leading to substantial cosmological consequences discussed below.

2.1. Field Configurations in Bubble Wall Collisions

Consider a field theory, in which the probability of the false vacuum decay per unit volume is equal to Γ and the energy difference of the false and true vacuum states in the unit volume is ρ_V.

The vacuum decay proceeds through the nucleation of bubbles with the new phase separated walls initially at rest. Following Hawking, Moss, Stewart (1982) and Watkins, Widrow (1992), suppose for simplicity, that the horizon size is much greater than the distance between the bubbles. Due to conversion of the false vacuum energy into its kinetic energy the velocity of the bubble wall grows rapidly up to the velocity of light,

$$v = c = 1 \ . \tag{449}$$

Let us discuss the dynamics of the collision of two bubbles being initially nucleated in points (r_1, t_1) and (r_2, t_2) and expanding in the false vacuum region. The successive stages of the process of are illustrated in Fig. 11.

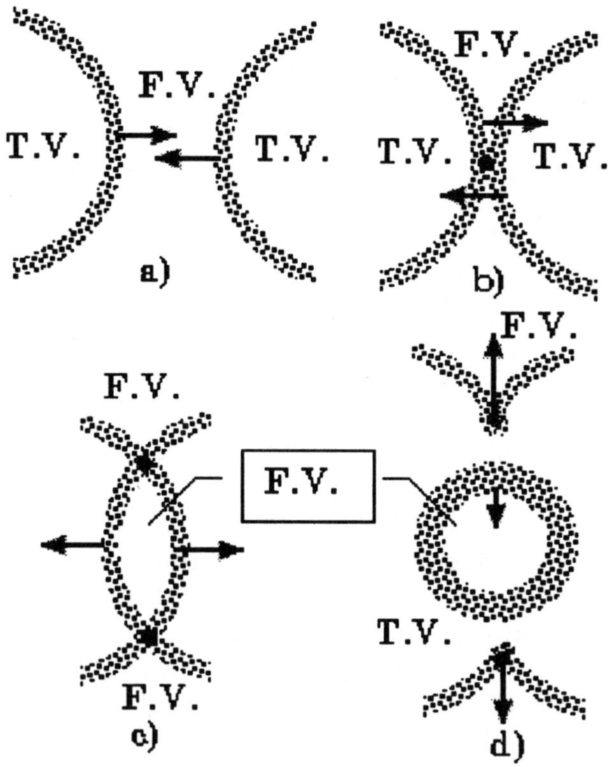

Fig.11. Formation of the false vacuum bag in the collision of two bubble walls.

Just after the collision, mutual penetration of walls at the distance, comparable with the wall width, is accompanied by the increase of the potential energy (Konoplich, 1980). Then the walls reflect and move backwards into the region of the true vacuum (Fig. 11c).

The space between them is filled by the field in the false vacuum state converting the kinetic energy of the wall back to the energy of the false vacuum state and slowing down the velocity of the walls.

Meanwhile the outer wall, that expands and accelerates outwards, absorbs the false vacuum in the outer area. Evidently, there is some moment, depending on the parameters of the theory, when the central region of the false vacuum is separated and a trapped false vacuum bag (FVB) is formed.

It was shown by Hawking, Moss, Stewart (1982) and Watkins, Widrow (1992), that the further evolution of FVB consists of the following stages:
1) the FVB area grows up to the definite size D_M until the kinetic energy of its wall vanishes;
2) the false vacuum volume begins to shrink down to the size D^*, comparable with the width of the wall;
3) the volume expands again and then contracts again, so that the increase of the false vacuum volume is followed by its decrease and so on.

Such a process of periodical mutual interchange of expansion and contraction, i.e. oscillation leads to the loss of the FVB energy, transforming into fluctuations of the classical scalar field. It was shown in papers by Watkins, Widrow (1992) and Belova, Kudryavtzev (1988) that only several oscillations take place.

On the other hand it is important that the secondary oscillations might occur only if the minimal size of the FVB appears to be larger than its gravitational radius,

$$D^* > r_g \ . \tag{450}$$

The opposite case

$$D^* < r_g \tag{451}$$

leads to the formation of black hole with the mass of the FVB.

The main idea of the indirect mechanism of BH formation in two-bubble collisions in the first-order phase transitions is to consider the contraction of the false vacuum bag down to its gravitational radius. As we will show later, the probability of the BH formation in this process is almost unity for a wide range of parameters of theories, predicting the first-order phase transitions in the Universe.

Note that the FVB does not possess spherical symmetry at the moment of its detachment from outer walls but the wall tension restores the symmetry during the first contraction of FVB.

2.2. Gravitational Collapse of FVB and BH Formation

Consider now the conditions for the FVB transformation into a black hole in more details.

The mass M of FVB can be expressed in terms of the parameters of considered theory and calculated in the coordinate system K', in which the colliding bubbles were nucleated simultaneously.

In this system, the radius of each bubble b' is equal at the first moment of collision to the half of the initial distance between the centers of bubbles. Apparently, the maximum size D_M of the FVB is of the same order as the size of a bubble, for it is the only length parameter at this scale, and is equal to

$$D_M = 2b'C \, . \tag{452}$$

The parameter

$$C \cong 1 \tag{453}$$

can be calculated for each theory, but its numerical value does not affect the conclusions significantly.

One can find the mass of the FVB that is formed in the collision of two bubbles of radius b':

$$M = \frac{4\pi}{3}(Cb')^3 \rho_V \, . \tag{454}$$

This mass is contained in the shrinking area of the false vacuum.

The minimal size of FVB D^* is of the order of or less than the wall width Δ.

The BH is created if the minimal size of FVB appears to be smaller than its gravitational radius. It means that at least at the condition

$$\Delta < r_g = 2GM \tag{455}$$

the FVB should transform into BH.

As an example, consider a quite simple model with the Lagrangian

$$L = \frac{1}{2}\left(\partial_\mu \Phi\right)^2 - \frac{\lambda}{8}\left(\Phi^2 - \Phi_0^2\right)^2 + \varepsilon\Phi_0^3(\Phi + \Phi_0) \, . \tag{456}$$

In the thin wall approximation the wall width can be expressed in terms of the parameters of Lagrangian as follows:

$$\Delta = 2\left(\sqrt{\lambda}\Phi_0\right)^{-1} \, . \tag{457}$$

Using the condition Eq. (455), one can easily find that FVB with the mass

$$M > \left(\sqrt{\lambda} G \Phi_0\right)^{-1} \tag{458}$$

should transform into a black hole of the mass M.

Note that this conclusion is valid provided that FVB is completely contained within the cosmological horizon, namely, that the condition

$$M_H > \left(\sqrt{\lambda} G \Phi_0\right)^{-1} \tag{459}$$

is valid.

At the moment of the phase transition, which may be related to the scale of the symmetry breaking as

$$t = \frac{m_{Pl}}{\Phi_0^2} \ , \tag{460}$$

the mass M_H within the cosmological horizon is

$$M_H \cong \frac{m_{Pl}^3}{\Phi_0^2} \ . \tag{461}$$

Thus, for the potential (456) and with the condition

$$\lambda > \left(\frac{\Phi_0}{m_{Pl}}\right)^2 , \tag{462}$$

the black hole formation from the false vacuum bag is possible.

It is easy to see that this condition is valid for any realistic model of particle theory.

Following Konoplich et al. (1998) consider now the distribution of the mass M and the velocity v of black holes formed in two-bubble collisions, assuming that their masses satisfy the inequality (455). These quantities depend on coordinates and times of nucleation of the colliding bubbles so that one obtains the mass and the velocity of black hole as the functions of these parameters:

$$M = M\left(\left|\mathbf{r}_2 - \mathbf{r}_1\right|, t_2 - t_1\right) \tag{463}$$

and

$$v = v\left(\left|\mathbf{r}_2 - \mathbf{r}_1\right|, t_2 - t_1\right) . \tag{464}$$

The probability dP of collision of two bubbles, which have been nucleated at the distance $r_2 - r_1$ from each other at the moments t_2 and t_1 , has the form

$$dP = dP_1 dP_2 dP_- , \tag{465}$$

where

$$dP_1 = \Gamma dt_1 d\mathbf{r}_1 \tag{466}$$

is the probability of the bubble nucleation in the space-time point with the coordinates (t_1, r_1)

$$dP_2 = \Gamma dt_2 4\pi \left|\mathbf{r}_2 - \mathbf{r}_1\right|^2 d\left|\mathbf{r}_2 - \mathbf{r}_1\right| \tag{467}$$

is the probability of the nucleation of the second bubble at the distance

$$\left|\mathbf{r}_2 - \mathbf{r}_1\right| \equiv 2b \tag{468}$$

from the first one. The integration over angles was performed here assuming the space isotropy.

The factor

$$P_- \equiv \exp(-\Gamma\Theta) \tag{469}$$

determines the probability that no other bubbles nucleate in the 4-dimensional region Θ, in which the collision of the two considered bubbles takes place.

In what follows, we consider the probability density of the false vacuum decay Γ being a free parameter. The region Θ will be calculated below.

After integration, assuming that the probability of false vacuum decay does not depend on time, we obtain (Konoplich et al., 1998)

$$\frac{dP}{V_c} = 32\pi\Gamma^2 \exp(-\Gamma\Theta) b^2 dt_1 dt_2 db\ . \tag{470}$$

Here V_c is the volume within the cosmological horizon at the moment of the phase transition.

In the following we shall consider the K' system mentioned above which moves with the velocity

$$\upsilon = \frac{(t_1 - t_2)}{2b}\ . \tag{471}$$

Evidently, the velocity υ is also the velocity of FVB and of the black hole it forms. The radius of the colliding bubbles is given by

$$b' = \frac{b}{\gamma}\ , \tag{472}$$

where

$$\gamma = \left(1 - \upsilon^2\right)^{-1/2}\ .$$

Using Eqs. (454) and (470), it is easy to obtain the FVB distribution in terms of new variables M, v, t, where M is the mass of FVB (or BH), that is formed in the bubble collision, v is its velocity, and

$$t = b + \frac{t_1 + t_2}{2} \tag{473}$$

is the first moment of the bubble contact. One finds that this distribution has the form (Konoplich et al., 1998)

$$\frac{1}{V_c}\frac{dP}{dv dM} = \int \frac{64\pi}{3}\Gamma^2 \exp(-\Gamma\Theta)\gamma^4 \left(\frac{M}{C\rho_V}\right)^{1/3} \frac{dt}{C\rho_V} \; . \tag{474}$$

The difficult problem of the exact determination of the four-dimensional area Θ can be considered in the reasonable approximation (Konoplich et al., 1998).

Namely, suppose that each bubble that has reached the sphere with radius b' and with the center in the point O (see Fig. 11b) at the moment of the first bubbles contact t' prevents the creation of FVB. With this assumption the area Θ can be determined quite easily and is given by

$$\Theta = \int_0^{t'} d\tau' \, d^3r' \, \theta(\tau' + r' - b' - t') = \frac{\pi}{3}\left[(b' + t')^4 - (b')^4\right] . \tag{475}$$

Here the parameter b' is connected with the mass M according to the Eq. (454) and the time t' in the K' system is equal to

$$t' = \gamma t \; . \tag{476}$$

Therefore, after integration over t, the mass and the velocity distributions of FVB have the form

$$\frac{1}{V_c}\frac{dP}{dvdM} = \frac{64\pi}{3}\Gamma^2 \exp\left(-\Gamma\frac{\pi}{3}\left(\frac{M}{C\rho_V}\right)^{4/3}\right)\gamma^4\left(\frac{M}{C\rho_V}\right)^{1/3}\frac{I}{C\rho_V}\ , \quad (477)$$

where

$$I = \int_{t_-}^{\infty} d\tau \exp\left(-\frac{\pi}{3}\Gamma\left[\left(\frac{M}{C\rho_V}\right)^{1/3} + \tau\gamma\right]^4\right) \quad (478)$$

and

$$t_- = (1+v)\gamma\left(\frac{M}{C\rho_V}\right)^{1/3}\ . \quad (479)$$

Let us compare the volume V_{bag} containing one FVB and the volume V_{bubble} of one bubble at the end of the phase transition.

After numerical integration of (477) we obtain

$$V_{\text{bag}} \approx V_{BH} \approx 3.9\Gamma^{-3/4}\ . \quad (480)$$

On the other hand, the average volume per one bubble is

$$V_{\text{bubble}} = \frac{4\pi}{3}\left(\frac{3}{\pi}\right)^{3/4}\Gamma^{-3/4} \approx 4.0\cdot\Gamma^{-3/4}\ , \quad (481)$$

where it was assumed, that the bubbles have the spherical form. The expected equality

$$V_{\text{bag}} = V_{\text{bubble}} \quad (482)$$

is fulfilled quite satisfactory to prove the above approximation.

Distribution (477) may be conveniently expressed in terms of the dimensionless variable μ , defined as

$$\mu \equiv \left(\frac{\pi}{3}\Gamma\right)^{1/4}\left(\frac{M}{C\rho_V}\right)^{1/3} . \tag{483}$$

In terms of this variable, this distribution has the form

$$\frac{1}{V\Gamma^{-3/4}}\frac{dP}{dv d\mu} = 64\pi\left(\frac{3}{\pi}\right)^{1/4}\mu^3 \exp\left(\mu^4\right)\gamma^3 J(\mu, v) . \tag{484}$$

Here

$$J(\mu, v) = \int_{\tau_-}^{\infty} d\tau \exp\left(-\tau^4\right) , \tag{485}$$

where

$$\tau_- = \mu\left(1 + \gamma^2(1 + v)\right) . \tag{486}$$

This distribution appears to be rather narrow: the number of black holes with the mass 30 times greater than the average one is suppressed by the factor of the order of 10^5.

The average value of the dimensionless "mass" is

$$\mu = 0.32 . \tag{487}$$

Using of this value Konoplich et al. (1998) related the average mass of the black hole and the average volume containing it in the moment of the phase transition:

$$\langle M_{BH} \rangle = 0.25C\mu^3\rho_V \langle V_{BH} \rangle \approx 0.012C\rho_V \langle V_{BH} \rangle \,. \quad (488)$$

Remind that the constant C here is the model dependent value of the order of unity and its exact value is not important for the further estimations.

2.3. First-Order Phase Transitions in the Early Universe

Inflational models, where inflation is finished by the first order-phase transition, occupy a notified position in the modern cosmology of the very early Universe (see for example La, Steinhardt, 1989; Holman, Kolb, Wang, 1990; Holman et al., 1991; Adams, Freeze, 1991; Copeland et al., 1994; Occhionero, Amendola, 1994; Amendola et al., 1996).

In particular, in the framework of such models it is possible to consider the observed large-scale voids as remnants of the primordial bubbles with the characteristic size of several tens of Mpc (Occhionero, Amendola, 1994; Amendola et al., 1996). A detailed analysis of cosmological first-order phase transitions in the context of extended inflation can be found in (Turner et al., 1992).

Hereafter we will concentrate our attention only on the final moment of the inflational stage when the phase transition is completed. Remind that the first-order phase transition is considered as completed immediately after the percolation regime in the true vacuum is established. The establishment of such regime roughly takes place when at least one bubble per unit Hubble volume is nucleated. Accurate calculation (Turner et al., 1992) shows that the first-order phase transition is completed if the condition

$$Q \equiv \frac{4\pi}{9}\left(\frac{\Gamma}{H^4}\right)_{t_{\rm end}} = 1 \quad (489)$$

is valid. Here Γ is the bubble nucleation rate, t_{end} denotes the time of the phase transition and H is the Hubble constant in this period.

In the framework of the inflational models the true vacuum fills all space in the result as the first-order phase transition due to collisions of bubbles, nucleated at the final moment of exponential expansion. The collisions between such bubbles occur when their co-moving spatial dimension l is less or equal to the effective Hubble horizon

$$l \le H_{\text{end}}^{-1} \tag{490}$$

at the transition epoch.

If we take

$$H_0 = 100h \frac{\text{km}}{\text{s} \cdot \text{Mpc}}$$

for the Universe with

$$\Omega = 1 \ ,$$

the co-moving size of these bubbles is approximately given by

$$l \sim 10^{-21} h^{-1} \text{ Mpc} \ . \tag{491}$$

It is generally believed that such bubbles are rapidly thermalized leaving no trace in the distribution of matter and radiation.

However, it has been shown in the previous Section that for realistic parameters of the underlying field theory the collisions between only two bubbles lead to the black hole formation with the probability of the order of unity. The mass of these black holes is given by (see Eq. (488))

$$M_{BH} = \gamma_1 M_{\text{bub}} \,, \tag{492}$$

where

$$\gamma_1 < 10^{-2} \tag{493}$$

and M_{bub} denotes the mass that could be contained within the bubble volume in the period of bubble collision, if the energy density of bubble walls is completely thermalized.

The mechanism, suggested by Konoplich et al. (1998) leads to a new possibility of the PBH formation at the epoch of reheating in the inflational models, assuming the first-order phase transition in the end of inflation.

In the previous discussion we have considered either the process of PBH formation in the course of the evolution of small initial density perturbations at the post-inflational dust-like stage, what can lead to rather large probability but takes long time, or rapid PBH formation from the Gaussian tails of the distribution of the fluctuations, which is suppressed exponentially in the case of small post-inflation initial perturbations.

Quite different situation takes place in the end of the inflational stage, if the first-order phase transition takes place.

Namely, in the percolation regime, the collisions between the bubbles, having the size of the order of the effective Hubble radius, lead to formation of black holes with the masses

$$M_0 = \gamma_1 M_{\text{end}}^{\text{hor}} \,, \tag{494}$$

where $M_{\text{end}}{}^{\text{hor}}$ is the mass within the Hubble horizon at the end of inflation. According to (488) the initial mass fraction of such PBH is given by the relation:

$$\beta_0 = \gamma_1 \exp(-1) . \tag{495}$$

The suppression factor

$$P_- = \exp(-1) \tag{496}$$

causing slight difference with the expression in Eq. (488) is connected with the condition, that no secondary bubble nucleation takes place inside FVB, while it collapses into black hole.

One can rewrite the expression for the PBH mass, given by Eq. (494), in terms of inflationary energy scale H_{end} in the form

$$M_0 = \frac{\gamma_1}{2} \frac{m_{Pl}^2}{H_{\text{end}}} . \tag{497}$$

For example, let us take in Eq. (497) the estimation for the inflationary scale

$$H_{\text{end}} \approx 4 \cdot 10^{-6} m_{Pl} , \tag{498}$$

which follows from the observations of the anisotropy of the thermal electromagnetic background. One obtains then, that the initial mass fraction

$$\beta_0 = 6 \cdot 10^{-3} \tag{499}$$

is contained in the Universe immediately after the first-order phase transition in the form of PBH with the mass

$$M_0 \approx 0.7 \text{ g} . \tag{500}$$

In the radiation-dominated stage, after reheating, the relative contribution of these PBH into the total cosmological density grows as the scale factor, and it means that at the moment

$$t_1 \approx \left(\beta_0^2 H_{\text{end}}\right)^{-1} \tag{501}$$

over 50% of the total density is contained in the primordial black holes.

Since the PBH behave as the dust-like matter, from the moment t_1 the equation of state of the Universe should have the form

$$p = 0 \ .$$

The PBH-dominated dust-like stage ends at the moment of the complete evaporation of PBH at the moment:

$$t_2 = \frac{1}{g}\left(\frac{M_0}{m_{Pl}}\right)^3 t_{Pl} \ , \tag{502}$$

where g is the effective number of degrees of freedom of massless particles at this time.

In general, at the PBH-dominated stage black holes with greater masses could be formed. However, the probability of formation of such more massive PBH will be negligibly small for non-ultraviolet spectra of density fluctuations. In the case when the initial spectrum does not grow to small scales, the amplitude of initial post-inflation density perturbations does not exceed

$$\delta \cong 10^{-6} \ . \tag{503}$$

For such amplitude, the probability of the direct black hole production at the dust-like stage is very small and the evolutional process of PBH formation takes much more time than the time scale of the initial small black hole evaporation.

There is a number of constraints on the maximum allowed mass fraction of PBH, covering various ranges of PBH mass (Zeldovich, Starobinsky, 1976; Naselsky, 1978; Miyama, Sato, 1978; Lindley, 1980; Zeldovich et al., 1977; Rothman, Matzner, 1981; MacGibbon, Carr, 1991; Hawking, 1974; see for review Polnarev, Khlopov, 1985 and Khlopov et al., 1984). Based on the astrophysical observations they use different signatures for the possible existence of PBH and can be divided into two major groups.

The first group of constraints is based on the analysis of the effect of the Hawking PBH evaporation, and the second uses the PBH gravitational effects only.

The PBH evaporation via thermal emission has potentially observable astrophysical consequences. Observations have established limits on the maximum fraction of PBH allowed during the period of their evaporation. The PBH with mass

$$M_{ev} \leq 5 \cdot 10^{14} \text{ g} \tag{504}$$

should evaporate before the present epoch.

For more massive PBH, the effect of evaporation is not significant, and they should be present in the modern Universe. The universal constraint on their number follows from the condition, that their density does not exceed the upper limit on the total density. In the case of simple inflational scenario that gives

$$\Omega_{PBH} \leq 1 \ . \tag{505}$$

It is generally assumed, that the evaporation proceeds until the PBH completely vanishes (Hawking, 1974), but there are various arguments against this assumption (Markov, 1984; Zeldovich, 1981; Barrow et al., 1992; Carr, Gilbert, 1994; Alexeyev, Pomazanov, 1997; Dymnikova, 1996).

If one supposes that the BH evaporation leaves a stable relic, normally assumed to have a mass of the order of

$$m_r = \kappa m_{Pl} \ , \tag{506}$$

where

$$\kappa \cong 1 \div 10^3 \tag{507}$$

one can derive the density of PBH relics from all PBH, evaporated to the present moment.

In particular, one can estimate the density of PBH relics from the small-mass PBH formed in the first-order phase transition after inflation.

Since the probability of the formation of such PBH was estimated to be large, they soon start to dominate in the Universe and to realise the early dust-like stage of their dominance, which is finished at the moment of PBH evaporation. The mass fraction of the Universe going to PBH evaporation relics is given by

$$\alpha_r = \kappa \left(\frac{m_{Pl}}{M_0} \right) \left(\frac{t_{eq}}{t_2} \right)^{1/2} \ , \tag{508}$$

where t_2 is the period of evaporation of PBH with the mass M_0 , and the beginning of the modern stage of matter dominance is

$$t_{eq} = 3.2 \cdot 10^{10} h^{-4} \ \mathrm{s} \ . \tag{509}$$

From the condition, that the matter dominance can not start at

$$t < t_{eq} \;, \tag{510}$$

one finds (Konoplich et al., 1998) that the inequality

$$\alpha_r < 1 \tag{511}$$

must be valid. With the use of Eqs. (497), (506) and (507) one can express this condition as the following constraint:

$$\frac{H_{end}}{m_{Pl}} < \frac{0.6\gamma_1}{g^{1/5}\kappa^{2/5}}\left(\frac{t_{Pl}}{t_{eq}}\right)^{1/5} \leq 10^{-11}\,\frac{h^{4/5}\gamma_1}{g^{1/5}\kappa^{2/5}} \;. \tag{512}$$

On the other hand, the restriction

$$\beta(M) < 10^{-8}\left(\frac{10^{11}\ \mathrm{g}}{M}\right) \tag{513}$$

(Zeldovich, Starobinsky, 1976), obtained from the condition that the PBH evaporation does not lead to the overproduction of the entropy of the Universe, implies that the evaporating PBH can represent the dominant fraction of the cosmological density in the period of their formation only provided that their mass does not exceed 10^3 g . In principle, it means that all observed entropy of the Universe can be produced in the evaporation of small-mass PBH. So PBH with the mass

$$M < 10^3\ \mathrm{g} \tag{514}$$

could be produced in the very early Universe with the probability of the order of unity without any contradiction with the observations. Moreover, one can ascribe the observed entropy of the Universe to the evaporation of small-mass PBH.

In terms of PBH, formed as the result of the first-order phase transition at the end of inflation, the condition, that the observed entropy is the result of PBH evaporation, implies the following lower limit on the inflation energy scale

$$\frac{H_{end}}{m_{Pl}} \geq 10^{-9}\gamma_1 \,. \tag{515}$$

The two conditions, given by Eqs. (512) and (515) are incompatible, if the hypothesis on the existence of stable relics of PBH evaporation is valid.

One may conclude, that it is difficult to make compatible the existence of relics of BH evaporation and the first-order phase transition at the end of inflation.

CHAPTER 5

PRIMORDIAL PARTICLES AND THE THEORY OF BIG BANG NUCLEOSYNTHESIS

1. New Particle Effects on Nucleosynthesis

The simplest way to use the observed abundance of light elements for tests of particle theory is to estimate the effect of hypothetical phenomena on the parameters of the theory of standard Big Bang nucleosynthesis (see Chapter 3).

In such an approach, the effects of new particles are reduced to parameters of Big Bang nucleosynthesis theory, so that the restrictions on possible properties of such particles follow from constraints on these parameters.

In order to restrict the parameters of Big Bang nucleosynthesis, one compares the observed abundance of light element, extrapolated to their primordial values, with the results of detailed calculations by Wagoner, Fowler, Hoyle (1967), Wagoner (1973) (see also Gamow, 1946; 1948; Hayashi, 1950; Alfer, Herman, 1953; Reeves et al., 1973; Yahil, Baude, 1976; Yang et al., 1979; Olive et al., 1981; Yang et al., 1984; Olive et al., 1990; Schramm, Copi, 1996).

Let us consider the main assumptions, parameters and results of these calculations.

Assumptions.

1. The expansion of the Universe is considered on the basis of the metric theory of gravitation;

2) There are no strong inhomogeneity and anisotropy during the period of nucleosynthesis in the Universe (On the effect of inhomogeneity on nucleosynthesis see: Zeldovich, 1975; Scherrer, Applegate, Hogan, 1987; Alcock, Fuller, Mathews, 1987; Fowler, Malaney, 1988; Kurki-Suonio et al., 1990; Alcock et al., 1990. On effects of anisotropy see: Rothman, Matzer, 1984; Yushkevich et al., 1985 and Refs. therein);
3) It is assumed, that there was a high-temperature stage in the evolution of the Universe with the temperature

$$T > 10^{10}\ \mathrm{K}\ . \tag{516}$$

The equilibrium between nucleons, neutrino-antineutrino, electron-positron pairs and radiation was established on such stage. In the course of successive expansion beta-processes went out of equilibrium and the freezing of neutron to proton ratio takes place.
4) All local processes of antinucleon annihilation had been finished before the beginning of the period of nucleosynthesis, and the effects of annihilation in nucleosynthesis were negligible (On the effects of annihilation, or non-equilibrium particles see below).

Parameters.

1) The baryon density ρ_b.

It is assumed, that cosmological expansion takes adiabatically (see Chapter 3), so that the specific entropy

$$s_b = \frac{T^3}{\rho_b} \propto \frac{n_\gamma}{n_b} = const \tag{517}$$

(compare with Eq. (198)). If there was no additional energy release, enlarging the number density of primordial photons (with the exception of electron-positron pair annihilation, being taken into account explicitly in calculations) the value of s_b is defined by the modern baryon to photon ratio.

2) The parameter ξ, determining the expansion rate at the stage when

$$p = \tfrac{1}{3}\varepsilon$$

as follows

$$\frac{dv}{vdt} = \frac{\ddot{a}}{\dot{a}} = -\xi\sqrt{24\pi G\rho_n} \ . \tag{518}$$

Here a is the scale factor, and ρ_n is the cosmological density in the period of Big Bang nucleosynthesis, accounting for all known types of particles (photons, electron-positron pairs, neutrinos and antineutrinos) only. The baryon relative contribution into the total cosmological density does not exceed

$$\frac{\rho_b}{\rho_{tot}} \leq 10^{-5} \ . \tag{519}$$

If the parameter $\xi > 1$, some new types of relativistic particles should be present in the Universe. (Effects of non-relativistic particles will be discussed further in this Chapter).

Shvartsman (1969) first obtained general restriction on the possible number of species of light weakly interacting particles, i.e. on the value of ξ from the results of Big Bang nucleosynthesis. In particular, assuming, that all such particles represent new types of neutrino; the maximal possible value of ξ imposes an upper limit on the number of neutrino species. According to Shvartsman (1969), this upper limit on the number of new neutrino species was

$$\Delta N_\nu < 8 \div 10 \, , \tag{520}$$

giving no practical use for neutrino physics.

However, in successive papers (Steigman, Schramm, Gunn, 1977; Yang et al., 1979; 1984) the limit was reduced to

$$\Delta N_\nu < 1 \, . \tag{521}$$

For long time, this number was the strongest restriction on the number of neutrino species (Schramm, 1990).

Severe constraints, obtained in these papers, correspond to the absence of new families of quarks and leptons, except for the three families,

$$\begin{pmatrix} \nu_e \\ e \\ u \\ d \end{pmatrix} \begin{pmatrix} \nu_\mu \\ \mu \\ c \\ s \end{pmatrix} \begin{pmatrix} \nu_\tau \\ \tau \\ t \\ b \end{pmatrix}$$

related to the three known types of light neutrinos ν_e , ν_μ and ν_τ.

The existence of extra families does not contradict this stringent limit, if the respective neutrinos are massive, since the contribution of heavy neutrinos or hypothetical heavy stable neutral leptons with the mass

$$m_\nu >> 1 \text{ MeV} \tag{522}$$

into the total density is small

$$\frac{\rho_\nu}{\rho_{tot}} \propto \left(\frac{3 \text{ MeV}}{m_\nu} \right)^2 \tag{523}$$

due to annihilation of the bulk of this particles in the period of their freezing out in the very early Universe at

$$T \le m_\nu \tag{524}$$

(see review in: Dolgov, Zeldovich, 1981, and Chapter 3).

In this case, putting the cosmological limit together with the measurements of the Z-boson width, one can exclude the existence of new heavy neutrinos with the mass, smaller than

$$m_\nu \le \frac{m_Z}{2} \ , \tag{525}$$

where m_Z is the mass of Z-boson.

On the other hand, the upper limit on the value of ξ imposes severe constraints on the particle models, predicting new types of light weakly interacting particles, related to new fundamental symmetries, arising in such models.

In the case of CP violation, the equivalence of left- and right-handed co-ordinate systems can be restored only due to the existence of invisible twins of the ordinary particles — their mirror partners (Lee, Yang, 1956; Kobzarev, Okun, Pomeranchuk, 1966; Okun, 1983).

Such particles should not have practically any interactions, except for gravitational interaction, with the ordinary particles, and their existence can be tested only by their astrophysical effects (Blinnikov, Khlopov, 1982, 1983; Dubrovich, Khlopov, 1989; Khlopov et al., 1991).

In this model, mirror electrons, positrons, photons, right-handed neutrinos and left-handed antineutrinos are predicted with the masses coinciding with the masses of corresponding ordinary particles, and they should be present in the Universe with the same concentrations as their ordinary twins.

Thus, the model of mirror particles predicts

$$\xi = \sqrt{2} \,, \tag{526}$$

what is equivalent to about eight additional types of neutrinos.

It means that the restriction (Yang et al., 1979, 1984) excludes completely the possibility of exact symmetry of ordinary and mirror matter in the Universe (Carlson, Glashow, 1987). The room is left for asymmetry in either mirror particle properties or in their cosmological evolution. The former case corresponds to shadow world with distinct difference of properties of ordinary and shadow particles, as in the case of the $E_8\mathrm{x}E_8'$ superstring model. The latter can take place for asymmetric mechanisms of inflation, suppressing the relative contribution of mirror particles into the cosmological density.

Note, however, that the severe constraints on the value of ξ follow from the comparison of predictions of Big Bang nucleosynthesis and the extrapolation of the observed abundance of light elements to their primordial values. The critical analysis of the observational data, given in Chapter 3, makes us to be more cautious, than it is usually accepted in the literature, concerning possible ambiguities and model dependence in such extrapolation.

The frozen neutron-to-proton ratio n/p is determined from the comparison of rates of expansion and beta-processes. In particular, the cross section of beta-reactions

$$\begin{aligned}
&\nu_e n \rightarrow ep \,, \\
&\bar{\nu}_e p \rightarrow e^+ n \,, \\
&ep \rightarrow \nu_e n \,, \\
&e^+ n \rightarrow \bar{\nu}_e p
\end{aligned} \tag{527}$$

is determined by the value of the weak transition matrix element, extracted from the data on the lifetime of neutrons.

In the calculations of Wagoner (1973), Wagoner, Fowler, Hoyle (1976), Yang et al. (1979, 1984), the free neutron lifetime relative to beta-decay was taken equal to

$$\tau = \frac{926 \text{ s}}{C} , \tag{528}$$

where the constant C took into account for the ambiguity in the experimental value of neutron lifetime.

Neutrino degeneracy in the period of nucleosynthesis (Yahil, Baude, 1976; Linde, 1979; Salam, 1981), CP violation effects in non-equilibrium neutrino oscillations to sterile neutrino states (Khlopov, Petcov, 1981; see below), the presence of non-equilibrium neutrino fluxes (Weiner, Naselsky, 1977; Dolgov, Kirilova, 1987) can influence the rate of beta-processes.

In the standard model of Big Bang nucleosynthesis, it is assumed that

$$C = \xi = 1 , \tag{529}$$

and the annihilation of electron-positron pairs is considered as the only source of energy release after nucleosynthesis.

Results.

The analysis (see, e.g., review by Khlopov, Chechethkin, 1987) of the dependence of light element abundance, predicted in the standard model with the parameters given by Eq. (529), on the baryon density ρ_b shows that the deuterium concentration is most sensitive to the value of ρ_b. The greater is ρ_b , the stronger is deuterium depletion due to its burning.

Thus, significant deficit of deuterium abundance $X_D{}^{th}$ is predicted for

$$\rho_b > 10^{-30} \frac{\mathrm{g}}{\mathrm{cm}^3} \tag{530}$$

as compared with the observed average value of this abundance $X_D{}^{th}$,

$$X_D^{th} < X_D^{obs} \sim 10^{-5} \div 10^{-4} \ . \tag{531}$$

The qualitative physical explanation of this result, which belongs to Ya.B.Zeldovich (private communication), is as follows.

In the chain of reactions of nucleosynthesis after the deuterium production in the reaction (see Chapter 3)

$$n + p \rightarrow D + \gamma \ ,$$

the most part of it converts into tritium in the reaction

$$D + D \rightarrow T + p \ .$$

Therefore, the main process of deuterium burning is the reaction (see Eq. (297))

$$D + T \rightarrow {}^4He + n \ ,$$

in which the relative concentration of deuterium is reduced according to the kinetic equation

$$\frac{dX_D}{dt} = -X_D X_T (n_b \sigma v) \ . \tag{532}$$

Since at sufficiently large baryon densities

$$\rho_b > 10^{-31}\,\frac{\text{g}}{\text{cm}^3} \tag{533}$$

the relative concentration of tritium X_T is virtually independent on baryon density, the frozen out concentration of deuterium depends exponentially on the baryon density:

$$X_D^{th} \propto \exp(-A\Omega_b)\ . \tag{534}$$

On the other hand, ^{4}He abundance has significant dependence on ρ_b at very small baryon densities only. At

$$\rho_b < 10^{-31}\,\frac{\text{g}}{\text{cm}^3} \tag{535}$$

the frozen n/p ratio is greater, than $\langle n/p \rangle$ in nuclei, produced as the result of nucleosynthesis, due to decay of the fraction of free neutrons.

The dependence of ^{4}He abundance on ξ (or on the number of neutrino species N_ν) was given in Chapter 3 for

$$\rho_b > 10^{-31}\,\frac{\text{g}}{\text{cm}^3}\ .$$

The existence of the symmetric mirror world corresponds to the weight concentration of ^{4}He

$$Y_{\text{prim}} \sim 0.29 \tag{536}$$

for

$$\rho_b > 10^{-31}\,\frac{\text{g}}{\text{cm}^3}\ .$$

If

$$Y_{\text{prim}} \le 0.25 \tag{537}$$

in accordance with the most severe constraints, the allowed value of the total number of light neutrino species N_ν can not exceed

$$N_\nu < 4\ . \tag{538}$$

1.1. Restrictions on the Total Cosmological Density

According to the standard Big Bang model, the weak beta-transitions, namely, the reactions

$$\begin{aligned} ep &\leftrightarrow \nu_e n \\ e^+ n &\leftrightarrow p\bar{\nu}_e \end{aligned} \tag{539}$$

at

$$T > 1\ \text{MeV} \tag{540}$$

maintain the equilibrium value of neutron-to-proton concentration ratio

$$\frac{n}{p} = \exp\left(-\frac{\Delta m_{np}}{T}\right), \tag{541}$$

where Δm_{np} is the difference of neutron and proton masses.

At

$$T < 1 \text{ MeV} \tag{542}$$

the rate of these transitions turns out to be smaller than the rate of cosmological expansion. Relative concentration of neutrons freezes out. The most part of neutrons goes after successive nuclear reactions into ^{4}He nuclei. Thus, the ^{4}He abundance is essentially determined by the value of the frozen n/p ratio and is very sensitive to the conditions of n/p freezing in the early Universe.

In the absence of significant electron neutrino (or electron antineutrino) excess in the Universe in the period of n/p freezing, the rate of beta reactions is determined only by the temperature T so that the corresponding time scale is given by

$$\tau(n \leftrightarrow p) = (n_\nu \sigma v)^{-1} = \left(AT^3 \frac{G_F^2 T^2}{\pi} \right)^{-1} , \tag{543}$$

where G_F is the Fermi constant of weak interactions and the numerical factor is equal to

$$A = 0.1 \; .$$

At the given temperature, the expansion rate depends on the equation of state. If this equation is relativistic,

$$p = \tfrac{1}{3} \varepsilon \; ,$$

the expansion rate is determined by the total density of all relativistic particles, that existed in the Universe in the considered period. The total density should account for the contribution of both known and unknown types of particles, and is equal to

$$\rho = \xi^2 \rho_n , \tag{544}$$

where ρ_n is the density of all known types of particles.

The greater is ξ , the higher is the expansion rate, the earlier (at higher T_f) fulfils the condition of freezing out

$$\tau(n \leftrightarrow p) = t \tag{545}$$

leading to

$$BG_F^2 T_f^5 = \xi \frac{T_f^2}{m_{Pl}} , \tag{546}$$

where B is constant.

The greater is the frozen ratio

$$\frac{n}{p} = \exp\left(-\frac{\Delta m_{np}}{T_f} \right) , \tag{547}$$

the more ^{4}He is produced.

At

$$\xi > 2 \tag{548}$$

the predicted amount of ^{4}He exceeds the amount of ^{4}He, observed in the modern Universe. It provides upper limit on the allowed contribution of light weakly interacting particles into the cosmological density in the period of n/p freezing out (Shvartsman, 1969).

Restriction of the concentration of primordial ^{4}He by the value (Steigman et al., 1977; Olive et al., 1979; Ellis et al., 1986; Schramm, Copi, 1996) (compare with the Eq. (537))

$$Y_{\text{prim}} \leq 25\% \tag{549}$$

rules out the existence of even one extra type of light two-component neutrino with the ordinary weak interaction. Because the restriction on the number of neutrino species is

$$N_{\nu} < 4$$

(see Eq. (538)).

In this case, the right-handed neutrinos (left-handed antineutrinos) can exist, provided that their cosmological abundance is significantly smaller, than the one of the ordinary left-handed neutrinos (right-handed antineutrinos). Such a suppression of concentration of the right-handed neutrinos can take place, if their interaction is much more weak, than the ordinary weak interaction, so that decoupling of the right-handed neutrinos has taken place earlier, than the ordinary neutrinos, at higher temperature

$$T > 100 \text{ MeV} , \tag{550}$$

when there were more relativistic species in the equilibrium with plasma and radiation (see Chapter 3).

The restriction on the primordial ^{4}He abundance, given by Eq.(549), provides in this case an upper limit on the cross section σ_R of right-handed neutrino interaction

$$\sigma_R \sim G_R^2 T^2 , \tag{551}$$

where G_R is the dimensional constant of four-fermion interaction of right-handed currents.

Assuming that this interaction is mediated by corresponding gauge bosons, so that

$$G_R \propto M_R^{-2} \ , \tag{552}$$

the upper limit on the cross section of interaction of right-handed currents, following from the upper limit on ^{4}He abundance, corresponds to the lower limit on the value of M_R.

If

$$G_R \propto G_F \left(\frac{M_W}{M_R} \right)^2 \ , \tag{553}$$

one obtains (Olive et al., 1979; Dolgov, 1981)

$$M_R > 10 M_W \ . \tag{554}$$

Similar restriction

$$M_{Z'} > 10 M_Z \tag{555}$$

can be obtained for neutral current Z' interaction of right-handed neutrinos, predicted in E_8 x E_8' superstring models (Ellis et al., 1986).

In the realisation of the original idea of Shvartsman (1969), such consideration can be applied to any form of relativistic matter, predicted in the framework of particle theory. For instance, Koikawa (1991) has considered the contribution of gravitational waves, emitted by the cosmic strings, into the cosmological density in the period of nucleosynthesis and imposed constraint on the parameters of cosmic strings.

Let us show now, that the observational upper limit on the ^{4}He abundance can also be used in order to lay on restrictions on the possible properties of massive non-relativistic particles, dominating in the early Universe during the period of freezing out of n/p ratio.

Consider particles with the mass m and relative frozen out concentration

$$r = \frac{n_m}{n_r} \ , \tag{556}$$

where n_m and n_r are the number densities of the frozen out and the relativistic particles, respectively. The density of considered particles ρ_m , given by

$$\rho_m = mn_m = rmn_r \ , \tag{557}$$

begins to exceed the density of relativistic particles ρ_r , which can be approximated as

$$\rho_r \approx 3Tn_r \ , \tag{558}$$

at the temperature

$$T_0 \approx \tfrac{1}{3} rm \tag{559}$$

what corresponds to the period (see Chapter 4)

$$t_0 \approx \frac{m_{Pl}}{T_0^2} \approx \left(\frac{3m_{Pl}}{rm}\right)^2 t_{Pl} \ . \tag{560}$$

At

$$t > t_0 \tag{561}$$

the frozen out non-relativistic particles dominate in the Universe, and that corresponds to the dust-like stage with the equation of state

$$p = 0 \ .$$

On this stage the temperature decreases with time as

$$T = \left(\frac{t_0}{t}\right)^{2/3} T_0 \tag{562}$$

and time is related to the temperature as

$$t = \frac{m_{Pl}}{T_0^{1/2} T^{3/2}} . \tag{563}$$

If the n/p ratio is frozen out at this stage at the freezing temperature

$$T_{fr} < T_0 , \tag{564}$$

the inequality

$$T_{fr} > T_n \tag{565}$$

is valid, where the temperature T_n, defined as

$$T_n = \left(\beta G_F^2 m_{Pl}\right)^{-1/3} \approx 1 \text{ MeV} \tag{566}$$

with numerical constant β, is the n/p ratio freezing temperature in the standard model with

$$\xi = 1 .$$

Indeed, using Eqs. (543), (545), (546) and (563), one obtains

$$T_{fr} = \left(\frac{T_0^{1/2}}{\beta G_F^2 m_{Pl}} \right)^{2/7} = \left(\frac{T_0}{T_n} \right)^{1/7} T_n \ . \tag{567}$$

With the account for Eq. (564) one obtains from (567) that the n/p ratio freezes out on $p = 0$ stage at higher temperature T_{fr} than on $p = \varepsilon/3$ stage in the standard model without unknown particles, so that the corresponding frozen ratio is higher.

Dicus et al. (1978) have studied the kinetics of nuclear reactions, leading to the ^{4}He formation on $p = 0$ stage after n/p freezing out. It was shown, that the predicted ^{4}He abundance exhibited no significant dependence on the equation of state, being determined mainly by the value of T_{fr} (or frozen out n/p).

It is interesting, that according to Dicus et al. (1978) and Polnarev and Khlopov (1982), for the fixed modern baryon density the predicted deuterium abundance is higher in the presence of $p = 0$ stage in the period of nuclear reaction, corresponding to the time interval

$$1 < t < 10^2 \text{s} \ . \tag{568}$$

The effect is enhanced in the presence of inhomogeneities (Zeldovich, 1975; Polnarev, Khlopov, 1982) together with some decrease of the ^{4}He abundance. So there is no contradiction with the observed helium abundance, provided that

$$T_0 < T_n \ , \tag{569}$$

and, consequently, the n/p ratio freezes out on $p=\varepsilon/3$ stage under the condition

$$T_{fr} = T_n \ . \tag{570}$$

If hypothetical massive particles start to dominate in the Universe before cosmological nucleosynthesis at the temperature T_0, satisfying the inequality

$$T_0 > 2T_n \ , \tag{571}$$

the freezing out of the n/p ratio on such stage with $p = 0$ results in the predicted ^{4}He abundance, exceeding its maximal observed value.

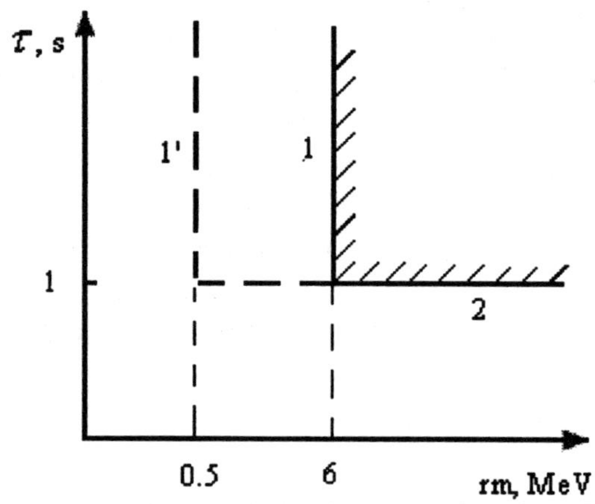

Fig.12. Constraints on *rm* and τ of heavy metastable particles from observed (lines 1 and 2) and extrapolated to primordial (line 1′) helium abundance

Thus, the existence of massive particles does not contradict the observational data on ^{4}He abundance, if either:

1) they start to dominate in the Universe at the temperature, T_0 , satisfying the condition

$$T_0 < 2T_n \tag{572}$$

that corresponds to the values of rm, given by

$$rm < 6 \text{ MeV} \tag{573}$$

and constrained by line 1 on Fig.12,

or

2) the $p = 0$ stage of massive particle dominance has finished before freezing out of the n/p ratio, i.e. the lifetime of massive particles relative to decay or to annihilation in inhomogeneities, formed at $p = 0$ stage, did not exceed 1 s. The respective constraint is given by line 2 on Fig.12.

According to more restrictive condition Eq. (549)

$$Y_{\text{prim}} \leq 25\% ,$$

one gets much stronger limits, presented also on Fig.12 (dashed lines).

The observed ^{4}He abundance provides the possibility to constrain the contribution of any form of hypothetical particles and fields at the first second of expansion. Such contribution can not exceed the contribution of the known particles (or even 1/6 of that), taken into account by the old standard theory of Big Bang nucleosynthesis.

This restriction is universal. It takes into account the energy density of new particles only. The corresponding contribution into the total density leads to the higher expansion rate and the higher temperature freezing of the n/p ratio.

However, the effect of hypothetical particles on the relationship between the rates of expansion and β-processes, resulting in the change of the frozen n/p ratio, can be induced not only by the increase of expansion rate.

The existence of new particles, fields or phenomena can give rise to a number of effects in the period of n/p freezing, influencing the rate of β-processes. All these effects, being different by their physical nature, influence directly the kinetics of nucleon reactions, changing the value of the n/p ratio.

1.2. The Shift of Equilibrium Rates of β-processes

Let us consider first the effects, changing the equilibrium rates of β-processes.

The theory of Big Bang nucleosynthesis has analysed possible effects of nonzero chemical potential of electron neutrinos on the rates of β-reactions (Fowler, 1970; Steigman, 1979; Linde, 1979; Salam, 1981). The dependence of ^{4}He abundance on the value of this chemical potential was determined in numerical calculations by Yahil, Baude (1976) and Fowler (1970).

This dependence can be easily understood qualitatively (see Fig.13), because the nonzero chemical potential μ corresponds to the suppression of either antineutrino, or neutrino density, depending on the sign of excessive lepton charge.

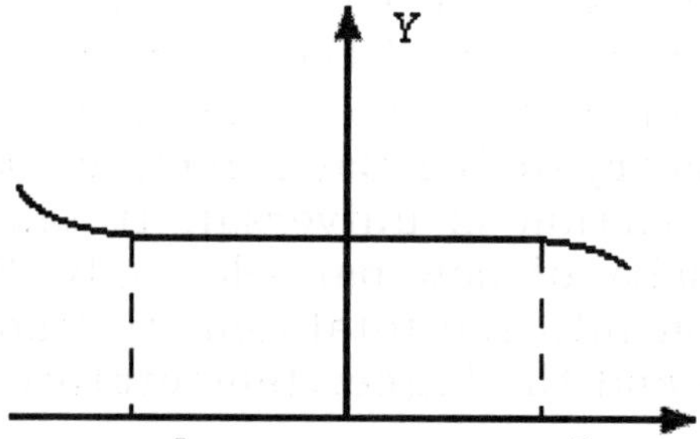

Fig.13. Large chemical potential of the electron neutrino leads to the decrease of helium abundance for $\varphi > 0$ and to the increase of this abundance at $\varphi < 0$.

In the former case, the neutrino excess is present, inducing the decrease of the frozen n/p ratio. In the latter case, the antineutrino excess leads to the increase of this ratio.

In the presence of chemical potential μ instead of Eq.(541) one obtains the frozen value of n/p ratio

$$\frac{n}{p} = \exp\left(-\frac{\Delta m_{np}}{T_f} - \varphi \right) . \tag{574}$$

Here the dimensionless parameter φ is introduced, defined as

$$\varphi = \frac{\mu}{T_f} . \tag{575}$$

In the framework of GUT models, the existence of lepton asymmetry by 8-10 orders of the magnitude larger, than the baryon asymmetry, seems rather improbable (Langacker, Pi, 1980; Kolb, Turner,1983). As a rule, in these models the value of lepton asymmetry, being generated in the very early Universe together with the baryon asymmetry, is of the same order of magnitude. Moreover, if there was a period in the very early Universe, when lepton non-conserving processes were in equilibrium, any initial lepton asymmetry has been washed out.

At high temperatures, of the order and higher than the electroweak scale, i.e. at

$$T \geq m_W ,$$

rapid processes due to nonperturbative electroweak effects (see Chapter 3) become possible, in which lepton (L) and baryon (B) numbers are not conserved separately, and only the combination $B - L$ is conserved (Kuzmin, Shaposhnikov, 1985). These processes result in vanishing of the sum $B + L$, so that the absolute value of the baryon asymmetry ΔB should be equal to the value of the lepton asymmetry ΔL.

Thus, on the basis of various theoretical arguments, the observed ΔB seem to exclude large value of ΔL in the period of freezing out of the n/p ratio. However, having in mind various possible modification of the scenario of very early Universe, the question on the dependence of the frozen n/p ratio on the value of lepton charge of the Universe still remains open.

Khlopov and Petcov (1981) have pointed out the possibility, that significant difference in electron neutrino and antineutrino concentrations can appear in the period of n/p freezing out even for the zero lepton number of the Universe.

The possibility is related to the existence of "sterile states" of right-handed neutrinos (and left-handed antineutrinos) and CP violation in non-equilibrium oscillations of electron neutrino into left-handed sterile antineutrino

$$\nu_e \leftrightarrow \bar{\nu}_L \, , \tag{576}$$

and of electron antineutrino into right-handed sterile neutrino

$$\bar{\nu}_e \leftrightarrow \nu_R \, . \tag{577}$$

Owing to superweak interaction of sterile neutrino with the ordinary particles, the concentration of sterile neutrinos in the Universe should be much smaller, than the concentration of ordinary neutrinos (see Chapter 3). The same is true, if mirror particles exist, interacting with the sterile neutrino states in the same manner as ordinary particles interact with ordinary neutrino, provided that there is no symmetry in the distribution of mirror and ordinary particles in the Universe.

Indeed, even if sterile neutrinos were in equilibrium with the other particles, their decoupling should have taken place much earlier (at much higher temperatures), than the decoupling of ordinary neutrinos due to much weaker interaction of sterile neutrinos, as compared to the ordinary weak interaction. Therefore, the decoupled sterile neutrinos should have the smaller concentration, than the ordinary neutrinos.

The concentration of sterile neutrinos should have been even more smaller, if there was no period in the very early Universe, when sterile neutrinos were in the equilibrium with the rest of particles (Dolgov, 1980).

If neutrino have both Majorana and Dirac masses, the oscillations between the ordinary and sterile neutrino states given by Eq. (576)

$$\nu_e \leftrightarrow \bar{\nu}_L$$

and by Eq. (577)

$$\bar{\nu}_e \leftrightarrow \nu_R$$

are possible, together with the oscillations of electron neutrino into ordinary left-handed neutrino

$$\nu_e \leftrightarrow \nu_L \tag{578}$$

and the oscillations of electron antineutrino into ordinary right-handed antineutrinos

$$\bar{\nu}_e \leftrightarrow \bar{\nu}_R \ . \tag{579}$$

If such transitions are "switched on" before the ordinary left-handed neutrino (right-handed antineutrino) decouple from radiation, there is an effective mechanism, bringing sterile left-handed antineutrino (right-handed neutrino) states into the thermodynamic equilibrium.

It takes place, if the oscillation length L_{osc} , being for the bulk of neutrino at the temperature T of the order of

$$L_{osc} \sim \frac{T}{\delta m^2} , \tag{580}$$

where δm^2 is the square mass difference of neutrino mass states, is smaller, than the mean free path l of ordinary neutrinos in plasma, given by

$$l = \left(n_l \sigma_{\nu l}\right)^{-1} \sim \left(G_F^2 T^5\right)^{-1} . \tag{581}$$

In Eq. (581), it is taken into account that at the temperatures 1-100 MeV the main effect of neutrino opacity is given by the neutrino interaction with relativistic ordinary neutrino-antineutrino and electron-positron gases, so that n_l is the number density of these gases, being at the temperature T of the order of

$$n_l \sim T^3 , \tag{582}$$

and $\sigma_{\nu l}$ is the corresponding weak interaction cross section

$$\sigma_{\nu l} \sim G_F^2 T^2 . \tag{583}$$

One obtains that the condition

$$L_{osc} < l \tag{584}$$

takes place at $T > 3\text{MeV}$, provided that the square mass difference for neutrino states with definite mass satisfies the inequality

$$\delta m^2 > 10^{-7}\,\mathrm{eV}^2 \ . \tag{585}$$

The secondary decoupling of sterile neutrinos occurs in this case, when the rate of weak left-handed neutrino (right-handed antineutrino) interactions with plasma turns out to be smaller than the rate of expansion. Sterile and ordinary neutrinos decouple simultaneously from the other particles. Hence the concentration of sterile neutrinos after their secondary decoupling turns out to be equal to the concentration of ordinary neutrinos.

Note, that low-energy neutrinos with the energy

$$E_\nu \ll T \tag{586}$$

begin to oscillate much earlier, than the condition

$$L_{osc} \sim l \tag{587}$$

is fulfilled for neutrino with

$$E_\nu \sim T \ .$$

However, low-energy neutrinos present a small fraction of all neutrinos, since the maximum of their energy distribution corresponds to

$$E_\nu \sim 3T \ . \tag{588}$$

That is why in order to estimate the effect of oscillations, it is reasonable to consider neutrinos with the energy near the maximum of the energy distribution. In this case there can be doubling of neutrino species in the period of freezing out of the n/p ratio due to equilibrium oscillations between ordinary and sterile neutrino states.

In principle, the condition (587)

$$L_{osc} \sim l$$

should be corrected by the factor, taking into account the amplitude of oscillations. However, Enqvist, Kanulainen and Maalampi (1990) shown, that in the case of negative square mass difference of neutrino mass states, lying in the interval

$$-10^{-5}\mathrm{eV} < \delta m^2 < 0 \tag{589}$$

resonant transition of MSW type (Mikheev, Smirnov, 1985; 1986; Wolfenstein, 1978; 1979) should take place in neutrino oscillations into sterile states. It should lead to the doubling of neutrino species for a wide range of δm^2 and neutrino mixing angles, satisfying the condition

$$\sin 2\theta > 10^{-4} \ . \tag{590}$$

Combining the prediction on doubling of neutrino species due to neutrino oscillations together with the stringent restriction on the primordial ^{4}He abundance

$$Y_{\mathrm{prim}} \leq 25\%$$

given by Eq.(549) one can exclude rather wide range of possible values of δm^2 and neutrino mixing angles (Dolgov, 1980; Enqvist, 1990).

Electron neutrino oscillations into sterile neutrino states

$$\nu_e \leftrightarrow \bar{\nu}_L$$

and

$$\bar{\nu}_e \leftrightarrow \nu_R$$

with the neutrino square mass difference, lying in the interval

$$10^{-10} < \delta m^2 < 10^{-8}\,\mathrm{eV}^2 \tag{591}$$

are "switched on" in the period

$$T \sim 1\ \mathrm{MeV} \tag{592}$$

that is just the period when the n/p ratio freezes out.

In the case of more than two different neutrino species phenomenology of neutrino oscillation (Bilenky et al., 1980; Kobzarev et al.,1980) allows CP violation effects in oscillations. Such effects can cause the difference in the "switching on" of

$$\nu_e \leftrightarrow \bar{\nu}_L$$

and

$$\bar{\nu}_e \leftrightarrow \nu_R$$

oscillations. Thus, in the period of "switching on" of these oscillations the significant excess of electron neutrino over electron antineutrino (or vice verse) can appear, comparable with their total concentration.

Due to existence of such an excess, the relationship between the rates of weak transitions

$$\nu_e n \rightarrow ep$$

and

$$\bar{\nu}_e p \rightarrow e^+ n$$

is changed, resulting in significant change in the frozen n/p ratio. In the case when oscillations cause substantial electron neutrino excess

$$n_{\nu_e} >> n_{\bar{\nu}_e} \; , \tag{593}$$

the n/p ratio decreases, whereas in the opposite case

$$n_{\nu_e} << n_{\bar{\nu}_e} \tag{594}$$

this ratio increases.

So the observed ^{4}He concentration can be used as a sensitive indicator of CP violation in neutrino oscillations with

$$10^{-10} < \delta m^2 < 10^{-8}\,\mathrm{eV}^2 \; .$$

Note, that the existence of such oscillations can give rise to season variations of fluxes of monochromatic electron neutrino from the Sun.

Even in the presence of a small magnetic moment μ_ν for neutrino with Dirac mass m_ν of the order of

$$\mu_\nu \sim G_F m_\nu \; , \tag{595}$$

where G_F is the Fermi constant, the existence of primordial magnetic field

$$B_{prim} > 4 \cdot 10^{-9}\,\mathrm{Gs}(1+z)^2 \left(\frac{1\mathrm{eV}}{m_\nu} \right) , \tag{596}$$

where z is the red shift, provides helicity flips of neutrino before they decouple from the cosmic plasma. Then the concentrations of ordinary and sterile neutrino turn out to be equal, what is excluded by "stringent" interpretation of the observational data on primordial ^{4}He abundance (Wasserman, Feinberg, 1980; Enquist et al., 1995).

The shift of the time moment of the n/p ratio freezing out can be induced by the time variation of fundamental constants due to effects of additional (compactified) dimensions.

In this case, the observed ^{4}He abundance puts an upper limit on the rates of time variation of the radius of compact dimensions and of fundamental constants, determining the rate of β processes (Kolb et al., 1986).

1.3. Non-equilibrium particles and ^{4}He abundance.

Possible existence of non-equilibrium particles in the Universe in the period of Big Bang nucleosynthesis is related to another class of effects influencing the value of n/p ratio.

So, Vainer and Naselsky (1977) have estimated the change in the value of n/p ratio due to interactions of nucleons with energetic neutrinos from evaporating primordial black holes (PBH). It is easily seen, that interaction of nucleons with equal fluxes of electron neutrino and electron antineutrino from PBH, evaporating at

$$t < 10^2 \mathrm{s}$$

(on the PBH evaporation see Chapters 4 and 6), increases the value of n/p ratio. It provides restrictions on possible neutrino fluxes and, consequently, on the concentration of PBHs, evaporating in this period.

The same argument can be used in order to restrict the concentration of any hypothetical type of massive particles, for which neutrino are predicted among the products of their decay at

$$t < 10^2 \mathrm{s}\ .$$

Depending on the mass of particles with neutrino among their decay products, the value of n/p ratio can be shifted either to higher, or to lower values.

For energetic non-equilibrium neutrino n/p ratio increases, since the interactions of energetic neutrino with nucleons tend to make equal concentrations of neutrons and protons

$$\frac{n}{p} \to 1 . \tag{597}$$

The existence of hypothetical sources of non-equilibrium neutrino with low energy

$$E_\nu < 5 \text{ MeV} \tag{598}$$

shifts down the value of the n/p ratio due to threshold effects in weak reactions

$$\begin{aligned} &\nu_e n \to ep , \\ &\bar{\nu}_e p \to e^+ n , \end{aligned} \tag{599}$$

in which the mass difference of neutron and proton and the mass of electron should be taken into account (Dolgov, Kirillova, 1987; Sato, 1987).

Note, that in the considered period

$$t < 10^2 \text{s}$$

electron-positron pairs were in equilibrium with plasma and radiation, and their concentration exceeds the concentration of nucleons by 8-10 orders of magnitude.

Thus, due to neutrino and antineutrino scattering on electrons and positrons, the Universe turns out to be opaque for neutrinos with the energy

$$E_\nu > E_{th} \sim 100\text{MeV}\left(\frac{t}{1\text{s}}\right) \tag{600}$$

and such energetic neutrinos can not virtually manage the interaction with nucleons. Thus, the value of n/p ratio can be influenced by the weak interaction of neutrinos with the energy, not exceeding

$$E_{th} \sim 100\text{MeV}\left(\frac{t}{1\text{s}}\right), \tag{601}$$

with nucleons.

It makes the ^{4}He abundance sensitive to such fluxes of non-equilibrium neutrino, which are only by 1-2 orders of magnitude less intensive, than the fluxes of equilibrium neutrino.

Relatively weak sensitivity of n/p ratio to the fluxes of non-equilibrium neutrino is related to the weakness of neutrino-nucleon interaction. The sensitivity should increase, if the sources of non-equilibrium particles are predicted, that give rise to stronger interactions with nucleons,.

Indeed, as it was shown by Zeldovich et al. (1977) on the example of evaporating PBH, the existence of sources of antinucleons in the Universe at

$$1 < t < 10^2\text{s} \tag{602}$$

should lead to much more noticeable influence of such sources on the value of n/p ratio.

In the case of primordial black hole evaporation, the production of nucleon-antinucleon pairs can be considered. The isotopic invariance of the process of PBH evaporation leads to equal amount of neutron-antineutron and proton-antiproton pairs, produced with the concentration

$$n_n = n_p = n_{\bar{n}} = n_{\bar{p}} = n_1 , \tag{603}$$

where the nucleon concentration n_1 is proportional to the relative contribution $\alpha(M)$ of PBHs into the total density in the period of their evaporation

$$n_1 = f_N \frac{\alpha(M)\rho_{tot}}{T_{PBH}} . \tag{604}$$

Here ρ_{tot} is the total density and T_{PBH} is the temperature of the PBH evaporation.

The production of equal amounts of neutrons and protons in the considered period of cosmological evolution leads to the growth of the value of n/p ratio. After anti-nucleon annihilation, the value of n/p turns out to be equal to

$$\frac{n}{p} = \left(\frac{n}{p}\right)_f \tag{605}$$

instead of the value, defined by its freezing out and denoted as

$$\frac{n}{p} = \left(\frac{n}{p}\right)_i = \frac{n_0}{p_0} . \tag{606}$$

We see that

$$\left(\frac{n}{p}\right)_f = \frac{n_0 + n_1}{p_0 + n_1} > \frac{n_0}{p_0} = \left(\frac{n}{p}\right)_i . \tag{607}$$

It gives the strongest upper limit on the possible concentration of PBHs, evaporating in the Universe (Zeldovich et al., 1977) in the period (602)

$$1 < t < 10^2 \mathrm{s}$$

from the observed ^{4}He abundance.

Same arguments give the strongest upper limit on the possible value of the product rm for hypothetical metastable particles with the mass m, having the frozen relative concentration

$$r = \frac{n_m}{n_r} \tag{608}$$

and decaying in this period having nucleon-antinucleon pairs among the products of their decay:

$$rm < 6.5\mathrm{eV}\frac{\Omega_b}{B_N}(1 - Y_{obs})^{-1}(1 + Y_{obs}/2)^{-1}(Y_{obs} - Y_0) \; . \tag{609}$$

Here Y_{obs} is the observed abundance of ^{4}He, the magnitude Y_0 is defined as

$$Y_0 = 2\left(\frac{n}{p}\right)_i , \tag{610}$$

Ω_b is the relative cosmological baryon density and B_N is the fraction of nucleon-antinucleon pairs, produced in decays of such particles. In Fig.14 the restriction (609) is shown together with other restrictions, following from the analysis of effects of hypothetical particles on the ^{4}He abundance.

In addition to CP-violating effects in neutrino oscillations, considered above, the ^{4}He abundance turns out to be sensitive to another possible effect, related with some mechanisms of CP violation, namely, with existence of antimatter domains, predicted in baryon-asymmetrical Universe in the models of inhomogeneous baryosynthesis, based, in particular, on the models with spatial variation of CP, such as the models of spontaneous CP violation (Chapter 3).

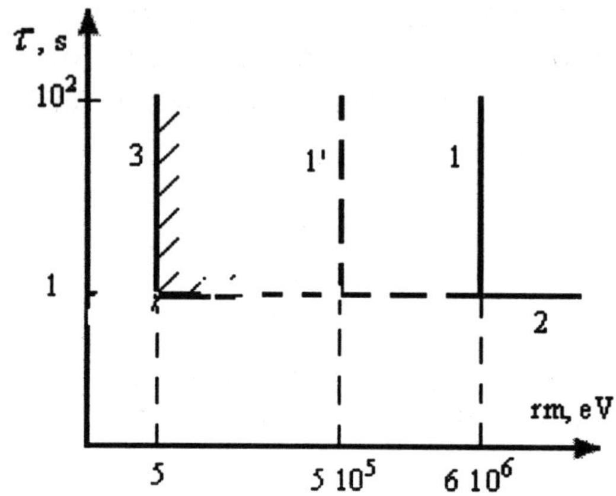

Fig.14. Constraints on metastable particles following from effects of nucleon–antinucleon pairs on helium abundance (line 3) in comparison with analogous constraints from "thermodynamic" effects (lines 1′ and 1).

Since after local nucleon-antinucleon annihilation further annihilation is possible only at the boundaries of domains, the domain structure of the matter-antimatter distribution provides conservation of significant amount of antinucleons in the periods, when antinucleons should be absent in the case of homogeneous matter-antimatter mixing. The time scale of domain annihilation is determined by the size of domain. The greater is domain, the greater is the time scale of its annihilation, and the greater is the period, when antinucleons are retained in the Universe.

The annihilation of antimatter domain in the period

$$1 < t < 10^2 \mathrm{s}$$

reduces the concentration of neutrons, because they diffuse more effectively, than protons, to the boundary of the matter domain (Steigman, 1976). As a result of such annihilation the n/p ratio decreases. The observed ^{4}He abundance imposes an upper limit on the averaged concentration of antinucleons in the Universe, annihilating in the considered period. The limit follows from the lower estimation of the primordial ^{4}He abundance.

Indeed, let N be the concentration of annihilated antinucleons. The ratio n/p is reduced due to neutron diffusion to the boundary of domain and annihilation and after annihilation turns out to be equal to

$$\left(\frac{n}{p}\right)_f = \frac{n_0 - N}{p_0} < \frac{n_0}{p_0} . \tag{611}$$

In this estimation, the difference of cross sections of nucleon-antinucleon annihilation in the states with isotopic spin 0 and 1 was not taken into account. However, this does not influence the result significantly.

The fraction of annihilated antinucleons

$$f = \frac{N}{p} , \tag{612}$$

is constrained by the following inequality

$$f < \frac{n_0}{p_0} - \left(\frac{n}{p}\right)^{obs}_{\min} . \tag{613}$$

The evolution of antimatter domains was discussed in the book by Zeldovich and Novikov (1975) and in the reviews of Steigman (1976) and Chechetkin et al. (1982). The new level of understanding of the problem of possible existence of antimatter in the Universe as a profound signature of the mechanism of matter generation is considered in the present book.

Thus, the observed ^{4}He abundance provides nontrivial restrictions on various effects of hypothetical particles and fields in the period of Big Bang nucleosynthesis, corresponding to

$$1 < t < 10^2 \text{s} .$$

In the next section we shall consider the observational data on the spectrum of the thermal electromagnetic background radiation as the source of information on the possible cosmological effects of the hidden sector of particle theory.

2. Metastable Particles and the BBBR Spectrum

2.1 Distortion of the thermal background spectrum

As we have already mentioned, the data on the thermal background spectrum provide important observational information on the possible physical conditions on the RD stage, that supplements the information obtained from the observed abundance of light elements. The data from COBE satellite (Smoot, 1990) show, that the spectrum of relic radiation corresponds to the Planck distribution with the temperature

$$T = 2.7\text{K} \tag{614}$$

with the accuracy of some tenths of percent. It means, that possible distortion of this thermal spectrum should lie within these observational errors. According to Zeldovich, Sunyaev (1969), Sunyaev, Zeldovich (1970), this provide restrictions on the character of out-of-equilibrium processes on the RD stage.

In the framework of the standard Big Bang scenario, some non-equilibrium processes are predicted.

They are: electron-positron annihilation, and recombination of hydrogen. The predicted distortions of the thermal spectrum due to these processes are by one-two orders of magnitude smaller, than the level of sensitivity, accessible in the modern observations.

However, some hypothetical non-equilibrium processes are predicted on the RD stage in the framework of various particle models. Decays of massive metastable particles, PBH evaporation, particle production as a result of cosmic string decay, annihilation of antimatter domains are the examples of such non-equilibrium processes.

Interaction of fluxes of non-equilibrium particles from these sources with the matter and radiation heats the matter, so that hot electrons appear. On sufficiently late stages of expansion, the interaction of these electrons with radiation can not provide the Planck spectrum of heated radiation, so that the thermal background spectrum is distorted (Zeldovich, Sunyaev, 1969; Sunyaev, Zeldovich, 1970; Zeldovich, Novikov, 1975).

The magnitude of distortion and, consequently, the possibility of its observation is determined by the energy release and by the period when it takes place.

The observations impose upper limit on the possible magnitude of distortions, and, consequently, restrict the possible time scale and the energy release of hypothetical sources (in particular, the lifetime, the concentration and the mass of hypothetical particles).

Let us give a brief discussion of the results of the theory of distortions of the thermal background spectrum. According to this theory (see Zeldovich, Novikov, 1975; Zeldovich, Sunyaev,1969; Sunyaev, Zeldovich, 1970; Gunn et al.,1977, Sunyaev et al.,1978), if the energy release takes place at the values of the red shift

$$z > 10^8 \Omega_b^{1/2} , \tag{615}$$

where Ω_b is the relative baryon density in the Universe, corresponding to the period

$$t < 10^3 \Omega_b^{-1} \text{ s} , \tag{616}$$

additional soft photons are produced in the reactions

$$\begin{aligned} &ep \to ep\gamma , \\ &\gamma e \to \gamma\gamma e . \end{aligned} \tag{617}$$

After the energy release the Planck spectrum is established.

On the other hand, if the energy release takes place at the red shift

$$z < 10^8 \Omega_b^{1/2} , \tag{618}$$

these reactions are ineffective for production of additional photons and the Planck form of the spectrum can not be established.

The theory (Zeldovich, Sunyaev,1969; Sunyaev, Zeldovich, 1970; Zeldovich, Illarionov, Sunyaev,1972) distinguishes two cases depending on the red shift z, corresponding to the period of energy release and satisfying the condition (618):

i) early energy release, corresponding to the red shift

$$4 \cdot 10^4 \Omega_b^{1/2} < z < 10^8 \Omega_b^{1/2} , \tag{619}$$

and

ii) late energy release at the red shift

$$z < 4 \cdot 10^4 \Omega_b^{1/2} . \tag{620}$$

In the case of the early energy release, the thermal equilibrium is established between hot electrons and cool photons under the condition of the fixed number of photons. If the energy release takes place in the period (619), the photon distribution established by photon-electron scattering differs from the Planck distribution, but is the equilibrium distribution of the fixed number of photons, interacting with matter, i.e. the Bose-Einstein distribution

$$F_{em}(\nu) = \frac{2h\nu^3}{c^2}\frac{1}{\exp\left(\frac{h\nu + \mu kT_e}{kT_e}\right) - 1} \quad . \tag{621}$$

Here ν is the frequency, and T_e is the electron temperature. The value of the dimensionless chemical potential μ defines the relative contribution of the energy release to the energy density of the thermal electromagnetic background radiation. It determines the difference of the spectrum (621) from the Planck form, and measures the distortion of the Planck thermal spectrum of the electromagnetic background radiation.

The observational data on the spectrum of the thermal background imposes the following restriction on the possible energy release in the period (619) (Sunyaev, Zeldovich, 1970; Khlopov, Chechetkin, 1987; Smoot, 1990):

$$\frac{\delta\varepsilon}{\varepsilon_\gamma} < \tfrac{1}{3}|\mu| < 1.1 \cdot 10^{-4} \tag{622}$$

for the ideal Bose-Einstein distribution, using the stringent COBE constraint on the value of μ (Mather et al., 1994):

$$|\mu| < 3.3 \cdot 10^{-4} \quad . \tag{623}$$

Here $\delta\varepsilon$ is the specific energy release, and ε_γ is the energy density of the thermal radiation.

The detailed theory of the early energy release spectrum distortions (Sunyaev, Zeldovich, 1970) accounts for the nonideal form of the distorted spectrum due to bremstrahlung and Compton scattering, what specifies the parts of the photon spectrum, most sensitive to distortion. From the COBE data, stringent limits on the energy release can be derived as a function of red shift (Wright et al., 1994). However, such constraints are strongly model-dependent and sensitive to very specified parameters of the hypothetical source.

If the energy release has taken place at rather late stage, corresponding to the period (620), the equilibrium Bose-Einstein spectrum (621) can not be established. The distortion of the thermal spectrum is determined by the kinetics of heating of the photon gas by hot electrons.

With the use of the COBE observational constraint on the parameter y, given by Mather et al. (1994),

$$|y| < 2.5 \cdot 10^{-5} \ , \tag{624}$$

it follows from the observational data, that the late energy release should not exceed

$$\frac{\delta\varepsilon}{\varepsilon_\gamma} < 12|y| < 3 \cdot 10^{-4} \tag{625}$$

(Zeldovich, Sunyaev, 1969; Khlopov, Chechetkin, 1987; Smoot, 1990; Mather et al., 1994; Wright et al., 1994) for Compton scattering induced distortions.

On the basis of restrictions (622) and (625) one can, in particular, put an upper limit on the contribution f to the total density of heavy metastable particles with the mass m and concentration n_m (relative to baryons)

$$f = \frac{m \cdot n_m}{m_p \cdot n_B} \tag{626}$$

decaying in the Universe in the period (618).

Indeed, the energy density of the thermal background ε_γ is given by (see Chapter 3)

$$\varepsilon_\gamma = 4 \cdot 10^{-13} \left(\frac{T_0}{2.7K} \right)^4 (1+z)^4 \frac{\text{erg}}{\text{cm}^3} , \tag{627}$$

where T_0 is the modern temperature of the black body background radiation. Assuming for simplicity, that particles decay instantaneously at

$$z = z_d , \tag{628}$$

one obtains the energy density, released in these decays,

$$\delta\varepsilon = a_{em} mc^2 n_m = a_{em} \frac{m \cdot n_m}{m_p \cdot n_b} \rho_b c^2 =$$

$$= 5 \cdot 10^{-9} a_{em} \Omega_b f (1+z)^3 \frac{\text{erg}}{\text{cm}^3} , \tag{629}$$

where

$$a_{em} \sim 0.5 \tag{630}$$

is the fraction of the rest energy of metastable particle, contributing into the heating of electrons.

At $z=z_d$ one obtains from the Eqs. (627) and (629) the expression for relative energy release

$$\frac{\delta\varepsilon}{\varepsilon_\gamma} = 6 \cdot 10^3 \Omega_b f (1+z)^{-1} \left(\frac{2.7K}{T_0}\right)^4 . \tag{631}$$

The upper limits (622) and (625) lead to the following constraints on the possible value of f:

$$f < 1.6 \cdot 10^{-4} \Omega_b^{-1} (1+z_d) \left(\frac{T_0}{2.7K}\right)^4 \cdot \\ \cdot \begin{cases} 3 \cdot 10^{-4}, z < 4 \cdot 10^4 \Omega_b^{1/2} \\ 1.1 \cdot 10^{-4}, 4 \cdot 10^4 \Omega_b^{1/2} < z < 10^8 \Omega_b^{1/2}. \end{cases} \tag{632}$$

It is seen from Eq. (632) that the restrictions on f are rather weak at

$$z_d > 10^5 \tag{633}$$

i.e. if the lifetime of particles is equal to

$$\tau < 10^9 \text{s} . \tag{634}$$

The constraint (632) for such particles is

$$f < 2 \cdot 10^{-7} (1 + z_d) \tag{635}$$

at

$$\Omega_b = 0.1 .$$

Relatively weak sensitivity of distortions of the thermal background spectrum to the possible properties of hypothetical particles, decaying on the RD stage, is explained by the dominance of the radiation energy density in the total cosmological density on this stage. Thus, significant effect on the spectrum of thermal background radiation can take place, if the hypothetical sources of non-equilibrium particles release the energy density, exceeding some tenths of percent of the total density, what strongly exceeds the matter density during earlier periods of the RD stage.

However, the restriction on the energy release at the red shifts (618) strictly excludes the possibility of complete "regeneration" of the thermal background in this period.

In particular, at these red shifts the stages of dominance of massive metastable particles can not end due to decay or annihilation of such particles, if the products of their decays or annihilations interact with plasma and radiation.

If relativistic particles, produced in such decays (or annihilations), do not interact with plasma and radiation, they do not influence directly on the spectrum of thermal radiation, but they can cause an influence on the development of gravitational instability and on the formation of the large scale structure of the Universe (see below).

Putting together restrictions on the possible effect of non-equilibrium particles, produced in the end of dust-like stage, with the restrictions on the existence of such stage in the period of the Big Bang nucleosynthesis, one can exclude long $p = 0$ stages after the first second of expansion.

On the other hand, the energy release at the red shifts

$$z > 10^8 \Omega_b^{1/2}$$

corresponding to the cosmological times

$$t < 10^3 \Omega_b^{-1} \text{ s} , \tag{636}$$

results in thermalisation and recovering of Planck spectrum of radiation. With the account for restriction on the non-relativistic matter dominance at

$$t < 1 \text{ s}$$

deduced in the Sec 1.1, one can conclude that there is not excluded the principal possibility (Doroshkevich, Khlopov, 1983; Polnarev, Khlopov, 1981) of relatively short dominance of massive metastable particles in the period

$$1 < t < 10^4 \text{s} . \tag{637}$$

2.2. Dominance of metastable particles in the period of nucleosynthesis.

The natural possibility of energy release in the period (637), corresponding to the temperature

$$30 \text{ keV} < T < 3 \text{ MeV} \tag{638}$$

is related to decay of hypothetical metastable neutrinos (e.g. τ-neutrino) with the lifetime τ in the interval

$$1 \text{ s} < \tau < 10^4 \text{ s} \tag{639}$$

and the mass m, lying in the interval

$$30 \text{ keV} < m < 30 \text{ MeV} . \tag{640}$$

For the considered period, the frozen concentration of massive left-handed neutrinos and right-handed antineutrinos relative to the number density of all the relativistic particles n_r is given by (Dolgov, Zeldovich, 1981; Lee, Weinberg, 1977; Vysotsky et al., 1977, see also Chapter 3):

$$r = \frac{\left(n_{\nu L} + n_{\bar{\nu}R}\right)}{n_r} = f(m) = \frac{1}{\kappa}\begin{cases} 1\,, & m < 3\text{ MeV} \\ \left(\dfrac{3\text{ MeV}}{m_\nu}\right)^3, & m \geq 3\text{ MeV}, \end{cases} \quad (641)$$

where the numerical factor κ , taking into account derelativisation of electron-positron pairs and their annihilation in the considered period, lies in the interval

$$5 < \kappa < 6\ . \quad (642)$$

Neutrinos with the mass in the considered interval begin to dominate at

$$t > t_0\ , \quad (643)$$

where t_0 is

$$t_0 = 1\text{ s}\left(\frac{rm}{1\text{ MeV}}\right)^{-2}\ . \quad (644)$$

The dominance of such neutrinos ν_H in the period

$$t_0 < t < \tau \quad (645)$$

corresponds to the dust-like equation of state $p = 0$.

The dominant decay mode of neutrino ν_H can be the process

$$\nu_H \to e^+ e^- \nu_e\ , \quad (646)$$

if the mass of ν_H is

$$m > 1\ \text{MeV}\ , \tag{647}$$

or the process

$$\nu_H \to \nu_L \gamma \tag{648}$$

(Kim, Uehara, 1981). Reasonable choice of parameters of such decays may lead to the values of τ, compatible with the condition (639).

Indeed, consider, for instance, the decay mode given by Eq. (646). This decay can occur due to the mixing of ν_H and ν_e in the ordinary weak interaction of charged currents

$$j_\mu^e = \overline{\Psi}_e \gamma_\mu (1 + \gamma_5) \Psi_{\nu_e} \tag{649}$$

and

$$j_\mu^{H+} = \overline{\Psi}_{\nu_H} \gamma_\mu (1 + \gamma_5) \Psi_e\ . \tag{650}$$

The Fermi constant G_F in the interaction of these currents is suppressed by the mixing factor U_{He} determined by the neutrino mass matrix, so that the Lagrangian of the current-current weak interaction is given by

$$L = \frac{G_F}{\sqrt{2}} U_{He} j_\mu^{H+} j_\mu^e\ . \tag{651}$$

Assuming that the mass of ν_H is

$$m = m_{\nu_H} >> m_e\ , \tag{652}$$

where m_e is the mass of electron, the probability of the considered decay is given by the expression, similar to the muon decay

$$W = \frac{G_F^2 |U_{He}|^2 m^5}{192\pi^3} . \tag{653}$$

The lifetime of ν_H is

$$\tau = 10^4 \mathrm{s} \left(\frac{1\mathrm{MeV}}{m} \right)^5 |U_{He}|^{-2} . \tag{654}$$

General restrictions on the value of mixing factor are given in the Review of Particle Physics (Barnett et al., 1996). The most stringent laboratory restrictions on this factor in the considered mass interval (640) follow from the data on the decay

$$\pi \to e\nu . \tag{655}$$

The deviations from the ordinary mode

$$\pi^+ \to e^+ \nu_e \tag{656}$$

induced by the neutrino mixing with the massive neutrino state and by the additional channel

$$\pi^+ \to e^+ \nu_H \tag{657}$$

were searched for (Bryman et al., 1981; Barnett et al., 1996). As a result of these experimental studies, the upper limit on the mixing factor was obtained:

$$|U_{He}|^2 < 2.5 \cdot 10^{-3} \left(\frac{1\mathrm{MeV}}{m} \right)^2 . \tag{658}$$

It follows from Eq.(654) that this experimental upper limit on the mixing factor imposes the lower limit on the life time of massive neutrino:

$$\tau > 4 \cdot 10^{3}\,\mathrm{s}\left(\frac{10\mathrm{MeV}}{m}\right)^{3} . \tag{659}$$

Hence, the existing experimental constraints do not exclude the possibility for existence of heavy neutrinos ν_H with the mass

$$10\ \mathrm{MeV} < m < 30\ \mathrm{MeV}\ , \tag{660}$$

which can dominate in the Universe at

$$t < 10^{4}\,\mathrm{s}\ . \tag{661}$$

Using the data on the direct measurement of the number of neutrino species from the measurement of Z-boson width (Barnett et al., 1996), given by

$$N_{\nu} = 3.09 \pm 0.13 \tag{662}$$

for neutrino with the ordinary weak interaction and the mass

$$m < \frac{m_Z}{2}\ , \tag{663}$$

and with the account for the experimental upper limit on the mass of ν_τ given by

$$m_{\nu_\tau} < 24\ \mathrm{MeV}$$

(Barnett et al., 1996), ν_τ can dominate in the considered period, if their mass lies in the interval

$$10 \text{ MeV} < m_{\nu_\tau} < 24 \text{ MeV} . \tag{664}$$

Electron-positron mode of ν_H decays at

$$1 \text{ s} < t < 10^4 \text{ s}$$

due to successive thermalization and annihilation of positrons and electrons lead to the increase of the photon concentration by the factor of

$$n_\gamma = \left(\frac{\tau}{t_0}\right)^{1/2} (n_\gamma)_0 \, . \tag{665}$$

The Planck form of the photon spectrum should be established after thermalization. As was shown in Section 1, ν_H dominance in the Universe at

$$t_0 > 1 \text{ s} \tag{666}$$

virtually does not influence on the value of the frozen n/p ratio. If the lifetime of these particles is

$$\tau > 10^3 \text{s} , \tag{667}$$

their dominance in the period of thermonuclear reactions of Big Bang nucleosynthesis leads at the fixed baryon density to small (< 1%) increase of the primordial ^{4}He abundance and to some increase of the primordial deuterium abundance. Both effects do not contradict the observational data. Much more serious problems for the considered schemes of massive unstable neutrino can arise in their confrontation with the data on Supernova explosions (Falk, Schramm, 1978; Hut, White, 1984).

In the models of GUT, SUSY GUT, superstrings, and other particle models considered above, various types of massive metastable particles are predicted. Since their existence follows from different physical motivations, virtually all of them may be present in the Universe.

Self-consistent treatment of these possibilities includes joint analysis of multicomponent scenarios, in which some particles may be predicted with the parameters, at which they can dominate in the Universe in the period (637).

The restriction (637) on the possible duration of the corresponding $p = 0$ stage imposes constraints on the predicted by each concrete model: frozen concentration r , mass m , and the lifetime τ of such hypothetical particles

$$10 \text{ keV} < rm < 3 \text{ MeV} , \tag{668}$$

$$\max(t_0, \ 1 \text{ s}) < \tau < 10^4 \text{ s} , \tag{669}$$

where t_0 in Eq.(669) is given by Eq.(644).

The discussed scenarios of short dust-like stages can result in the change (decrease) of the predicted relative concentration of all other frozen particles, since after the end of dominance of massive metastable particles photon concentration grows, and the concentration of frozen primordial stable particles has the same value.

Consider some stable frozen particles present in the Universe in the period under discussion and having after the 1 s of expansion the frozen concentration relative to photons

$$r = \frac{n_m}{n_\gamma} \ . \tag{670}$$

The maximal possible decrease of relative concentration for all such frozen particles after the short dust-like stage is finished can be reached at

$$\tau \sim t_{max} \sim 10^4 \text{ s} \tag{671}$$

for $\Omega_b = 0.1$ and

$$t_0 \sim 1 \text{ s} . \tag{672}$$

Remind that the upper limit on the duration of dust-like stage follows from the condition that the Planck spectrum is recovered after this stage is finished, and the lower limit follows from the Big Bang nucleosynthesis constraints on the possible contribution of massive particles into the cosmological density.

With the account for these constraints, the maximal decrease of relative concentration after the end of dust-like stage is

$$r = \left(\frac{n_m}{n_\gamma}\right)_{t>\tau} = \left(\frac{\tau}{t_0}\right)_{max}^{-1/2} \left(\frac{n_m}{n_\gamma}\right)_{t<1s} . \tag{673}$$

In the case, when the dust-like stage is provided by the massive neutrino dominance, the experimental bounds, given by Eqs. (658), lead to the minimal time moment, when the dust-like stage can start,

$$t_0 \sim 10^2 \text{s} , \tag{674}$$

corresponding to the mass of unstable neutrino

$$m_\nu \sim 10 \text{ MeV} . \tag{675}$$

At these parameters possible dominance of massive unstable neutrino can take place in the limited time interval,

$$t_0 \sim 10^2 \text{s} < t < 10^4 \text{ s} . \tag{676}$$

The maximal possible decrease of relative concentration of primordial stable particles can not exceed in this case the factor of 10.

The true scenario of very early Universe can contain various particle physics effects and their possible combinations. It makes the considered effect of short-term dust-like stages only one of the possible factors to be accounted for in the prediction for the relic concentration of particles, which were created, frozen out or decoupled in the very early Universe.

However, such effect may cause quantitative change in the cosmology of massive neutrino, generally considered as quantitatively well established one.

In the case of primordial neutrinos, the standard Big Bang model seems to predict unambiguously the concentration of primordial neutrinos on the basis of quantitatively definite relationship with the observed concentration of relic photons (see Chapter 3). The numerical factors in this relationship given by Eq. (285),

$$\frac{n_\nu + n_{\bar{\nu}}}{n_\gamma} = \frac{3}{4} \cdot \frac{4}{11} \tag{677}$$

are determined by the thermodynamic equilibrium of relativistic boson and fermion gases, and the specific entropy conservation in the equilibrium mixture of radiation and electron-positron pairs.

The account for a possible stage of dominance of non-relativistic particles induces an additional factor in this relationship, which is inversely proportional to the relative increase of photon concentration as a result of the electromagnetic energy release in the end of the dust-like stage. The dilution factor for the primordial radiation is given by

$$\frac{(n_\gamma)_{in}}{(n_\gamma)_f} = \left(\frac{t_0}{\tau}\right)^{1/2} . \qquad (678)$$

Taking into account this dilution factor due to possible dust-like stage after 1 s of expansion, one can obtain the generalised form for the neutrino photon ratio

$$\frac{n_\nu + n_{\bar{\nu}}}{n_\gamma} = \frac{3}{11} \cdot \left(\frac{t_0}{\tau}\right)^{1/2} . \qquad (679)$$

Note, that in a number of papers there were given arguments (Lindley, 1980, 1984; Sarkar, Cooper, 1984; Hut, White, 1984) against possible dominance in the Universe of particles with the mass

$$5 \text{ MeV} < m < 10 \text{ MeV} \qquad (680)$$

decaying at

$$\tau > 10^2 \text{ s} \qquad (681)$$

with the production of photons or electron-positron pairs. These arguments were based on rather detailed calculation of deuterium photodesintegration by the products of these decays (Lindley, 1984).

In particular, these results exclude the possibility for ν_τ dominance in the early Universe, and, being put together with the bounds on the mixing and lifetime of ν_τ, they completely exclude the possibility for the mass of ν_τ, exceeding 1 MeV (Sarkar, Cooper, 1984).

However, these calculations do not take into account, that the neutron, produced as the result of deuterium destruction in the reaction

$$\gamma D \rightarrow np \ ,$$

can further be captured by proton in the inverse reaction

$$np \rightarrow \gamma D \ .$$

This possibility, first pointed out by Zeldovich et al. (1977) in the analysis of the PBH evaporation and then considered in more general case of various sources of non-equilibrium particles (see Chechetkin et al., 1982; Khlopov, Chechetkin, 1987) recovers the effects of deuterium photodesintegration. One can easily estimate (Zeldovich et al., 1977), that at

$$t < 10^5 \mathrm{s}\ \Omega_b^{2/3} \tag{682}$$

the time scale of neutron capture by protons, given by

$$t_{np} = \left(n_p \sigma v(np \rightarrow \gamma D)\right)^{-1} \ , \tag{683}$$

is smaller, than the neutron lifetime $\tau_n \sim 10^3$ s.

Thus, neutrons can form deuterium before their decay. Moreover, as we shall see later, in the most cases the decay products of metastable particles can destruct ^{4}He nuclei, resulting in the increase of deuterium and ^{3}He abundance.

CHAPTER 6

ANTIPROTONS IN THE UNIVERSE AFTER THE BIG BANG NUCLEOSYNTHESIS

In the baryon asymmetrical Universe, the baryon-antibaryon pairs, being in the equilibrium with plasma and radiation at earlier stages, should annihilate after the first microsecond of expansion.

The important consequence of the baryon asymmetry is the exponentially small amount of antiprotons, which can survive after this annihilation.

In the standard baryon asymmetrical cosmology, the exponential suppression of antiproton abundance is predicted from the first microsecond of expansion up to the period of galaxies formation, when the interaction of cosmic rays with the matter can lead to the production of antiproton component of cosmic rays.

One may conclude, that the appearance of non exponentially small amount of antiprotons and sizeable effects of their annihilation from the first microsecond to the period of galaxy formation serves as the profound signature of new cosmological phenomena, to be related with the new physics, which are not taken into account in the standard Big Bang scenario.

On the basis of the observational data, one can use this signature to constrain the allowed parameters of the hypothetical sources of antiprotons on the RD stage after the first microsecond of expansion, and to impose restrictions on the parameters of a particle model, predicting such sources as its cosmological consequence.

The particle theory leads to a number of nontrivial possibilities for antiprotons to appear in the Universe after the annihilation of equilibrium nucleon-antinucleon pairs and even after the period of Big Bang nucleosynthesis (Khlopov, Chechetkin, 1987; Chechetkin et al., 1982).

Primordial antinucleons may retain in antimatter domains, predicted in the framework of some particle models, in particular, in the models of spontaneous CP violation.

Another possibility is related to the production of nucleon-antinucleon pairs by some sources of non-equilibrium particles at the RD stage. Such sources are: a) evaporation of PBH, formed in the very early Universe (see Chapter 4), b) decays of metastable particles with the mass, exceeding 1 GeV, c) residual annihilation of frozen massive stable particles.

Let us consider first the PBH evaporation as a source of nucleon-antinucleon pair production.

1. The PBH evaporation as a source of nucleon-antinucleon pairs.

As it was already discussed in Chapter 4 the possibility of black hole evaporation, first discovered by Hawking (1975), is the most important property of PBH from the point of view of their possible effects on the RD stage.

The black hole emission is described as emission from the surface of a black body with the temperature T_{PBH}, given by Eq. (443)

$$T_{PBH} = \frac{1}{4\pi r_g} = \frac{m_{Pl}^2}{8\pi M} \ .$$

Here M is the mass of black hole and r_g is its gravitational radius, given by Eq. (442). According to Eq. (447), the time scale of evaporation is

$$t_{\rm evp} = 10^{-27}\,{\rm s}\left(\frac{M}{1\ {\rm g}}\right)^3 .$$

The fluxes of particles, produced in the PBH evaporation, can cause the same influence on the physical processes in the Universe as the products of decay of metastable particles.

Consider now, following Chechetkin et al. (1982) and Khlopov, Chechetkin (1987), only one particular channel of the PBH evaporation, namely, the antinucleon production by evaporating PBH.

Let us estimate the average number of antinucleons F, evaporated by the PBH with the mass M. It follows from Eq. (443), that PBH with the mass

$$M < M_{\bar{p}} = 10^{13}\,{\rm g} \tag{684}$$

have the surface temperature, exceeding 1 GeV, so in the process of evaporation of such PBH the antinucleon production is possible.

The modern theory of strong interactions (see Chapter 2) assumes that quarks and gluons are produced first in hadronic processes at high energies. Then their fragmentation into hadrons (hadronization) takes place.

The same seems to be true for the evaporating PBH, emiting gluons and quarks. The distribution of these particles, corresponding to thermodynamic equilibrium at

$$T = T_{PBH} ,$$

is realised through the set of individual processes of emission. Each individual process is the process of creation of a pair of particles in strong gravitational field. One of created particles falls to singularity of PBH, and the other goes to infinity.

If a pair of coloured particles (quark-antiquark pair or a pair of gluons) is created, their fragmentation into hadrons may be considered similar to description of hadron production in electron-positron annihilation (or in hard hadron processes).

Namely, when the separation l between two colour charges reaches the scale of confinement

$$l \sim \frac{1}{\Lambda_{QCD}} \sim 10^{-13} \text{ cm} \tag{685}$$

hadronization should take place, and the emitted quark or gluon transforms into hadronic jet.

The estimation of antiproton content of such jet can be based on the arguments of Azimov, Dokshitzer, and Khoze (1983), which considered antiproton production in electron-positron annihilation. One obtains

$$F = \frac{\left(\dfrac{dN_{\bar{p}}}{dt}\right)}{\left|\dfrac{dM}{dt}\right|} = \frac{\sum\limits_{i=q,\bar{q},g} \int dp_{\bar{p}} \int D_i^{\bar{p}}\left(p_{\bar{p}}, p_i\right) f(p_i) dp_i S}{\sum\limits_i \int f(p_i)\varepsilon_i dp_i S}, \tag{686}$$

where

$$f(p_i) = f(p_i, T_{PBH}) \tag{687}$$

is the momentum distribution of i-type elementary particles, emitted by PBH,

$$S = 4\pi r_g^2 \tag{688}$$

is the "surface" of PBH,

$$D_i^{\bar{p}}\left(p_{\bar{p}}, p_i\right)$$

is the probability for fragmentation of quark, antiquark or gluon with momentum p_i into antiproton, ε_i is the energy of the particle of type i.

With the account for fragmentation of gluons in antinucleons, which is more effective, than the quark (antiquark) one (Azimov et al., 1983), one has

$$F = f_{\bar{p}} = \left\langle N_{\bar{p}} \right\rangle = \frac{1}{7}\left\langle N_{\pi} \right\rangle \ , \tag{689}$$

where the average multiplicity of pions is related to the gluon multiplicity:

$$\left\langle N_{\pi} \right\rangle = \left(\frac{T_{PBH}}{\mu} \right)^{1/2} \left\langle N_{g} \right\rangle \tag{690}$$

with

$$\mu \sim 1 \text{ GeV} \ . \tag{691}$$

Since the PBH mass decreases in the course of evaporation, antiprotons can be produced on late stages of evaporation of PBH with the mass

$$M > M_{\bar{p}} = 10^{13}\,\text{g} \tag{692}$$

when PBH mass decreases down to 10^{13} g.

In this case, the antinucleon multiplicity is equal to

$$F = f_{\bar{p}} \left(\frac{M_{\bar{p}}}{M} \right) . \tag{693}$$

Relativistic antiprotons, produced by the PBH evaporation at

$$t < 10^2 \text{ s} \tag{694}$$

when the temperature of plasma is

$$T > 10^2 \text{ keV} , \tag{695}$$

slow down before annihilation, because the presence of electron-positron pairs in the equilibrium with the radiation makes the rate of the Coulomb deceleration of antiprotons greater, than the rate of their annihilation.

After positron annihilation, i.e. at

$$t > 10^2 \text{ s} \tag{696}$$

the value of the electron number density decreases to the value of the baryon number density, and the energy losses of antiprotons due to their Coulomb interaction with plasma are small.

However, at high energy, the annihilation channel is suppressed in antiproton interaction with protons. The annihilation cross section σ_{ann} is of the order of the difference of total cross sections of antiproton and proton interactions

$$\sigma_{ann} \sim \sigma^{tot}_{\bar{p}p} - \sigma^{tot}_{pp} , \tag{697}$$

which is decreasing with the energy.

The cross section of inclusive inelastic reactions

$$\bar{p} + p \to \bar{p} + anything \tag{698}$$

contributes about one half into the total cross section, and relativistic antiprotons lose energy in these reactions.

So, even at

$$t > 10^2 \text{ s} ,$$

a significant fraction of antiprotons from evaporating PBHs slows down. At the RD stage, the annihilation rate for relativistic antiprotons exceeds the rate of expansion. Therefore, the bulk of antiprotons, produced in this period, should annihilate.

Let us estimate the number density of antiprotons, produced at

$$t = t_e = 10^{-27} \text{s} \left(\frac{M}{1\text{g}}\right)^3 \tag{699}$$

in the process of evaporation of PBH with the mass M. The total number of antinucleons N_a, produced by one PBH, is equal to

$$N_a = \int_0^{\min\left(M, M_{\bar{p}}\right)} \left(\frac{dN_{\bar{p}}}{dM}\right) dM = \int_0^{\min\left(M, M_{\bar{p}}\right)} f_a \, dM , \tag{700}$$

where crude estimation for

$$f_a \equiv f_{\bar{p}}$$

gives

$$f_a = \frac{dN_{\bar{p}}}{dM} = \frac{g_g}{g_{tot}} \cdot \frac{\left(\frac{\langle N_{\bar{p}} \rangle}{\langle N_g \rangle}\right)}{3T_{PBH}} = \frac{g_g}{g_{tot}} \cdot \frac{\left(\frac{T_{PBH}}{\mu}\right)^{1/2}}{21T_{PBH}} =$$

$$= \frac{g_g}{21 g_{tot}} \left(\mu T_{PBH}\right)^{-1/2} = \frac{g_g}{21 g_{tot}} \left(\frac{8\pi M}{\mu}\right)^{1/2} \frac{1}{m_{Pl}} . \tag{701}$$

In Eq. (701) one assumes that the ratio of gluon and total multiplicities of particles, emitted by PBH, is equal to the ratio of their statistical weights g,

$$\frac{\langle N_g \rangle}{\langle N_{tot} \rangle} = \frac{g_g}{g_{tot}} \tag{702}$$

and that the PBH emits particles with the mean energy

$$E = 3T_{PBH} \ . \tag{703}$$

The number density of antiprotons, produced in the PBH evaporation averaged over the Universe, is given by

$$n_{\bar{p}} = n_{PBH}(M) \cdot N_a \ , \tag{704}$$

where $n_{PBH}(M)$ is the number density of PBH with the mass M in the Universe in the period of their evaporation

$$n_{PBH}(M) = \frac{\rho_{PBH}(M)}{M} = \alpha(M)\frac{\rho_{tot}}{M} \ . \tag{705}$$

Here

$$\alpha(M) = \frac{\rho_{PBH}(M)}{\rho_{tot}} \tag{706}$$

is the relative contribution of the PBH with the mass M into the total cosmological density ρ_{tot} . One obtains from Eqs. (686)-(706)

$$n_{\bar{p}} = \frac{4}{63} \cdot \frac{g_g}{g_{tot}} \cdot \left(\frac{2\pi}{\mu}\right)^{1/2} \cdot \frac{1}{m_{Pl}} \cdot$$

$$\cdot \min\left(M^{1/2}, \frac{M_{\bar{p}}^{3/2}}{M} \right) \cdot \alpha(M) \cdot \rho_{tot} \ . \qquad (707)$$

Primordial black holes, evaporating on the RD stage, have masses

$$M < M_{\bar{p}} = 10^{13}\ \mathrm{g}\ . \qquad (708)$$

In this case,

$$n_{\bar{p}} = \frac{g_g}{g_{tot}} \cdot \frac{4\left(\dfrac{2\pi M}{\mu}\right)^{1/2}}{63 m_{Pl}} \cdot \alpha(M) \cdot \rho_{tot} \ . \qquad (709)$$

Here the total cosmological density is

$$\rho_{tot} = \frac{m_{Pl}^2}{t^2} \ .$$

In the numerical estimations we also used the expression

$$\rho_{tot} = 5 \cdot 10^5 \ \frac{\mathrm{g}}{\mathrm{cm}^3} \left(\frac{1\mathrm{s}}{t}\right)^2 \ .$$

From Eq. (447) one can obtain the following relationship between the mass of the PBH M and the time of its evaporation

$$M = 10^9 \text{g} \left(\frac{t_e}{1 \text{ s}}\right)^{1/3} . \tag{710}$$

Substituting the expression for ρ_{tot} into Eq. (707), one obtains the estimation for the number density of antiprotons from evaporating PBH

$$n_{\bar{p}} = \frac{g_g}{g_{tot}} \cdot \frac{4}{63} \left(\frac{2\pi M}{\mu}\right)^{1/2} \cdot \alpha(M) \cdot \frac{m_{Pl}}{t^2} =$$
$$= 6 \cdot 10^{21} \text{ cm}^{-3} \frac{g_g}{g_{tot}} \left(\frac{M}{1\text{g}}\right)^{1/2} \alpha(M) \left(\frac{1 \text{ s}}{t_e}\right) =$$
$$= 2 \cdot 10^{26} \text{ cm}^{-3} \frac{g_g}{g_{tot}} \alpha(M) \left(\frac{1 \text{ s}}{t_e}\right)^{11/6} . \tag{711}$$

at time moment

$$t = t_e . \tag{712}$$

Using of the relation

$$n_\gamma = 2.5 \cdot 10^8 \Omega_b^{-1} n_b = 5 \cdot 10^{31} \text{cm}^{-3} \left(\frac{1\text{s}}{t}\right)^{3/2} \tag{713}$$

one obtains

$$n_{\bar{p}} = 10^3 \frac{g_g}{g_{tot}} \alpha(M) \left(\frac{1\text{s}}{t_e}\right)^{1/3} \Omega_b^{-1} n_b . \tag{714}$$

Since virtually all antiprotons, produced in the PBH evaporation on RD stage, annihilate, one obtains from Eq.

(714) the averaged amount of annihilated antiprotons per baryon

$$r_a = \frac{n_{\bar{p}}}{n_b} = 10^3 \frac{g_g}{g_{tot}} \alpha(M) \left(\frac{1 \text{ s}}{t_e} \right)^{1/3} \Omega_b^{-1} . \tag{715}$$

Eq. (714) is valid for the PBH, evaporating on the RD stage. Primordial black holes with the mass

$$10^{13} \text{ g} < M < 10^{15} \text{ g} \tag{716}$$

evaporate on the matter-dominant stage at

$$t > 10^{12} \text{ s} . \tag{717}$$

It follows from Eq. (707), that the averaged number density of antiprotons, evaporated from such PBH is equal to

$$\begin{aligned} n_{\bar{p}} &= \frac{g_g}{g_{tot}} \cdot \frac{4}{63} \left(\frac{2\pi}{\mu} \right)^{1/2} \cdot \frac{M_{\bar{p}}^{3/2}}{m_{Pl} M} \cdot \alpha(M) \cdot \rho_{tot} = \\ &= \frac{g_g}{g_{tot}} \cdot \frac{4}{63} \left(\frac{2\pi}{\mu} \right)^{1/2} \cdot \frac{M_{\bar{p}}^{3/2} m_{Pl}}{M} \cdot \alpha(M) \cdot t_e^{-2} = \\ &= \frac{g_g}{g_{tot}} \cdot \frac{4}{63} \left(\frac{2\pi}{\mu} \right)^{1/2} \cdot \frac{M_{\bar{p}}^{3/2} m_p}{m_{Pl} M} \cdot \alpha(M) \cdot \Omega_b^{-1} n_b , \end{aligned} \tag{718}$$

where the total density is taken equal to the critical density.

The relative concentration of antiprotons, produced in the PBH evaporation, is

$$r_a = \frac{n_{\bar{p}}}{n_b} = \frac{g_g}{g_{tot}} \cdot \frac{4}{63}\left(\frac{2\pi}{\mu}\right)^{1/2} \cdot \frac{M_{\bar{p}}^{3/2} m_p}{m_{Pl} M} \cdot \alpha(M) \cdot \Omega_b^{-1} =$$

$$= 4 \cdot 10^{-2} \frac{g_g}{g_{tot}} \left(\frac{M_{\bar{p}}}{M}\right) \Omega_b^{-1} \alpha(M) \ . \qquad (719)$$

Note, that for the homogeneous baryon distribution

$$n_b(x) = \langle n_b(x) \rangle \qquad (720)$$

on the MD stage, the magnitude (719) does not coincide with the fraction f of antiprotons, participating annihilation, since for relativistic antiprotons the time scale of their annihilation

$$t_{ann} = \left(n_b (\sigma v)_{ann}\right)^{-1} \qquad (721)$$

exceeds the cosmological time scale at

$$t_e > t_{\bar{p}} = 2 \cdot 10^{14} \mathrm{s}\ \Omega_b \qquad (722)$$

so that the rate of their annihilation is smaller, than the rate of expansion, and the most part of relativistic antiprotons do not participate annihilation. For nonrelativistic antiprotons, the effects of the Coulomb attraction are significant in the rate of antiproton-proton annihilation given by

$$(\sigma v)_{ann} = 6.5 \cdot 10^{-17} \frac{\mathrm{cm}^3}{\mathrm{s}} \left(\frac{c}{v}\right), \qquad (723)$$

where v is relative velocity (Morgan, Hudges, 1970; Cohen et al., 1997). Thus, for non-relativistic antiprotons the probability for their annihilation is enhanced, and at the low velocity, satisfying the condition

$$\frac{v}{c} < 10^{-4}\Omega_b \tag{724}$$

the rate of annihilation exceeds the expansion rate even in the modern Universe.

After the galaxies are formed at

$$t > t_s \sim 10^{16} \text{ s}, \tag{725}$$

neither matter, nor PBH are distributed homogeneously. Primordial black holes, being in this case one of the forms of dark matter, are concentrated in halos of galaxies (see below), and their evaporation may be the local (galactic) source of antiproton component of cosmic rays (Carr, 1981; Chechetkin et al., 1982).

Antiprotons, produced in the period

$$t_{\bar{p}} < t < t_s , \tag{726}$$

experience red shift, but such a red shift can not move relativistic and semi-relativistic antiprotons, produced in PBH evaporation, to low-velocity conditions of complete annihilation, as well as can not provide their condensation in galaxies. The latter can take place only for very slow antiprotons, having the velocity in the period of evaporation t_e, smaller, than

$$\frac{v}{c} < 10^{-3}\left(\frac{t_s}{t_e}\right)^{2/3} , \tag{727}$$

but the annihilation rate for such slow antiprotons is of the order of the expansion rate, so that their significant fraction should annihilate. More energetic antiprotons should form isotropic background of antiprotons, homogeneously distributed in the Universe.

Consider now the possible effects of antinucleon annihilation on the RD stage. First let us discuss the antinucleon annihilation with nucleons.

As we have mentioned earlier, antiprotons should slow down in the cosmic plasma before their annihilation. As a result of annihilation of slow antiproton with nucleon, pions are produced with the averaged multiplicity about 5 (see Chesquire, 1974; Chechetkin et al., 1982; Cohen et al., 1997).

One can easily estimate that at

$$t > 10^2 \mathrm{s}\ \Omega_b^{2/3} \tag{728}$$

the time scale of pion-nucleon interaction

$$t_{\pi N} \sim \left(n_b \sigma_{\pi N} c\right)^{-1} \tag{729}$$

exceeds the lifetime of charged pions

$$\tau_{\pi^+} = \tau_{\pi^-} \approx 2 \cdot 10^{-8} \mathrm{s}\ . \tag{730}$$

Hence, both neutral and charged pions from nucleon-antinucleon annihilation should decay. The annihilation of slow antiprotons in the Universe on the RD stage leads to a cascade of pion and muon decays. As a result, there are finally produced neutrinos, carrying out about 50% of the energy released in annihilation, γ-quanta, carrying out about 34% of this energy, and electron-positron pairs, carrying out the rest 16% of the energy.

Neutrinos, produced in pion and muon decay cascades from nucleon-antinucleon annihilation, have energies E_0, not exceeding several hundred MeV. Such neutrinos practically do not interact with the matter and they should retain in the Universe, having experienced the red shift down to the energies

$$E = E_0(1 + z_{ann})^{-1} , \tag{731}$$

where z_{ann} is the red shift, corresponding to the period of annihilation. The fluxes of these neutrino with the energy, not exceeding the value

$$E \sim 10^{-2} \text{ MeV} \tag{732}$$

for

$$z_{ann} > 10^5 \tag{733}$$

seem to be not detectable.

Gamma quanta, electrons and positrons from annihilation interact with electrons of the cosmic plasma, as well as with positrons, born in annihilation process, annihilate further with cosmic electrons. The energy release in these processes leads to distortions of the thermal background radiation spectrum.

However, the existing restrictions on such distortions (Eq. (632)) are not very sensitive to antinucleon annihilation on the early RD stage. The effect in the energy release should be compared with the total energy density of the thermal radiation background. The significant effect implies the amount of annihilating antiprotons of the order of the total amount of baryons in the Universe.

As we shall show, much stronger restrictions on antiproton sources follow from the analysis of the astrophysical effects of antiproton annihilation with ^{4}He nuclei.

2. Effects of antiproton annihilation with ^{4}He nuclei on D and ^{3}He abundance.

Helium ^{4}He is the most prevalent after hydrogen element in the Universe. Its weight concentration is equal to

$$Y \equiv X_{^4\mathrm{He}} = \frac{m_{^4\mathrm{He}} n_{^4\mathrm{He}}}{\sum_{\mathrm{A}} m_{\mathrm{A}} n_{\mathrm{A}}} \cong \frac{4n_{^4\mathrm{He}}}{\left(n_{\mathrm{H}} + 4n_{^4\mathrm{He}}\right)} \cong 0.25 \ , \qquad (734)$$

whereas the concentration of other light elements is much smaller. For example, deuterium abundance is estimated as

$$X_{\mathrm{D}} = \frac{m_{\mathrm{D}} n_{\mathrm{D}}}{\sum_{\mathrm{A}} m_{\mathrm{A}} n_{\mathrm{A}}} = 2.5 \cdot 10^{-5} \ , \qquad (735)$$

and the abundance of ^{3}He is given by

$$X_{^3\mathrm{He}} = 4.2 \cdot 10^{-5} \ . \qquad (736)$$

Annihilation of antiprotons with ^{4}He nuclei can lead to the production of deuterium and ^{3}He in the following reactions

$$\bar{p} + {}^4He \rightarrow D + p(n) + N_1\pi \ , \qquad (737)$$

$$\bar{p} + {}^4He \rightarrow {}^3He + N_2\pi \ , \qquad (738)$$

$$\bar{p} + {}^4He \rightarrow T + N_3\pi \ , \qquad (739)$$

where N_1 , N_2 , N_3 are the numbers of pions.

If there were antiprotons in the Universe after the Big Bang nucleosynthesis, their annihilation should inevitably result in production of additional amounts of deuterium and ^{3}He.

Comparison of ^{4}He and D, ^{3}He abundances shows that the destruction of even a small fraction of ^{4}He (~10^{-4}) due to interaction with antiprotons can lead to the creation of all nowadays observed abundance of deuterium and/or ^{3}He.

Note that in the early Universe deuterium can be produced not only directly in reactions (737). The destruction of ^{4}He nuclei in the process of antiproton annihilation is the source of free neutrons (Chechetkin et al., 1982; Zeldovich et al.,1977).

If the matter density in the Universe is sufficiently high in the period of annihilation, neutrons can participate the reaction

$$n + p \rightarrow D + \gamma \,, \tag{740}$$

thus forming additional amount of deuterium.

So, there are two ways to produce deuterium in antiproton annihilation with ^{4}He nuclei after the Big Bang nucleosynthesis:

1) in the direct reactions (737)

and

2) indirectly, due to the interaction of protons with neutrons, produced after antiproton-induced destruction of ^{4}He nuclei.

The indirect mechanism is effective, if protons can capture neutrons before neutron decay, i.e. if the time scale of the reaction (740) does not exceed the neutron lifetime

$$\tau_n \sim 10^3 \text{ s} \,. \tag{741}$$

This condition has the form:

$$t_{np} = \left(n_p \sigma_{np} v\right)^{-1} \leq \tau_n \ . \tag{742}$$

Here n_p is the proton concentration and $\sigma_{np}v$ is the rate of the neutron capture reaction. The estimations by Zeldovich et al. (1977) show that the indirect mechanism is not effective after

$$t_D = 10^5 \Omega_b^{2/3} \text{ s} \tag{743}$$

since

$$t_{np} > \tau_n \tag{744}$$

at

$$t > t_D \ . \tag{745}$$

In the period (745), the direct mechanism is the only source of deuterium production in the early Universe after the Big Bang nucleosynthesis.

Let us discuss the restrictions on f from the analysis of the influence of antiproton interaction with ^{4}He nuclei on light element abundance. Due to antiproton annihilation with ^{4}He, an additional amount of deuterium

$$\Delta n_{\rm D} = \begin{cases} n_{^4\rm He}(f_{\rm D} + f_n)f & \text{for } 10^2 \text{ s} \leq t \leq t_D \ , \\ n_{^4\rm He} f_{\rm D} f & \text{for } t > t_D \ , \end{cases} \tag{746}$$

can be produced, where $n_{\rm A}$ with A = ^{4}He is ^{4}He concentration and $f_{\rm D}$ and f_n are the average multiplicities of D nuclei and neutrons, produced in the result of annihilation, respectively. Similarly, one obtains for ^{3}He

$$\Delta n_{^3\mathrm{He}} = n_{^4\mathrm{He}} f_{^3\mathrm{He}} f \,. \tag{747}$$

Assuming that Δn_{D} should not exceed the observed D abundance X_{D}, one obtains the following upper limit on the value of f:

$$f \le \begin{cases} \dfrac{2X_{\mathrm{D}}}{Y(f_{\mathrm{D}} + f_n)} & \text{for } 10^2 \text{ s} \le t_{ann} \le t_D \,, \\ \dfrac{2X_{\mathrm{D}}}{Y f_{\mathrm{D}}} & \text{for } t_{ann} \le t_D \,, \end{cases} \tag{748}$$

where Y is the observed ^{4}He abundance.

As to the ^{3}He production, note that ^{3}He nuclei are also the products of decay of tritium, produced in the reaction (739). So both reactions (738) and (739) contribute into the ^{3}He production.

With the use of Eq. (747), one can find the restriction on f similar to the one given by Eq. (748) from the observed ^{3}He abundance

$$f \le \frac{4}{3} \cdot \frac{X_{^3\mathrm{He}}}{Y f^{eff}_{^3\mathrm{He}}} \,, \tag{749}$$

where

$$f^{eff}_{^3\mathrm{He}} = f_{^3\mathrm{He}} + f_{\mathrm{T}} \,. \tag{750}$$

It follows from Eqs. (748) and (749), that the upper limit on f depends essentially on the values of f_{A}, where A = D, n, T and ^{3}He. To obtain more reliable information on the possible sources of antiprotons at the RD stage (Chechetkin et al., 1982), special experimental studies of antiproton interaction with ^{4}He nuclei were undertaken (Balestra et al., 1980; Batusev et al., 1984) in the experiment PS179 in LEAR. The value

$$f_{^3\mathrm{He}} = 0.102 \pm 0.008 \tag{751}$$

was obtained in this experiment (Batusev et al., 1984).

With these experimental data, the relative probability of tritium production was also estimated:

$$f_{\mathrm{T}} = (1.32 \pm 0.05) f_{^3\mathrm{He}} \ , \tag{752}$$

giving the value for the effective ^{3}He production constant

$$f^{eff}_{^3\mathrm{He}} = 0.237 \pm 0.014 \ . \tag{753}$$

It follows from Eq. (749) that the maximal ^{3}He abundance allowed from the observations,

$$X_{^3\mathrm{He}} = 1.2 \cdot 10^{-4} \tag{754}$$

corresponds to the restriction on f given by

$$f < 2.5 \cdot 10^{-3} \ . \tag{755}$$

The data of Batusev et al. (1984) show, that the magnitude f_{A} for A = ^{3}He exhibits rather weak dependence on antiproton momentum in the range 200-600 MeV/c. It confirms the theoretical arguments of Kondratyuk and Shmatikov (1983) in favour of weak dependence of this magnitude on the antiproton energy. So, in the first approximation, one can assume this magnitude to be independent on the energy of antiproton. In this approximation the predicted concentration of ^{3}He, produced in antiproton annihilation with ^{4}He nuclei, is independent on the spectrum of antiprotons and on their behaviour in the cosmic plasma (in particular, on the mechanisms of deceleration in plasma in the case of energetic antiprotons).

In this approximation the upper limit (755) is valid for all hypothetical antiproton sources on RD stage. It provides reliable test for the possible physical nature and, as we shall see in the next section, makes it possible to obtain some nontrivial information on the hidden sector of particle theory, as well as on some phenomena, probing the physical mechanisms for inflation, baryosynthesis and dark matter.

Note, that decays of metastable particle and PBH evaporation are the sources of not only antiprotons but also of other particles, namely, of protons, mesons and hyperons, electron-positron pairs, gamma quanta and neutrinos. Moreover, for any source of antiprotons, the proton-antiproton annihilation is also the source of γ-quanta with the energy about 100 MeV (see above).

The interaction with ^{4}He nuclei of protons and γ-quanta with the energy, exceeding 20 MeV, can also lead to destruction of ^{4}He nuclei and D and ^{3}He production (Ellis, Nanopoulos, Sarkar, 1984; Dimopoulos et al., 1987; Levitan et al., 1988). Analysis of these mechanisms involves detailed study of the evolution of energetic nucleons and γ-quanta in cosmic plasma with the use of numerical simulations. Such analysis should take into account all mechanisms of energy losses and interactions of non-equilibrium particles, as well as all specific properties of the considered hypothetical source of non-equilibrium particles (in particular, spectra of particles, emitted by the source).

In the first approximation, the analysis of effects of antiproton-^{4}He interaction on light element abundance avoids the consideration of all these details, giving reliable universal minimal estimation of the effect of antiproton sources. Additional production of D and ^{3}He due to ^{4}He destruction by energetic particles strengthens the effect and corresponding restrictions on the parameters of antiproton sources on the RD stage.

Such an enhancement of restriction (755) due to additional ^{3}He production in ^{4}He destruction by energetic particles of nucleon cascade in the early Universe, induced by antinucleon with initial energy E_0, was calculated by

Levitan et al. (1988) in Monte Carlo simulations of development of nucleon cascade (see Fig.15).

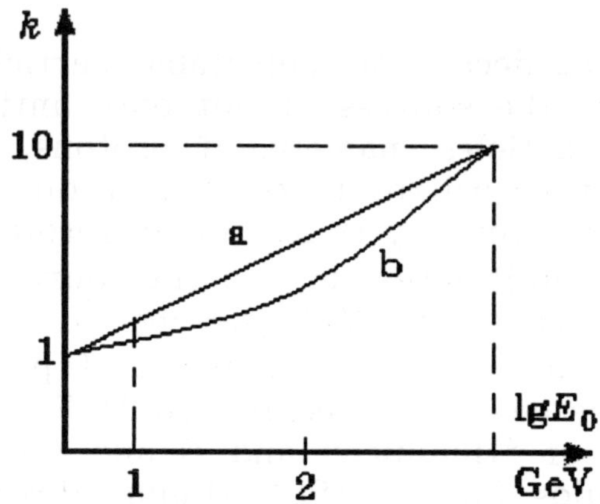

Fig.15. The energy dependence of nucleon cascade enhancement factor for the constraint on antinucleon sources from ^{3}He abundance.

Detailed analysis of the predicted ratio of light element abundance, corresponding to each specific source of non-equilibrium particles, makes it possible to distinguish effects of different sources and to test their existence and possible parameters in the confrontation with the observed chemical composition of the Universe.

In conclusion, let us consider another aspect of the relationship between the abundance of light elements and parameters of antiproton sources.

In the case of strict equality in the Eq. (748), the existence of observed deuterium may be ascribed to non-equilibrium processes of ^{4}He destruction on RD stage (Zeldovich et al., 1977). New mechanism of deuterium production arises, in which the deuterium abundance does not depend on the baryon density, as it is in the case of the standard theory of Big Bang nucleosynthesis. Even in the case $\Omega_b = 1$, when the theory of standard Big Bang nucleosynthesis predicts

$$X_D < 10^{-8} \tag{756}$$

(see Chapter 3) the ^{4}He destruction on the post-nucleosynthesis RD stage can give rise to the prediction of the amount of deuterium, consistent with its observed abundance.

Such a possibility is of special interest for the problem on the physical nature of the dark matter in the Universe (see below), since the standard theory of Big Bang nucleosynthesis favours $\Omega_b = 0.1$ thus providing serious argument for nonbaryonic nature of the dark matter, dominating in the Universe and making the total density equal to the critical density, in accordance with the simplest inflationary models.

So, the possibilities of non-equilibrium cosmological nucleosynthesis and, in particular, of antiproton-^{4}He mechanism of deuterium production deserve special investigation. Such an investigation generally involves detailed consideration of possible properties of hypothetical sources of non-equilibrium particles on the RD stage.

However, if the antiproton-^{4}He mechanism dominates in the non-equilibrium cosmological deuterium production, one can test its viability independent on the nature of antiproton source (Chechetkin et al., 1982).

To probe experimentally the viability of this mechanism, it is sufficient to compare the values of f_A for A = D and A = ^{3}He.

If in the experimental measurement of these quantities it will be found that

$$f_D < 0.1 f_{^3He} \tag{757}$$

The antiproton annihilation with helium should result in the deuterium abundance, which is of the order of the magnitude smaller, than the abundance of ^{3}He.

Ascribing all the observed deuterium abundance to the antiproton-^{4}He mechanism, one comes to overproduction of ^{3}He, since with the account for all the uncertainties in the observed deuterium and ^{3}He abundance they can not differ by an order of magnitude.

Hence, one can experimentally disprove the antiproton-^{4}He mechanism of deuterium production at the RD stage in the measurement of f_A ratio for A = D and A = ^{3}He.

3. Restriction on antiproton sources.

Let us consider now the relationship between antiproton sources on the RD stage and the parameters of particle models. We shall see, how the restrictions on the effects of antiproton annihilation on the RD stage impose constraints on these parameters.

3.1. Constraints on PBHs and Physics of Their Origin

In the Chapter 4, it was demonstrated that the spectrum of PBHs, formed on the stage of dominance of superheavy metastable particles (or scalar fields), can be related to the parameters of these particles (or fields). The upper bound on the amount of antiprotons from PBHs with the mass

$$10^{10} < M < 10^{13} \text{ g} \tag{758}$$

evaporating on the RD stage, imposes the restriction on the possible contribution, $\alpha(M)$, of such PBHs into the cosmological density in the period of their evaporation.

It follows from Eq. (715), that the fraction of the total density, corresponding to PBHs with the mass

$$M < 10^{13} \text{ g}$$

in the period of their evaporation t_e, can be related with the quantity f by the equation

$$\alpha(M) = 10^{-3} f \Omega_b \left(\frac{t_e}{1\text{s}} \right)^{1/3} k_{\bar{p}}^{-1} , \tag{759}$$

where

$$\alpha(M) = \frac{\rho_{PBH}(M)}{\rho_{tot}} \tag{760}$$

and

$$k_{\bar{p}} = \frac{g_g}{g_{tot}} \tag{761}$$

is the fraction of antiprotons in the spectrum of evaporating PBH.

It follows from Eq. (759) that the restriction given by Eq. (755) provides the upper limit on $\alpha(M = 10^{11} \text{ g})$, evaporating at

$$t_e \sim 10^6 \text{ s} . \tag{762}$$

In Eq. (759), one can take $\Omega_b = 0.1$ and

$$\frac{g_g}{g_{tot}} = \frac{8}{\frac{21}{4} N_q + \frac{7}{4} N_l + 9 + \frac{7}{8} N_\nu + N_{hs}} , \tag{763}$$

where $N_{q(l)}$ is the number of quark (lepton) flavours with the mass

$$m < T_{PBH} \ , \tag{764}$$

N_ν is the number of neutrino species, and N_{hs} is the effective two-component species of particles, predicted in the hidden sector of particle theory. Then, putting

$$N_{hs} = 0 \tag{765}$$

i.e. neglecting the effect of superweakly interacting particles, one obtains the upper limit

$$\alpha(M) < 10^{-4} \ . \tag{766}$$

To deduce from this upper limit the restriction on the probability W_{PBH} of PBH formation, one should take into account the succession of dust-like and radiation-dominant stages, which follows the period of the PBH formation.

Consider, for definiteness, the case, when PBH are formed at the dust-like stage of superheavy particles dominance, so that this probability defines the fraction of the particles, forming PBHs. When the particles decay or annihilate within the inhomogeneities, at

$$t \sim t_f \tag{767}$$

early dust-like stage comes to end, and the magnitude W_{PBH} defines the contribution of PBHs into the total cosmological density at

$$t \geq t_f \ . \tag{768}$$

During the successive stage of the dominance in the Universe of relativistic particles, produced after the thermalization of decay products (or annihilation) of superheavy particles, with the equation of state

$$p = \tfrac{1}{3}\varepsilon \, .$$

The relative contribution of PBHs into the cosmological density increases as

$$\frac{\rho_{PBH}(M)}{\rho_{tot}} \propto \frac{M}{T} \propto a(t) \propto \left(\frac{t}{t_f}\right)^{1/2} . \tag{769}$$

Hence, in the period of PBH evaporation on the RD stage, i.e. at

$$t_e = \left(\frac{M}{m_{Pl}}\right)^3 t_{Pl} \, , \tag{770}$$

the relative contribution of PBHs into the cosmological density $\alpha(M)$ is

$$\alpha(M) = W_{PBH}\left(\frac{t_e}{t_f}\right)^{1/2} = W_{PBH}\left(\frac{M}{m_{Pl}}\right)^{3/2}\left(\frac{t_{Pl}}{t_f}\right)^{1/2} . \tag{771}$$

If the particle model predicts one or more additional types of superheavy metastable particles, so that at

$$t > t_f$$

one or more stages of dominance of non-relativistic metastable particles take place, one should put into the right-hand side of Eq.(771) an additional factor

$$A = \prod_i \left(\frac{t_{fi}}{t_{0i}} \right)^{1/2} , \tag{772}$$

where t_{0i} and t_{fi} correspond to the beginning and the end of the i-th stage. The factor A takes into account that relative contribution of PBHs into the total density does not grow during dust-like stages. But the possibility of such a succession of dust-like stages $p = 0$ is constrained from the observed baryon asymmetry of the Universe, since after each stage the specific entropy S increases by the factor

$$S \propto \left(\frac{t_{fi}}{t_{0i}} \right)^{1/2} , \tag{773}$$

and the initial baryon asymmetry is suppressed by the factor

$$\frac{n_b}{n_\gamma} \propto A^{-1} . \tag{774}$$

The weaker are restrictions on the probability of PBH formation, the stronger are constraints on the mechanisms of baryogenesis.

The upper limit (766) on the magnitude $\alpha(M)$ imposes the upper limit on the probability W_{PBH}, depending on the amplitude of density perturbations δ. It provides tests for superheavy particles and fields, predicted by the particle theory.

Indeed, the spectrum of masses of PBHs, formed during the stages of superheavy particles dominance, is restricted (see Chapter 4) within the mass interval from

$$M_0 = m_{Pl}\left(\frac{m_{Pl}}{rm}\right)^2 \tag{775}$$

up to the maximal mass

$$M_{max} = m_{Pl}\frac{\tau}{t_{Pl}}\delta^{3/2}\ , \tag{776}$$

and the spectrum in this interval is determined by the spectrum of initial perturbations $\delta(M)$. In the framework of inflational models, the shape and the amplitude of this spectrum are related to the parameters of the inflaton potential.

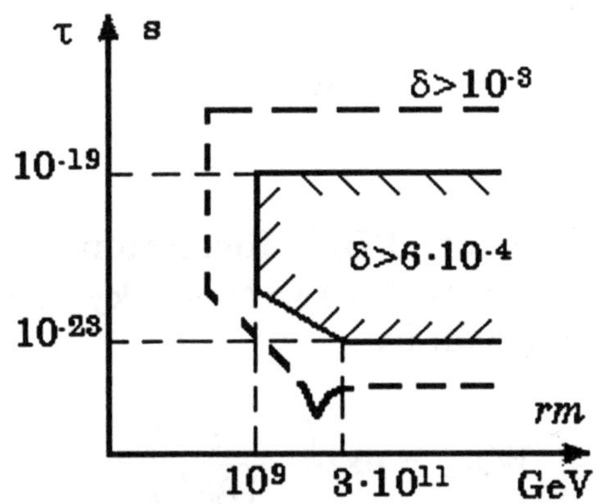

Fig.16. Constraints on τ and rm of metastable particles, forming PBH in the period of their dominance in the very early Universe.

Therefore, the parameters of superheavy fields and particles determine the lower bound of the PBH spectrum, given in Chapter 4. The condition, that the lower bound of the PBH spectrum does not contradict the upper limit on this spectrum, given above, provides nontrivial constraints on the masses, concentrations and lifetimes of superheavy metastable particles, predicted by particle theory at super-high energy scales. Such restrictions on the values of *rm* and τ, depending on

$$\delta(M) = \delta = const \tag{777}$$

are given in Fig.16. Some examples of such restrictions were presented in Chapter 4.

Note, that according to Carr (1975), Zabotin and Naselsky (1982) the PBH may form on

$$p = \tfrac{1}{3}\varepsilon$$

stages due to the high amplitude tail in the Gaussian distribution of amplitudes of perturbations. In this case, PBHs with the mass M are formed at

$$t \sim \left(\frac{M}{m_{Pl}}\right) t_{Pl} \tag{778}$$

and the probability of PBH formation is determined only by the value of δ (Zabotin, Naselsky, 1982; Khlopov et al., 1984) and is given by

$$W_{PBH} \sim \exp\left(-\frac{1}{18\delta^2}\right) . \tag{779}$$

Thus, the upper limit on $\alpha(M)$ due to the influence of PBH evaporation on primordial chemical composition via the antiproton interaction with ^{4}He provides strong constraints on the inhomogeneity of the early Universe at

$$t < 10^{-28} \text{ s} . \tag{780}$$

Since the relative contribution of PBHs into cosmological density increases as

$$\frac{\rho_{PBH}(M)}{\rho_{tot}} \propto \frac{M}{T} \tag{781}$$

during the RD stage, the upper limit on $\alpha(M)$, given by Eq. (766), imposes the restriction on the probability of PBH formation at

$$t = \frac{M}{m_{Pl}} t_{Pl} , \tag{782}$$

given by

$$W_{PBH} \sim \exp\left(-\frac{1}{18\delta^2}\right) = \beta(M) =$$

$$= \frac{m_{Pl}}{M}\alpha(M) < 4 \cdot 10^{-20}\left(\frac{10^{10}\,\mathrm{g}}{M}\right) , \tag{783}$$

thus providing the upper limit on the possible amplitude of perturbations in small scales:

$$\delta < 3.5 \cdot 10^{-2}\left(1 + \frac{1}{45}\ln\left(\frac{M}{10^{10}\,\mathrm{g}}\right)\right)^{-1/2} . \tag{784}$$

This restriction imposes constraints on possible parameters of self-interaction of scalar fields, being introduced in inflational models (Khlopov et al., 1984).

On the other hand, it was pointed out in Chapter 4 that relatively long stages of coherent scalar field oscillations can be predicted in the framework of inflational models (Linde, 1984; Starobinsky, 1980; for review see Dymnikova, 1998). The PBH formation is possible on these stages (Khlopov et al., 1985; 1984). After the end of such stages (Starobinsky, 1980) the decay products thermalize due to decay of the scalar inflaton field and the Universe is reheated up to the temperature T_R, determined by the duration τ of the coherent field oscillation stage (see Chapter 4):

$$T_R \sim \left(\frac{m_{Pl}}{\tau}\right)^{1/2} . \tag{785}$$

In the simplest case, the duration of this stage, as well as the probability of the PBH formation are defined by the parameters of the same inflationary model and by the inflaton scalar field self-couplings. Constraints on these parameters follow from the upper limit on the probability of PBH formation (see Fig.17).

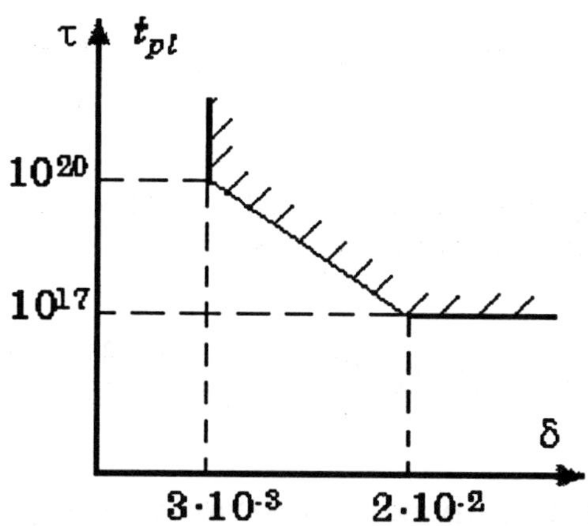

Fig.17. Constraints on the duration and inhomogeneity of post-inflational dust-like stage from upper limits on the spectrum of primordial black holes.

Note that the restriction on the value of τ imposes lower limit on the reheating temperature after inflation T_R for

$$\delta > 6 \cdot 10^{-4} \ . \tag{786}$$

These lower limits are closely related with the restrictions on the parameters of local supersymmetric models (Balestra et al., 1984; Khlopov, Linde, 1984; Ellis et al., 1984; see Chapter 2), predicting the existence of gravitino with the mass

$$m_G \sim 100 \text{ GeV} \tag{787}$$

(Ellis et al., 1983).

3.2. The Problem of Primordial Gravitino

Gravitino plays a special role in the supersymmetric models due to its extremely weak interaction with other particles. The cross section of the processes of gravitino absorption or emission is of the order of

$$\sigma \sim \frac{\alpha}{m_{Pl}^2} \sim 10^{-66} \text{cm}^2 \ , \tag{788}$$

where

$$\alpha = \frac{g^2}{4\pi} \sim 10^{-2} \tag{789}$$

and g is the gauge constant. On that reason, in the presence of other more light supersymmetric particles, the gravitino turns out to be a particle with the very long lifetime, given by

$$\tau_G = \left(\frac{m_{Pl}}{m_G}\right)^3 t_{Pl} \tag{790}$$

(Weinberg, 1982; Nanopoulos et al., 1983). For gravitino with the mass (787) predicted in many minimal supersymmetric models (Nanopoulos et al., 1983), the lifetime exceeds

$$t \sim \tau_G \sim 10^8 \text{s} . \tag{791}$$

Thus, supersymmetric models (see Chapter 2) predict the existence of a massive particle, being either absolutely stable (if it is the lightest supersymmetric particle and the R-parity is conserved), or metastable one with the lifetime, exceeding 1 s at

$$m_G < 10^5 \text{GeV} . \tag{792}$$

This prediction of supersymmetric models caused troubles in cosmology (Weinberg, 1982).

Indeed, in the case of gravitino decoupling from other particles at

$$T \sim m_{Pl} \tag{793}$$

the standard arguments on the decoupling of weakly interacting particles in the early Universe (see Chapter 3) give the primordial gravitino concentration of the order of

$$r_G = \frac{n_G}{n_\gamma} \sim \frac{1}{N} , \tag{794}$$

where N is the effective number of relativistic species, being in the equilibrium in the Universe at

$$T \sim m_{Pl} \ .$$

Such an abundance of stable gravitinos with the mass $m_G \sim 100$ GeV corresponds to their density in the modern Universe

$$\rho_G = m_G n_G = 10^2 m_p \cdot \frac{1}{N} \cdot n_\gamma \sim 3 \cdot 10^7 \rho_{cr} \tag{795}$$

being by seven orders of magnitude larger, than the upper limit on the modern cosmological density (see Weinberg, 1982).

The contradiction with observations is not removed in the case of metastable gravitino, decaying on the RD stage. The products of gravitino decays interact with the plasma and radiation, and that should induce distortion in the Planck form of the spectrum of the thermal electromagnetic background.

Indeed, the energy density ε_G released in gravitino decays at $t \sim \tau_G \sim 10^8$ s is given by

$$\varepsilon_G = m_G n_G \sim 100 \text{ MeV } \cdot n_\gamma \cdot \left(\frac{m_G}{100 \text{ GeV}}\right), \tag{796}$$

whereas the energy density of the thermal background in this period is only

$$\varepsilon_\gamma \sim 100 \text{ eV } \cdot n_\gamma \ . \tag{797}$$

One obtains the energy release by six orders of magnitude exceeding the thermal background energy density for the period, in which no more, than cents of percent of the total radiation energy density are allowed to be released (see above).

The attempt to resolve the contradiction was made on the basis of inflational models (Ellis et al., 1983).

In these models no equilibrium between gravitino and other particles takes place, since there was no Planck temperatures during inflation. The abundance of primordial gravitinos is not determined in this case by the equilibrium concentration in the period of their decoupling.

However, according to Nanopoulos et al. (1983), even in this case the non-vanishing abundance of gravitino is predicted. There is small but nonzero probability of gravitino production in collisions of particles with their supersymmetric partners.

Following Nanopoulos et al. (1983), the main source of gravitinos after inflation should be related to the process

$$X + V \to X + G\,, \tag{798}$$

where X are scalar particles, interacting with the gaugino V and gravitino G. The cross section for this process is of the order of

$$\sigma \sim \frac{\alpha}{m_{Pl}^{2}}\,, \tag{799}$$

where

$$\alpha = \frac{g^2}{4\pi} \sim 10^{-2}$$

and g is the gauge constant. Owing to this process, the relative concentration of gravitino grows after the end of inflation with the rate, given by

$$\frac{dr_G}{dt} \sim \alpha \cdot N(T) \cdot \frac{T^3}{m_{Pl}} \tag{800}$$

(Khlopov, Linde, 1984). Here

$$N(T) \sim 10^2 \tag{801}$$

is the effective number of degrees of freedom, defined as the number of species of particles with masses $m < T$ at the temperature T with the account of their statistical weights.

It follows from Eq. (800), that if the Universe after inflation was reheated up to the temperature

$$T = T_R \ , \tag{802}$$

the gravitino abundance grows from virtually zero values to the value

$$r_G \sim \frac{\alpha\sqrt{N(T)}T_R}{m_{Pl}} \sim 10^{-1}\frac{T_R}{m_{Pl}} \ . \tag{803}$$

Consider for definiteness the minimal N = 1 supergravity (Cremmer et al., 1979; 1983) coupled to the matter, in which photino and gluino are usually more light, than gravitino, so that gravitino decays

$$G \to \gamma\tilde{\gamma} \tag{804}$$

and

$$G \to g\tilde{g} \tag{805}$$

are possible.

The gravitino lifetime is in this case equal to

$$\tau_G = \left(\frac{m_{Pl}}{m_G}\right)^3 t_{Pl} \ . \tag{806}$$

If

$$m_{\tilde{\gamma},\tilde{g}} << m_G \tag{807}$$

one has the relation between the branching ratios for the considered modes of gravitino decays

$$\frac{Br(G \to \gamma\tilde{\gamma})}{Br(G \to g\tilde{g})} = \frac{1}{8} \; . \tag{808}$$

The concentration n_a of antinucleons produced in the gravitino decays is

$$n_a = n_G \cdot Br(G \to g\tilde{g}) \cdot \int dp_a \int dp D_g^a(p_a, p) \cdot \delta\left(p - \frac{m_G}{2}\right), \tag{809}$$

where $D_g{}^a(p_a,p)$ is the probability of fragmentation of gluon with momentum p to the antinucleon with momentum p_a , and δ is the Dirac delta-function.

One can estimate the fraction of antinucleons in gravitino decays similarly to our previous treatment of the antiproton production by evaporating PBHs. In this case this quantity defined as

$$f_a = \frac{n_a}{n_G} \tag{810}$$

is of order of

$$f_a \sim 1 \tag{811}$$

(c.f. with Khlopov, Linde, 1984; Balestra et al., 1984).

According to Balestra et al. (1984) and Khlopov, Linde (1984), the experimental data of Batusev et al.

(1984), leading to Eq. (755), correspond to the restriction on the value of relative gravitino concentration

$$r_G = \frac{n_G}{n_\gamma} < 10^{-12} \ . \tag{812}$$

Comparison of Eq. (812) with the theoretical estimation of the primordial gravitino concentration gives rather stringent upper limit on the possible maximal reheating temperature of the Universe after inflation:

$$T_R < \ 10^8 \ \mathrm{GeV} \ . \tag{813}$$

This restriction leads to serious problems in the explanation of the observed baryon asymmetry of the Universe. The bosons, predicted in GUTs (see Chapter 3), can not be present in the Universe in any sizeable amount at such low temperatures. With the account of Eq. (813), a similar problem arises for the particles with the masses 10^{10}-10^{11} GeV, predicted by supersymmetric models and considered as a source of cosmological baryon excess (Linde, 1984).

We see that the serious problem of primordial gravitinos appears in local supersymmetric GUT models. The analysis of the effect of gravitino decays on the primordial chemical composition implies such a low reheating temperature T_R that the problems with baryosynthesis arise.

Besides, the low reheating temperature T_R corresponds to a long stage of coherent scalar field oscillations, so that PBH, evaporating on RD stage, should form. Upper limits on the concentration of such PBH put additional constraints on the parameters of the theory, determining the spectrum of small-scale density perturbations.

Consider in conclusion restrictions on another source of antinucleon annihilation on RD stage, namely, on antimatter domains, predicted on the base of particle models, in particular, of the model of spontaneous CP violation.

3.3. Constraints on Antimatter Domain Structure

The analysis of dissipation of antimatter domains (for details see Chapter 3 and Chechetkin et al., 1982; Zeldovich, Novikov, 1975; Steigman, 1976) shows, that the domains, containing the average baryon number N_A, dissipate in the period

$$10^3 < t < 10^{12} \text{ s} , \tag{814}$$

if N_A is in the interval

$$10^{37} f\Omega_b < N_A < 10^{59} f\Omega_b . \tag{815}$$

Here f is the fraction of antibaryons within such domains.

The data on antiproton annihilation with ^{4}He nuclei make it possible to restrict f on the base of analysis of its effect on the abundance of light elements. One obtains

$$f < 2.5 \cdot 10^{-3} . \tag{816}$$

The analysis of possible distortions of the thermal electromagnetic background spectrum gives much more weak restrictions on such domains, since Eq. (632) gives

$$f < 2 \cdot 10^{-2} \left(\frac{1+z}{10^5} \right) \tag{817}$$

at the red shifts

$$z > 10^5 \tag{818}$$

corresponding to the period

$$t < 10^9 \text{s} . \tag{819}$$

Based on GUT models, the scale of domains is determined by the character of the phase transition in the early Universe (Kuzmin et al., 1981; Chechetkin et al., 1982), and in the models with spontaneous CP violation the magnitude f can be expressed through the phases of soft (spontaneous) φ_s and hard φ_h CP violation

$$f = \frac{\varphi_s - \varphi_h}{\varphi_s + \varphi_h} . \tag{820}$$

Then Eq. (816) imposes stringent constraint on the mechanisms of CP violation.

If

$$\varphi_s > \varphi_h \tag{821}$$

and the formation of antimatter domains of the considered scales is possible, φ_s should be very close to φ_h, so that the following condition is satisfied

$$\frac{\varphi_s - \varphi_h}{\varphi_s + \varphi_h} < 2.5 \cdot 10^{-3} . \tag{822}$$

Such restriction can not be obtained in the analysis of possible distortion of thermal electromagnetic background spectrum.

So, tracing only one effect of hypothetical sources of non-equilibrium particles on the RD stage, namely the effect of antiproton interaction with ^{4}He on the abundance of light elements, one can perform rather effective test of particle models, giving rise to sources of antiproton annihilation at

$$t > 10^3 \text{ s} .$$

The abundance of light elements turns out to be a much more sensitive detector of possible effects of non-equilibrium particles, than the spectrum of the thermal electromagnetic background.

It is clear, that more complete analysis of effects of non-equilibrium particles on the RD stage may serve as even more sensitive indicator of the presence of hypothetical sources of such particles in the early Universe.

CHAPTER 7

NON-EQUILIBRIUM EFFECTS AS THE PROBE FOR NEW PHYSICS

Decays of metastable particles, evaporation of primordial black holes, fragmentation of cosmic strings into small loops (Smith, Vilenkin, 1987) with successive decay of such loops on high energy particles are the sources of hadronic, electromagnetic and neutrino cascades of particles with the energies, exceeding significantly the mean thermal energy of equilibrium particles on the RD stage.

These processes are also the sources of considerable amount of particles, which are not in the equilibrium, or which equilibrium concentration is negligible.

The examples for the latter case are given by the sources of antiprotons in the baryon asymmetric Universe at $t > 10^{-4}$ s (see Chapter 6) or the sources of positrons after the first 100 s of expansion.

It is evident, that the physical conditions on the RD stage should change in the presence of hypothetical sources of non-equilibrium particles.

Moreover, the example of antiproton-He annihilation has demonstrated, that the astrophysical data are much more sensitive to non-equilibrium effects on the RD stage, than to the effects, shifting the equilibrium parameters on that stage.

Let us turn to general analysis of possible effects of non-equilibrium particles on the RD stage. Based on such analysis, the sensitivity of the astrophysical data to the presence of hypothetical sources of non-equilibrium particles in the early Universe can be studied, thus refining the probes for cosmological and physical phenomena underlying these sources.

1. The Non-Equilibrium Cosmological Nucleosynthesis.

Let us consider the succession of processes, induced by the source of non-equilibrium particles on the RD stage.

Elementary particles (quarks, leptons, photons, gluons,...) are the primary products of a source, such as massive particle decays or PBH evaporation. The sources of superhigh-energy particles can produce massive particles, e.g., W- and Z-bosons, among its products.

Successive decays of unstable particles and fragmentation of quarks, antiquarks and gluons into hadrons reduce the net output of the source to the fluxes of non-equilibrium photons, neutrino-antineutrino, electron-positron and nucleon-antinucleon pairs. In the RD stage, after the 1 s of expansion, all other particles decay before their interaction with plasma and radiation takes place. The interaction of energetic non-equilibrium particles with plasma and radiation, being in the equilibrium on the RD stage, decelerate the non-equilibrium particles.

Provided that the source may be considered as a stationary one, the stationary spectrum of non-equilibrium particles is formed as a result of particle emission by the source and their interaction with the medium.

Interaction of fluxes of non-equilibrium particles with nuclei – the products of the Big Bang nucleosynthesis – results in the fluxes of non-equilibrium secondary nuclei, whose deceleration in plasma is accompanied by reactions with nuclei from the equilibrium plasma.

The analysis of the chain of nuclear transformations, induced by non-equilibrium particles, as a function of the parameters of their source, is the subject of the theory of non-equilibrium cosmological nucleosynthesis.

The complete theory is not developed yet. Only some elements of this theory and the analysis of some special cases can be found in the literature. But the pattern of the future theory can be pictured rather definitely.

1.1. Kinetic Equation for Non-Equilibrium Particles

The two physically distinct types of the sources of non-equilibrium particles in the Universe are of interest.

The data on the spectrum of the thermal background (see Chapter 5) impose upper limit on the possible energy release $\delta\varepsilon$ at

$$t > 10^4 \text{ s} \tag{823}$$

given by

$$\frac{\delta\varepsilon}{\varepsilon_\gamma} < 1.1 \cdot 10^{-4} , \tag{824}$$

where ε_γ is the energy density of the black-body background radiation.

In this period, the sources of non-equilibrium particles can not dominate in the Universe and the existence of these sources practically does not influence the law of cosmological expansion, except for pure of weakly interacting particles, for example, pure sources of neutrino.

The sources of non-equilibrium particles, giving negligible contribution to the cosmological density, will be called weak sources.

At

$$t < 10^4 \text{ s} \tag{825}$$

the density of plasma is sufficiently high for effective transformation of the energy of non-equilibrium particles and formation of the Planck spectrum of the background radiation.

The value of $\delta\varepsilon$ may exceed the value of ε_γ in this period, what implies the change of the law of expansion in the period, when non-equilibrium particles appear in the Universe. So, strong sources of non-equilibrium particles may, in principle, exist in the Universe at $t < 10^4$ s .

The existence of such sources drastically changes the standard picture of the Big Bang nucleosynthesis. It influences both on the thermodynamic conditions for nuclear reactions between the equilibrium particles and on the kinetics of the processes, induced by non-equilibrium particles.

Unstable primordial gravitinos, discussed in the Chapter 6, are the example of the weak source of non-equilibrium particles.

Unstable massive τ neutrinos, considered in the Chapter 5, are the example of strong source of non-equilibrium particles.

Let us come to the quantitative description of non-equilibrium cosmological nucleosynthesis. For simplicity, following Sedelnikov, Filippov and Khlopov (1995) we shall consider the case of uniformly distributed sources of non-equilibrium particles in the Universe.

This approach is valid, when the inhomogeneity of the distribution of the sources is significant at the scales, much smaller than the scale of the modern inhomogeneities, since in the case of small scale strongly inhomogeneous source distribution, the net effect of sources, being averaged over large scales, is reduced to redefinition of parameters of uniformly distributed individual sources.

More complicated redefinition of parameters makes it possible to use this approach even in the case of the inhomogeneous early Universe on the RD stage or of inhomogeneous baryon distribution on this stage (Zeldovich, 1975; Polnarev, Khlopov, 1981).

In the considered case of the uniform medium at the moment t the momentum p_i distribution of particles of the

i-th type, being in the thermal equilibrium at the temperature $T(t)$, is given by the distribution function $f_i(p_i,T(t),t)$. Such distribution of non-equilibrium particles of the i-th type is given by the distribution function $\varphi_i(p_i,t)$. The number density of all particles of the i-th type is determined by both equilibrium and non-equilibrium particles, i.e. by the quantity

$$F_i = f_i + \varphi_i \, , \tag{826}$$

and is equal to

$$n_i(t) = \int F_i(p_i)dp_i \, . \tag{827}$$

The source of non-equilibrium particles is characterised by its intensity $Q_i(p_i,t,\tau)$, defined as the number density distribution of particles of the i-th type emitted with the momentum p_i per unit time at the moment t. Here τ is the time scale of the source. If the decays of metastable particles are the sources of non-equilibrium particles, τ means their lifetime. In the case of evaporating PBHs τ is the time of evaporation.

In the analogy with the case of photons in the expanding Universe (see, e.g. Zeldovich and Novikov, 1975) one can deduce the system of kinetic equations for distribution functions $F_i(p_i,t)$ (Filippov et al., 1993; Sedelnikov, Filippov, Khlopov, 1995)

$$\frac{\partial F_i}{\partial t} + 2HF_i - Hp_i \frac{\partial F_i}{\partial p_i} = I^+ - I^- + Q_i \, . \tag{828}$$

Here $H(t)$ is the Hubble constant at the moment t, and the collision terms I^+ and I^- determine the income and outflow of particles of the i-th type in the spherical layer of the radius p_i in the momentum space with the width Δp_i

due to their interaction with the other particles. The structure of collision terms has the form

$$I^{+}(p_i) = \begin{pmatrix} \int \sum\limits_{j} F_j \dfrac{d\langle \Gamma_j \rangle}{dp_i} dp_j + \\ + \iint \sum\limits_{j,k} F_j F_k \dfrac{d\langle \sigma^{jk} \rangle}{dp_i} dp_j dp_k + \ldots \end{pmatrix} A_i \tag{829}$$

and

$$I^{-}(p_i) = F_i \left(\langle \Gamma_i \rangle + \int \sum_j F_j \langle \sigma^{ij} \rangle dp_j + \ldots \right), \tag{830}$$

where dots stay for omitted terms, corresponding to n-particle collisions with $n > 3$, the occupation factor

$$A_i = 1 \pm \frac{F_i(p_i)}{4\pi p_i^2} \tag{831}$$

accounts for the effects of quantum statistics (the sign "+" for bosons corresponds to induced absorption and the sign "−" for fermions takes into account the Pauli exclusion principle) and the brackets <...> mean the average over the final states with the account for their occupation factors

$$\langle S_i \rangle = \prod_j \int d^3 p_j \frac{dS_i}{dp_j} A_j \ . \tag{832}$$

The system of equations (828) together with the system of Friedman-Robertson-Walker cosmological equations can be considered as the most general basis for the analysis of non-equilibrium processes in the homogeneous and isotropic Universe.

To analyse the influence of the sources of non-equilibrium particles on the light element abundance it is useful to pass from Eqs. (828)–(830) to more simplified description. It takes into account the difference in the effects of equilibrium and non-equilibrium components and sufficiently rapid (as compared with the expansion rate) formation of the spectrum of non-equilibrium particles.

If

$$H\tau_i \ll 1 \; , \tag{833}$$

where τ_i is the time scale of the i-th process of particle interaction (or decay), one can neglect in the left-hand side of Eq. (828) the effects of expansion and of the red shift, being proportional to H.

In the opposite case

$$H\tau_i \gg 1 \; , \tag{834}$$

the effects of particle interactions are insignificant, and one can neglect the contribution of collision terms into the right-hand side of Eq. (828).

To analyse the effects of non-equilibrium fluxes on the light element abundance, one can neglect the presence of low energy non-equilibrium charged particles with

$$E \ll 10 \text{ MeV} \; , \tag{835}$$

photons and neutrinos. We shall also neglect the effects of occupation numbers, putting them equal to

$$A_i = 1 \; . \tag{836}$$

1.2. Kinetic Equation for Weak Sources on RD Stage.

The source of non-equilibrium particles is weak, when the energy density of its products is small as compared with the energy density of the thermal background. We can formulate this condition as

$$\sum_i \int \varepsilon_i Q_i(p_i, t, \tau) dp_i \Delta t << \varepsilon_\gamma \ , \tag{837}$$

with

$$\Delta t \sim t \tag{838}$$

for all t.

In the presence of weak sources, the equilibrium conditions after thermalization of non-equilibrium particles do not change significantly. So, one can assume the expansion law

$$a(t) \propto t^{1/2}$$

and study effects of non-equilibrium particles on the background of the processes with equilibrium particles, considered by the standard Big Bang scenario on the RD stage.

If the condition (833) is valid, the system of equations (828)–(830) is reduced to the system of kinetic equations for the processes with non-equilibrium particles, taking place in the stationary equilibrium conditions. One obtains for the distribution function φ_i of non-equilibrium particles of the i-th type the equation

$$\frac{\partial \varphi_i}{\partial t} = \sum_{j,k} \int \varphi_j F_k \frac{d(\sigma v)^i_{jk}}{dp_i} dp_j dp_k + \sum_j \varphi_j \frac{d\Gamma_j}{dp_i} -$$
$$- \varphi_i \left(\Gamma_i + \sum_j \left(n_j (\sigma v)^{ij}(p_i) + \int \varphi_j (\sigma v)^{ij} dp_j \right) \right) + Q_i(p_i, t, \tau) \ . \tag{840}$$

In the case of equilibrium particles one can use the number density of particles $n_i(T)$ at the temperature T instead of their distribution function f_i. One has

$$n_i(T) = \int f_i(p_i, T, t)dp_i \ , \tag{841}$$

where $T(t)$ is determined by the law of expansion. For equilibrium particles we obtain the equation

$$\frac{\partial n_i(T)}{\partial t} = -n_i\left(\Gamma_i + \sum_j n_j \langle\sigma v\rangle_{ij} + \sum_j \int \varphi_j(p_j)\langle\sigma v\rangle_{ij} dp_j\right) + \\ + \sum_{j,k} n_j n_k \langle\sigma v\rangle^i_{jk} + \delta n_i(T) + \sum_j n_j \Gamma^i_j \ . \tag{842}$$

Here Γ_i is the total probability of decay of unstable particles of the i-th type, Γ_j^i is the relative probability of production of particles of the i-th type in the decays of the particles of the type j, and the term

$$\delta n_i(T) = \int_0^{p_{\max}} \varphi_i(p_i)dp_i \tag{843}$$

accounts for the rapid deceleration of nonrelativistic charged particles in the cosmic plasma and for the energy thresholds of reactions.

Consider now in more details the succession of the processes, induced in the Universe by a weak source of non-equilibrium particles.

In general, lepton-antilepton and quark-antiquark pairs, photons and gluons are the primary products of such a source. More massive fundamental particles, e.g. W-,Z-bosons, may be produced by a superhigh energy particle source.

Decays of unstable particles and quark, antiquark and gluon fragmentation into hadrons result in the formation of fluxes of non-equilibrium neutrino-antineutrino pairs, electrons, positrons, photons, nucleons and antinucleons and, possibly, charged pions and muons. These fluxes interact with the equilibrium particles.

Such interaction with nuclei, formed in the Big Bang nucleosynthesis, results in the fluxes of non-equilibrium nuclei. Deceleration of the latter in the cosmological plasma is accompanied by their reactions with the equilibrium nuclei.

Hence, the non-equilibrium particles can be direct products of the source, as well as the result of the interaction of some other non-equilibrium particles with the plasma and radiation.

For detailed analysis, the most significant particles and processes should be specified. Stable particles (photons, neutrinos and antineutrinos, electrons and positrons, nucleons and antinucleons) and the processes induced by them are definitely under consideration.

The metastable particles with the lifetime $t_i(p_i)$ should be taken into account in the processes with particles of the type j only provided that the following condition is valid:

$$\left(n_j\langle\sigma v\rangle_{ij}\right)t_i(p_i) > 1\ , \tag{844}$$

where

$$t_i(p_i) = t_i(0)\frac{\left(p_i^2 + m_i^2\right)^{1/2}}{m_i} \tag{845}$$

and $t_i(0)$ is the lifetime of a particle of the type i at rest. This condition means that the rate of particle interaction is greater then the probability of their decay.

For instance, charged pions, having the lifetime $t_\pi(0) \sim 10^{-8}$ s and the cross section of the interaction with nucleons of the order of $\sigma_{\pi N} \sim 10^{-26}$ cm^2 are significant in nucleon and nuclear reactions when

$$\Omega_b \left(\frac{H_0}{50 \frac{\text{km}}{\text{s} \cdot \text{Mpc}}} \right)^2 \left(\frac{2.7\text{K}}{T_0} \right)^3 \left(\frac{E_\pi}{m_\pi} \right) \left(\frac{1\text{s}}{t} \right)^{3/2} > 1 \,, \tag{846}$$

where

$$\Omega_b = \frac{\rho_b}{\rho_{cr}} \,,$$

ρ_b, H_0, T_0 are the modern density of baryons, the Hubble constant and the temperature of the thermal electromagnetic background, respectively, and t is the cosmological time. The above estimation shows, that charged pions with the energy

$$E_\pi < 100 \text{ GeV} \tag{847}$$

decay before their interaction with nucleons and nuclei at

$$t > 10^{-3} \text{ s} \,. \tag{848}$$

It follows from the definition (837) of the weak source, that all types i of non-equilibrium particles should satisfy the condition

$$\int \varepsilon_i \varphi_i dp_i << \varepsilon_\gamma \,. \tag{849}$$

In particular, restricting ourselves with the case

$$\omega_\gamma >> kT \,, \tag{850}$$

where ω_γ is the energy of non-equilibrium photon, and T is the temperature of the thermal background radiation, we obtain from the condition (849), that the number density of non-equilibrium photons $n_\gamma{}^{non}$ satisfies the inequality

$$n_\gamma^{non} = \int \varphi_\gamma d\omega_\gamma << \left(\frac{T}{1\ \text{MeV}}\right) n_\gamma \sim \left(\frac{1\ \text{s}}{t}\right)^{1/2} n_\gamma \ , \qquad (851)$$

where t is the cosmological time and n_γ is the number density of equilibrium photons. It means that at

$$t >> 1\ \text{s} \qquad (852)$$

the number density of energetic non-equilibrium photons is much smaller, than the density of equilibrium photons. Therefore, in each case, when the process can go due to thermal photons, the contribution from non-equilibrium photons into this process can be neglected.

On the other hand, the condition (849) allows for the number densities of non-equilibrium electrons, positrons, nucleons and antinucleons to be

$$n_{e(e^+)}^{non} = \int \varphi_{e(e^+)} dp_{e(e^+)} >> n_e \qquad (853)$$

for electrons (positrons) at $t > 10^2$ s , and

$$n_{N(\overline{N})}^{non} = \int \varphi_{N(\overline{N})} dp_{N(\overline{N})} >> n_N \qquad (854)$$

for nucleons (antinucleons) at $t > 10^{-3}$ s , where n_e and n_N are the electron and nucleon number densities in the equilibrium plasma.

In these cases, non-equilibrium electrons (positrons) and nucleons (antinucleons) play the dominant role in the development of electromagnetic and nucleon cascades, respectively.

To formulate the general problem of a weak source of non-equilibrium particles we present the structure of the set of Eqs. (840) explicitly

$$\frac{\partial \varphi_\gamma}{\partial t} = \left\langle \varphi_\gamma n_{e^\mp} \to \gamma \right\rangle + \left\langle \varphi_\gamma \varphi_{e^\mp} \to \gamma \right\rangle + \left\langle n_\gamma \varphi_{e^\mp} \to \gamma \right\rangle -$$
$$- \varphi_\gamma \left(\left\langle n_{e^\mp} \to \right\rangle + \left\langle n_\gamma \to \right\rangle + \left\langle \varphi_{e^\mp} \to \right\rangle + \left\langle \varphi_\gamma \to \right\rangle \right)$$
$$- \varphi_\gamma \left(\left\langle n_p \to \right\rangle + \left\langle n_4 \to \right\rangle \right) + Q_\gamma \left(p_\gamma, \tau, t \right)$$

$$\frac{\partial \varphi_{e^\mp}}{\partial t} = \left\langle \varphi_{e^\mp} n_\gamma \to e^\mp \right\rangle + \left\langle \varphi_\gamma \varphi_{e^\mp} \to e^\mp \right\rangle + \left\langle n_{e^\mp} \varphi_\gamma \to e^\mp \right\rangle +$$
$$+ \left\langle n_p \varphi_\gamma \to e^\mp \right\rangle + \left\langle n_4 \varphi_\gamma \to e^\mp \right\rangle - \varphi_{e^\mp} \left(\left\langle n_p \to \right\rangle + \left\langle n_4 \to \right\rangle \right) -$$
$$- \varphi_{e^\mp} \left(\left\langle n_{e^\mp} \to \right\rangle + \left\langle n_\gamma \to \right\rangle + \left\langle \varphi_{e^\mp} \to \right\rangle + \left\langle \varphi_\gamma \to \right\rangle \right) + Q_{e^\mp} \left(p_{e^\mp}, \tau, t \right)$$

$$\frac{\partial \varphi_{N,\overline{N}}}{\partial t} = \left\langle \varphi_{N,\overline{N}} n_p \to N, \overline{N} \right\rangle + \left\langle \varphi_{N,\overline{N}} n_4 \to N, \overline{N} \right\rangle +$$
$$\left\langle \varphi_{N,\overline{N}} \varphi_{N,\overline{N}} \to N, \overline{N} \right\rangle - \varphi_{N,\overline{N}} \left(\left\langle n_p \to \right\rangle + \left\langle n_4 \to \right\rangle + \left\langle \varphi_{N,\overline{N}} \to \right\rangle \right) -$$
$$- \varphi_{N,\overline{N}} \left\langle \ \right\rangle_{\text{Coul}} + Q_{N,\overline{N}} \left(p_{N,\overline{N}}, \tau, t \right)$$

$$\frac{\partial \varphi_d}{\partial t} = \left\langle \varphi_{N,\overline{N}} n_4 \to d \right\rangle + \left\langle \varphi_\gamma n_4 \to d \right\rangle + \left\langle \varphi_n n_p \to d \right\rangle +$$
$$\left\langle \varphi_n \varphi_p \to d \right\rangle - \varphi_d \left(\left\langle n_p \to \right\rangle + \left\langle n_4 \to \right\rangle + \left\langle \varphi_\gamma \to \right\rangle + \left\langle n_\gamma \to \right\rangle \right) -$$
$$- \varphi_d \left\langle \ \right\rangle_{\text{Coul}}$$

$$\frac{\partial \varphi_{T,{}^3He}}{\partial t} = \left\langle \varphi_{N,\overline{N}} n_4 \to T, {}^3He \right\rangle + \left\langle \varphi_\gamma n_4 \to T, {}^3He \right\rangle -$$
$$- \varphi_{T,{}^3He} \left(\left\langle n_p \to \right\rangle + \left\langle n_4 \to \right\rangle \right) - \varphi_{T,{}^3He} \left\langle \ \right\rangle_{\text{Coul}}$$

$$\frac{\partial \varphi_4}{\partial t} = \left\langle \varphi_{N,\overline{N}} n_4 \to {}^4He \right\rangle - \varphi_4 \left(\left\langle n_p \to \right\rangle + \left\langle \varphi_{N,\overline{N}} \to \right\rangle \right) - \varphi_4 \left\langle \ \right\rangle_{\text{Coul}}. \quad (855)$$

Here brackets < > stand for respective collision term, φ_i and n_j denote the distribution function of non-equilibrium particles of the type i and the number density of equilibrium particles of the type j, respectively, $< >_{\mathrm{Coul}}$ stands for Coulomb deceleration in plasma, φ_4 (n_4) is the distribution function (number density) of non-equilibrium (equilibrium) ^{4}He nuclei.

Due to strong Coulomb deceleration in plasma, in the first approximation one can neglect non-equilibrium fluxes of nuclei with the charge $Z > 2$, and consider the changes of their concentrations under the condition that these nuclei are in the thermal equilibrium with plasma.

Neutrino (antineutrino)-induced reactions and the evolution of neutrino cascade are not taken into account in Eqs. (855) and will be considered below.

1.3. Non-Equilibrium Nucleosynthesis. Some Special Cases

The full set of equations (855) is rather complicated, and it was not studied in detail so far even under the simplified assumptions, mentioned above. The results of simplified studies of some specific processes of non-equilibrium cosmological nucleosynthesis in electromagnetic and nuclear cascades induced by several hypothetical sources of non-equilibrium particles are only available now.

Following Burns, Lovelace (1982), Aharonian, Vardanian (1985), Lindley (1985), the reaction

$$\gamma + \gamma \to e^+ + e^- \tag{856}$$

is important in the development of electromagnetic shower by high-energy photons in astrophysical conditions and, especially, in the conditions of the RD stage. The pair production induces a sharp upper cut off in the non-equilibrium photon spectrum given by

$$E_{\max} \cong \frac{m_e^2}{25T} \ln(15\Omega_b) \ , \tag{857}$$

where the baryon density $\Omega_b \sim 0.1$, and m_e is the mass of electron.

Dimopoulos et al. (1987) expressed the non-equilibrium photon spectrum in terms of the energy distribution function $\xi_\gamma(E)$, defined as the photon number density in the unit energy interval per two body decay of a hypothetical superheavy particle X with the mass M , in which high-energy γ-quantum is emitted. They obtained

$$\xi_\gamma(E) = \begin{cases} \dfrac{M_X}{2E_{\max}^{1/2}} \cdot \dfrac{1}{E^{3/2}} \ , & E < E_{\max} \\ 0 \ , & E > E_{\max} \ , \end{cases} \tag{858}$$

where $E_{\max}$ is given by Eq. (857)

Dimopoulos et al. (1987a,b) have also reduced the set of equations (855) to the set of equations for reaction rates, accounting for non-equilibrium fluxes (Dimopoulos et al., 1987a), and found a simple analytic treatment, being in good agreement with the numerical solutions of this set of equations and giving an interesting possibility for light element production.

The analytical treatment by Dimopoulos et al. (1987b) was based on the numerically calculated (Dimopoulos et al., 1987a) quantities ξ_i being defined as the number of i-type nuclei per baryonic X decay and determined by the chain of nuclear reaction induced by nucleons and antinucleons from X decay.

It was assumed that ^{4}He abundance decreased dominantly due to baryodestruction, and the contribution of photodestruction is neglegible. The variation with time of the relative concentration of ^{4}He nuclei, defined as

$$f_4 = \frac{n_4}{n_\gamma} , \tag{859}$$

where n_4 and n_γ are the number densities of ^{4}He nuclei and photons, respectively, is

$$\frac{\partial f_4}{\partial t} = f_X^0 \Gamma_X \exp(-\Gamma_X t) r_b^* \xi_4 , \tag{860}$$

where

$$f_X^0 = \frac{n_X}{n_\gamma} \tag{861}$$

is the frozen relative concentration of superheavy particles X,

$$\tau_X^{-1} = \Gamma_X \tag{862}$$

is the rate of X decay, r_b* is the effective branching ratio of X decays to final states containing baryons and ξ_4 is negative.

The equations for D, ^{3}He, ^{6}Li and ^{7}Li abundance are derived under the assumption, that the dominant production occurs inside the hadronic showers and involves the relevant ξ_i and the dominant destruction is by photodissociation. These equations have the form:

$$\frac{\partial f_i}{\partial t} = f_X^0 \Gamma_X \exp(-\Gamma_X t) r_b^* \xi_i - n_i \int_{Q_i}^{E_{\max}(t)} f_\gamma(E) \sigma_{\gamma i}(E) dE , \tag{863}$$

where Q_i is the photodissociation threshold, n_i is the number density and $\sigma_{\gamma i}$ is the photodissociation cross section of the nuclear species i , and $E_{\max}(t)$ is given by Eq. (857)

with the account for the standard Big Bang time-temperature dependence on the RD stage

$$T \propto t^{-1/2} .$$

The nonthermal photon spectrum $f_\gamma(E)$ is given by its fixed point value, i.e. by the solution of the equation

$$\frac{\partial f_\gamma(E)}{\partial t} = f_X^0 \Gamma_X \exp(-\Gamma_X t) \xi_\gamma(E) - n_e \sigma_c(E) f_\gamma(E) \tag{864}$$

at

$$\frac{\partial f_\gamma(E)}{\partial t} = 0 . \tag{865}$$

One obtains

$$f_\gamma(E) = \frac{\xi_\gamma(E) f_X^0 \Gamma_X \exp(-\Gamma_X t)}{n_e \sigma_c(E)} . \tag{866}$$

Assuming, that equations (863) reach their fixed points

$$\frac{\partial f_i}{\partial t} = 0 , \tag{867}$$

Dimopoulos et al. (1987b) obtained for D, ^{3}He, ^{6}Li and ^{7}Li abundances the values

$$f_i = 30 \xi_i Q_i^{1/2} E_{\max}^{1/2}(t_{fi}) \frac{r_b^*}{M_X} f_b , \tag{868}$$

where f_b is the baryon-to-photon number density ratio

$$f_b = \frac{n_b}{n_\gamma} = 7.5 \cdot 10^{-9} \Omega_b \left(\frac{H_0}{50 \frac{\text{km}}{\text{sMpc}}} \right)^2 \left(\frac{2.7\text{K}}{T_0} \right)^3 , \tag{869}$$

the cross section ratio is

$$\frac{\sigma_{\gamma i}(E)}{\sigma_c(E)} = \frac{1}{30} \tag{870}$$

and t_{fi} is the freeze out time, when both photodissociation and baryoproduction come to end. The condition

$$E_{\max}(t_{fi}) > Q_i \tag{871}$$

and the constraint on ^{4}He photodissociation lead to rather narrow interval of the lifetimes

$$4 \cdot 10^5 < \tau_X < 8 \cdot 10^5 \text{ s} \tag{872}$$

when the fixed point values (868) are valid. That gives

$$\text{D} : {}^3\text{He} : {}^6\text{Li} : {}^7\text{Li} = 1 : 1 : 10^{-5} : 10^{-6} \tag{873}$$

determined predominantly by the ratio of ξ_i . The absolute values of these abundances correspond to the observed ones, if

$$2 \cdot 10^4 < \frac{M_X}{r_b^*} < 10^5 \text{ GeV} . \tag{874}$$

Assuming significant depletion of ^{6}Li abundance Dimopoulos et al. (1987b) show the principal possibility of interpretation of the observed abundance of the light elements in terms of the keV-era nucleosynthesis.

Such an approach removes the strong dependence of predicted abundance on the number of neutrino species or on the baryon-to-photon ratio but involves the hypothetical particle X with very specific properties.

The allowed intervals for the lifetime τ and mass-to-baryon branching ratio M_X/r_b^* are very narrow, and they can vanish after more detailed information on the nuclear processes involved will be available. Anyhow, it illustrates the possibilities of nuclear cosmoarcheology, relating the light element abundance to the relic "footprints" of hypothetical physics in the early Universe.

Non-equilibrium cosmological nucleosynthesis establishes new type of relations between the cosmological and nuclear data.

The results of the standard Big Bang nucleosynthesis are dominantly determined by the thermodynamic conditions, so that a slight change in thermodynamic parameters compensate any drastic error in nuclear reaction rates.

The results of non-equilibrium cosmological nucleosynthesis are directly determined by particular properties of nuclear reactions, namely, of their total and differential cross sections.

It turns out that sometimes the corresponding nuclear data are absent.

Experimental nuclear cosmoarcheology formulates the experimental tasks to fill the gap and by respective experimental measurements completes the link in the cosmoarcheological chain.

To prove that the source of non-equilibrium particles is responsible for the observed light element abundance, we need detailed self-consistent treatment of all channels of light element production in all nonthermal particle cascades, induced by the considered source of non-equilibrium cosmological nucleosynthesis. Such a treatment shows what the parameters of the hypothetical source **should be**.

However, we can use only one particular trace of the source to find what the parameters of this source **should not be**, i.e., to exclude their values, corresponding to overproduction of some elements.

The very existence of antinucleons in the Universe after the Big Bang nucleosynthesis is a profound signature of a source of nonthermal particles.

We have seen above that, whatever the source of antinucleons is, their annihilation in the Universe on the RD stage is severely restricted by the observed D and/or ^{3}He abundance. The only fact, that antiprotons were present in the Universe and annihilated with ^{4}He, was sufficient for the restrictions on antiproton sources, given in Chapter 6.

However, the sources of nucleon-antinucleon pairs on the RD stage, such as gravitino decays or PBH evaporation, produce high-energy nucleons and antinucleons, inducing nuclear cascades. One can expect, that the account for the development of non-equilibrium particle cascade should increase the sensitivity of cosmoarcheological analysis.

Levitan et al. (1988) analysed with Monte Carlo simulations the D and ^{3}He production in nucleon cascades, induced by energetic nucleons and antinucleons from hypothetical sources of nucleon-antinucleon pairs.

There are several factors, leading to the enhancement of D and ^{3}He production in such cascades as compared to slow antiproton annihilation with ^{4}He, considered in Chapter 6.

Due to suppression of relativistic antinucleon annihilation, the main mechanism of D and ^{3}He production by high energy antiprotons are inelastic inclusive processes

$$\bar{p} + {}^4\mathrm{He} \rightarrow \bar{p} + \mathrm{D} + anything \tag{875}$$

$$\bar{p} + {}^4\mathrm{He} \rightarrow \bar{p} + {}^3\mathrm{He} + anything \tag{876}$$

$$\bar{p} + {}^4\mathrm{He} \rightarrow \bar{p} + \mathrm{T} + anything \tag{877}$$

and the dominant mechanism of relativistic antiproton energy loss is related to inelastic scattering on nuclei (dominantly on the most abundant hydrogen nuclei, i.e. protons), in the process

$$\bar{p} + p \rightarrow \bar{p} + p + anything \ . \tag{878}$$

Hence, the relativistic antiproton, before it is slowed down to non-relativistic energies, can participate more than one collision, giving rise to D, ^{3}He and T production.

The next factor is related to recoil protons, acquiring after antiproton interaction with proton or antiproton destruction of helium the energy, exceeding the threshold of proton-induced ^{4}He destruction. Destruction of ^{4}He by recoil protons is the additional source of D, ^{3}He, T production in the case of high-energy primary antiproton.

The last, but not least, factor is related to interactions of primary protons, produced by the source in the pairs with antinucleons. Such protons, as well as the secondary protons, produced in the nucleon cascade, also give rise to D, T and ^{3}He production.

The quantitative estimation of relative significance of all these factors is obtained in numerical calculations by Levitan et al. (1988).

One finds that in the case of antiproton-induced D(^{3}He) production the dominant contribution is given by antiproton at its maximal energy. The second maximum at low energies is due to antiproton annihilation at rest. The minimum of D(^{3}He) production at low kinetic energies

$$E \leq 0.1 \text{ GeV} \tag{879}$$

of slowed down antiproton with the initial energy

$$E_0 > E \tag{880}$$

are due to strong Coulomb deceleration of non-relativistic charged particles in the cosmic plasma. The effective cross section of Coulomb scattering is proportional to

$$\sigma_{\text{Coul}} \sim \upsilon^{-4} , \tag{881}$$

where υ is the velocity of a charged particle.

We see that the probability for charged particle (proton and antiproton) to destruct ^{4}He before the particle is slowed down to the thermal energies by Coulomb interaction with plasma is strongly suppressed.

The energy dependence of the enhancement factor for the constraint on antiproton sources as well as the sensitivity of this factor to the data on the energy dependence of D(^{3}He) yield in antiproton interactions with ^{4}He was given on Fig.15 in Chapter 6.

Levitan et al. (1988) found simple analytical formulae, providing good approximation to their numerical results.

Consider, for example, the probability $S_D{}^A$ of D formation in antiproton interactions. The rate of antiproton slowing down exceeds significantly the rate of the change in the parameters of the medium or the rate of the evolution of the source. So the probability distribution $\varphi_a(E,E_0)$ for antiproton with the initial energy E_0 to have the energy E is stationary. This probability is determined by the equality of the rate of slowing of antiprotons with the energies

$$\varepsilon > E \tag{882}$$

down to the energy E and of the further slowing of antiproton with the energy E down to lower energies

$$\int_{E}^{E_0} \varphi_a(\varepsilon, E_0) \frac{d(\sigma v)^{in}(\varepsilon)}{dE} n_p d\varepsilon = n_p (\sigma v)^{tot} \varphi_a(E, E_0) \ . \tag{883}$$

At high energies one can use the approximation

$$\sigma^{tot}(\varepsilon) = \text{const} \ , \tag{884}$$

$$\sigma^{ann} << \sigma^{tot} \ , \tag{885}$$

$$\frac{d(\sigma v)^{in}(\varepsilon)}{dE} = \frac{\sigma^{in}}{\varepsilon} = \frac{2}{3} \cdot \frac{\sigma^{tot}}{\varepsilon} \ , \tag{886}$$

giving for the solution of Eq. (883)

$$\varphi_a(E, E_0) \propto E^{-2/3} \ . \tag{887}$$

The probability for deuterium formation in antiproton interactions is given by

$$S_{\rm D}^{A} = f_{\rm D}^{a} \int_{0}^{E_0} \varphi_a(E, E_0) n_{\rm He} \sigma_{\bar{p}{\rm He}}^{tot}(E) v \tau dE \ , \tag{888}$$

where $f_{\rm D}{}^{a}$ is relative yield of deuterium in antiproton-helium interaction, assumed to be equal

$$f_{\rm D}^{a} = \text{const} \tag{889}$$

and

$$\tau = \left(n_p \sigma_{pp}^{tot} v\right)^{-1} \tag{890}$$

is the time scale of cascade development. One finds from the above equations, that the energy dependence of $S_D{}^A$ is

$$S_D^A \propto E_0^{1/3} \, . \tag{891}$$

The examples, given by Dimopoulos et al. (1987), Levitan et al. (1988) exhibit the direct relationship between the data on particle interactions with nuclei, measured in experimental nuclear physics, and the light element abundance, measured in astronomical data.

Putting together the respective experimental and astronomical data one can make rather definite conclusions on the possibility for the sources of non-equilibrium particles to exist in the early Universe, thus providing complex indirect test of the underlying super high energy physics, unachievable by direct laboratory means.

The new type of relations between astronomical, experimental nuclear and theoretical research, arising in the context of nuclear cosmoarcheology, may be illustrated by the ASTROBELIX project, considered in the next section.

2. ASTROBELIX project.

2.1. AstroNuclear Experiment ASTROBELIX

The main motivation for the ASTROBELIX project follows from the fact, that no antiprotons are present in the baryon-asymmetrical Universe from the end of the local nucleon-antinucleon annihilation at $t < 10^{-3}$ s to the period of galaxy formation at $t > 10^{16}$ s when the interaction of cosmic rays can produce antiprotons.

Thus, the antiproton annihilation in the early Universe after the Big Bang nucleosynthesis is the pronounced signature of some new cosmological phenomena, related to the physics beyond the standard model.

Indeed, the nature of sources of antiproton annihilation can be related to cosmological consequences of particle models (Chechetkin et al., 1982b), so that the analysis of possible effects of antiproton annihilation provides constraints on possible parameters of these models.

Since different particle models can provide inflation, predict the same observed baryon asymmetry of the Universe and offer candidates for dark matter, one can use additional effects, related to cosmological consequences of such models to distinguish them and to choose among them. The sources of antiproton annihilation after the Big Bang nucleosynthesis turn out to be such additional effects, related to a wide class of physical mechanisms for inflation, baryosynthesis and dark matter.

From the phenomenological viewpoint, there are two principal possibilities of appearance of antiprotons in the baryon-asymmetrical Universe after the end of the local nucleon-antinucleon annihilation: a) antiprotons can survive in domains of the antimatter and b) they can be produced in nucleon-antinucleon pairs by some source of energetic particles.

Concerning the possibility (a), one can find that virtually each realisation of baryosynthesis (see Chapter 3) can naturally lead under some conditions to the generation of antibaryon excess.

In the mechanisms of baryosynthesis, related to baryon non-conserving non-equilibrium processes, the excess of baryons implies the specific choice of the CP-violating phase. The possibilities of the spatial variation of the value and the sign of this phase gives, the excess of antibaryons in some regions in the same mechanism (see Chapters 3 and 6).

In the mechanisms of Affleck, Dine (1985), Linde (1985) (see Chapter 3), the existence of baryon asymmetry is provided by the independence of the supersymmetric potential on the baryon charge of the condensate of scalar quarks, that leads naturally to the possibility of spatial variation of the baryon charge density in such a condensate.

The possibility (b) is related to the existence of new (approximately) conserving charges in virtually all extensions of the standard model. The lightest particles, having such charge, are metastable, and should be present long after they freeze out in the very early Universe.

The existence of primordial metastable particles leads directly to the sources of energetic particles on the RD stage owing to their decays.

In the case of gravitino (see Chapter 6) with the mass $m_G \sim 100$ GeV, gravitino decays at $t \sim 10^8$ s to gluino and gluon (805) resulting in hadronic jets, containing nucleon-antinucleon pairs, due to gluon fragmentation to hadrons.

Superheavy metastable particles may have insufficient lifetime to survive during the period of the Big Bang nucleosynthesis. However, they can be the cause of non-equilibrium particles on the RD stage indirectly through the primordial black holes (PBH) formation on stages of their dominance (see Chapter 4).

PBH evaporation on the RD stage results in creation of nucleon-antinucleon pairs among the products of evaporation (see Chapter 6).

The influence of antiproton annihilation with ^{4}He nuclei on the abundance of light elements turns out to be the most sensitive probe for the existence of hypothetical sources of antiproton annihilation on the RD stage (see Chapter 6).

In the earlier studies of nuclear cosmoarcheology, hypothetical sources of antiproton annihilation on the RD stage were probed by the analysis of D and ^{3}He production induced by antiproton annihilation with ^{4}He nuclei.

The gap in the logical chain linking the analysis of the relationship of antiproton sources to the existence of PBHs, antimatter domains and gravitino, predicted on the base of GUT, SUSY etc. particle models, to the observed D and ^{3}He abundances was filled by the data of CERN LEAR experiment PS 179, undertaken with the special aim to measure D and ^{3}He yields in antiproton annihilation with ^{4}He. As a result of this experiment, it was shown, that the predicted overproduction of ^{3}He in the antiproton-^{4}He annihilation puts the most restrictive constraints on the possible amount of antiprotons, annihilating on RD stage and, in particular, causes serious troubles for the simplest versions of local supersymmetric models (see Chapter 6).

Since Li, Be and B are by 6 orders of the magnitude less abundant, than ^{3}He, the possibility of the chain of nuclear processes

$$\bar{p}+{}^4\mathrm{He}\to{}^3\mathrm{He}+\ldots \quad\Rightarrow\quad {}^3\mathrm{He}+{}^4\mathrm{He}\to{}^7\mathrm{Be}+\gamma \qquad (892)$$

$$\bar{p}+{}^4\mathrm{He}\to\mathrm{D}+\ldots \quad\Rightarrow\quad \mathrm{D}+{}^4\mathrm{He}\to{}^6\mathrm{Li}+\gamma \qquad (893)$$

$$\bar{p}+{}^4\mathrm{He}\to\mathrm{T}+\ldots \quad\Rightarrow\quad \mathrm{T}+{}^4\mathrm{He}\to{}^7\mathrm{Li}+\gamma \qquad (894)$$

$$\bar{p}+{}^4\mathrm{He}\to\mathrm{T}+\ldots \quad\Rightarrow\quad \mathrm{T}+{}^4\mathrm{He}\to{}^6\mathrm{Li}+n \qquad (895)$$

provides more sensitive tool to probe the existence of antiproton annihilation on the RD stage by comparing Li, Be, B abundance, predicted in this chain, with the observed one.

However, this logical chain tracing the cosmological consequences of particle theory to astronomical observations of light element abundance is incomplete, since there is no information on the high momentum distribution of ^{3}He, T, D, produced in antiproton annihilation with ^{4}He as well as the pregalactic Li, Be and B abundance is poorly known.

To complete the chain, the ASTROBELIX project proposed (Khlopov, 1990, 1992):

a) to measure the momentum distribution of D, T, ^{3}He in the OBELIX studies of antiproton-^{4}He interactions in LEAR, and

b) to elaborate the methods of measurements of primordial chemical composition with the use of the narrow-band distortions in the Rayleigh-Jeans part of the relic radiation.

Though each source of antiproton annihilation is simultaneously the gamma source, γ-^{4}He interaction can not provide significant Li, Be, B production since

i) the cross sections of reactions of ^{4}He photodesintegration

$$\gamma + {}^4\mathrm{He} \to \mathrm{T} + p \tag{896}$$

$$\gamma + {}^4\mathrm{He} \to {}^3\mathrm{He} + n \tag{897}$$

are peaked near the threshold, and

ii) the γ-spectrum of electromagnetic cascade, formed in plasma on RD stage is also decreasing with the energy E of γ quanta,

$$f_\gamma(E) \propto E^{-3/2} . \tag{898}$$

Thus, contrary to the case of ^{3}He production, the photodesintegration-induced chain of nuclear reactions

$$\gamma + {}^4\mathrm{He} \to \mathrm{T} + p \;\Rightarrow\; \mathrm{T} + {}^4\mathrm{He} \to {}^6\mathrm{Li} + n \tag{899}$$

$$\gamma + {}^4\mathrm{He} \to \mathrm{T} + p \;\Rightarrow\; \mathrm{T} + {}^4\mathrm{He} \to {}^7\mathrm{Li} + \gamma \tag{900}$$

$$\gamma + {}^4\mathrm{He} \to {}^3\mathrm{He} + n \;\Rightarrow\; {}^3\mathrm{He} + {}^4\mathrm{He} \to {}^7\mathrm{Be} + \gamma \tag{901}$$

can not compete with (antiproton + ^{4}He)-induced Li, Be, B production.

The full body of ASTROBELIX project contained:
1) theoretical investigation of relation between parameters of particle models and sources of antiproton annihilation;
2) numerical simulation of antiproton annihilation in plasma on the RD stage and of the kinetics of corresponding reactions of non-equilibrium cosmological nucleosynthesis, clarifying the loopholes in the data on

antiproton-helium interaction and astronomical observations,
3) experimental measurement of cosmological relevant parameters of antiproton-helium interaction by the OBELIX collaboration in LEAR and
4) elaboration of radioastronomical methods of primordial chemical composition measurement via the searches of narrow-band distortion in the electromagnetic thermal background spectrum.

The project exhibited the cross-relations between the main components of cosmoparticle studies: the theoretical, the numerical, the experimental and the observational ones. In particular, the new, astronuclear, level of relations between nuclear physics and astrophysics is achieved in the context of cosmoparticle studies.

While nuclear astrophysics applies the well proved laws of nuclear physics to the problems of astrophysics, astronuclear experiments combine nuclear and astrophysical research in the studies of unknown physical laws, hidden at superhigh-energy scale.

The full-body realisation of ASTROBELIX programme turned out to be impossible. Experimental specifics of the OBELIX experiment was not adequate for proper measurements of secondary nuclei yields in antiproton-helium interaction. But the cosmoarcheological studies, undertaken in the framework of ASTROBELIX project, clarified the importance of their further development and have given rise to “Astroparticle Studies of Dark, Decaying and Annihilating Matter in the Universe” (ASTRODAMUS project), now being under way.

2.2. Li and Be Abundance Constraint on Antiproton Annihilation on the RD Stage

Among the results of cosmoarcheological studies, initiated by ASTROBELIX, let us discuss briefly the theoretical proof of high sensitivity of Li, Be, B abundance to parameters of antiproton sources as compared with the case of deuterium and helium-3.

It was shown by Khlopov et al. (1994), that even in the absence of direct experimental data rather definite cosmoarcheological conclusions can be done on the basis of observational data on Li abundance. Using the Monte Carlo simulations, it was shown, that the existing uncertainties in the exact form of the momentum distribution of secondary nuclei produced in antiproton interaction with ^{4}He cause rather small uncertainty in the relation between the number densities of antiprotons and of Li, Be nuclei, originated from antiproton induced cascade.

Relative stability of this relation followed from the fact, that virtually all existing nuclear physics models for distributions of secondary nuclei in destruction of ^{4}He by antiprotons, predict the maximum of the distribution at energies, exceeding the threshold of the successive nuclear processes. Thus, in the first approximation, the amount of produced Li is determined by the integral amount of secondary nuclei, which is much less ambiguous.

Based on the observed ^{6}Li abundance, Khlopov et al. (1994) found the constraint on the relative amount of antimatter, given by

$$f < 1.1 \cdot 10^{-4} \left(\frac{1 \text{ GeV}}{E_0} \right)^{1/2}, \tag{902}$$

where E_0 is the initial energy of antiproton. The constraint (902) imposes the upper bound on the gravitino concentration (see Chapter 6)

$$\frac{n_G}{n_\gamma} < 3.8 \cdot 10^5 \, \frac{n_{^6\text{Li}}}{n_\gamma} \; . \tag{903}$$

This upper limit was obtained for gravitino with the mass $m_G \sim 100$ GeV with the account for the relation between the mean energy of antiprotons and the energy of gluon E_g, given by

$$E_{\bar{p}} \sim \left(1 \text{ GeV} \cdot E_g\right)^{1/2} . \tag{904}$$

If the mass of gluino in the gravitino decay (805) is small as compared with the mass of gravitino

$$m_{\tilde{g}} << m_G , \tag{905}$$

the initial gluon energy is

$$E_g = \frac{m_G}{2} = 50 \text{ GeV} , \tag{906}$$

therefore the initial energy of antiproton is

$$E_0 \approx 7 \text{ GeV} . \tag{907}$$

Since the primordial gravitino concentration is related to the reheating temperature T_R as

$$\frac{n_G}{n_\gamma} = 0.1 \frac{T_R}{m_{Pl}} , \tag{908}$$

one obtains finally the upper limit on the temperature of reheating from the observed ^{6}Li abundance

$$T_R < 3.8 \cdot 10^6 \text{ GeV} . \tag{909}$$

The comparison with the corresponding limit, obtained in Eq. (813) from the observed ^{3}He abundance, shows that the account for antiproton-induced nucleosynthesis results in the more than the order of magnitude stronger constraint.

This example shows the importance of the non-equilibrium nucleosynthesis theoretical network as the sensitive tool in the cosmoarcheological equipment.

3. Cosmological Backgrounds of Non-Equilibrium Particles

3.1. High Energy Neutrino from the RD stage.

The weakness of neutrino interaction makes non-thermal neutrino cascades virtually ineffective in non-equilibrium cosmological nucleosynthesis, so we can hardly use abundance of light elements to test the presence of nonthermal neutrino sources on the RD stage.

The same reason makes impossible to use the thermal background radiation spectrum as a calorimeter for neutrino sources.

However, due to the weakness of neutrino interactions, direct experimental probes for high-energy neutrino sources are possible. Since nonthermal neutrinos can survive to the present time, they can be searched for in neutrino observatories.

The higher is the neutrino energy, the larger is the probability to detect them and/or to test the presence of their hypothetical source in the early Universe.

There are several possibilities for the source of energetic particles to produce high-energy neutrinos.

Neutrinos can be produced directly, as in the case of evaporating PBHs or superheavy metastable particles decaying on the channel

$$M \to \nu + anything\ . \tag{910}$$

The source can produce high-energy charged pions and kaons, having neutrinos among their decay products.

The development of nucleon cascade is accompanied by inelastic nucleon interactions, giving rise to pions and K-mesons, successively decaying on neutrinos.

The existence of the Big Bang thermal neutrino background imposes an upper limit on the energies of nonthermal neutrinos.

Due to the neutrino-neutrino weak interaction, neutrinos with the energy

$$E > E_{max} \tag{911}$$

should lose their energy in neutrino-neutrino scattering on the thermal neutrino background. One can estimate the value of E_{max} as (Khlopov, Chechetkin, 1987)

$$E_{max} \sim 1 \text{ GeV}\left(\frac{t}{1 \text{ s}}\right), \tag{912}$$

which is determined by the opacity of the thermal neutrino background for energetic neutrinos.

If the source of energetic neutrinos creates neutrinos with the energy $E > E_{max}$, neutrino-neutrino interaction induces neutrino cascade, since the scattering of energetic neutrino with

$$E >> E_{max} \tag{913}$$

on a thermal neutrino with the energy

$$E_T \sim 3T , \tag{914}$$

where T is the thermal neutrino temperature, results in the average energy of recoil neutrino

$$E' \sim \frac{E}{2} . \tag{915}$$

So, the "multiplication" of energetic neutrinos can also take place.

Superposition of all these processes forms the spectrum $F(E)$ of energetic neutrinos, induced by the source of superhigh-energy particles.

In general, the form of the spectrum can be determined only by detailed numerical simulations. The detailed information on all particle cascades, produced by the source, is needed. However, some principal features of the spectrum and the possibilities of its detection can be estimated on the base of such general properties of the source, as its time scale τ , relative contribution of its superhigh-energy products into the total cosmological density and the maximal energy E_s of energetic particles, produced by the source.

Consider, for definiteness, the case of hypothetical superheavy particles with the mass m, relative concentration $r = n_m/n_\gamma$ and the lifetime τ. The quantity $\xi_\nu(E)$ is the number of neutrinos with the energy E (in the energy interval with the width dE in the vicinity of E) per decay of this particle.

The spectrum $F(E)$ is to be cut at the high energies

$$E > \min(E_{\max}, m) \; . \tag{916}$$

To estimate the sensitivity of the existing or planned cosmic neutrino detectors to such a source of superhigh-energy neutrinos we should compare first the maximal modern energy of neutrinos from the considered source with the threshold energy of neutrino detectors.

Consider, for definiteness, the case

$$E_{\max} < m \; . \tag{917}$$

The maximal energy of nonthermal neutrinos is redshifted in the modern Universe down to

$$E_{max}^{mod} = E_{max}(1+z_d)^{-1} \ , \tag{918}$$

where z_d is the red shift, related to the period of particle decays at $t \sim \tau$ by the relation

$$\frac{t_U}{\tau} = (1+z_d)^{3/2} \begin{cases} 1 \ , & \text{for } \tau > t_{MD} \\ \left(\dfrac{1+z_d}{1+z_{MD}}\right)^{1/2}, & \text{for } \tau < t_{MD} \end{cases} . \tag{919}$$

Here

$$z_{MD} = 10^4 \left(\frac{H_0}{50 \frac{\text{km}}{\text{s Mpc}}} \right)^2 \tag{920}$$

for $\Omega = 1$, H_0 is the modern Hubble constant,

$$t_U = 4 \cdot 10^{17} \text{s} \tag{921}$$

is the age of the Universe and

$$t_{MD} = 4 \cdot 10^{11} \text{s} \tag{922}$$

is the beginning of the modern nonrelativistic matter dominant stage in the Universe.

Eq. (919) should be modified, accounting for possible recent decays of dark matter particles after the formation of the large scale structure of the Universe, what take place in unstable dark matter scenarios of large scale structure formation (see below). However, this modification is not significant for the further discussion.

It follows from the Eq. (919), that the modern value of E_{max} in the case of superheavy particles, decaying on the RD stage, i.e. at

$$t \sim \tau < t_{MD} \tag{923}$$

is given by

$$E_{\max}^{\text{mod}} = E_{\max}(1 + z_{MD})^{-1/4}\left(\frac{\tau}{t_U}\right)^{1/2}. \tag{924}$$

Using the estimation of $E_{\max}$ (924), one obtains that the modern maximal energy of nonthermal neutrinos, induced by decays of superheavy metastable particles with the mass satisfying the condition (917) at the time defined by Eq. (923) is given by (Khlopov, Chechetkin, 1985)

$$E_{\max}^{\text{mod}} = 2 \cdot 10^{-10}\ \text{GeV}\left(\frac{\tau}{1\ \text{s}}\right)^{3/2}. \tag{925}$$

It can be easily found from Eq. (925) that at

$$\tau > 10^5\ \text{s} \tag{926}$$

the maximal energy of nonthermal neutrinos exceeds the energy threshold for detection of such neutrinos in neutrino observatories.

The sensitivity of neutrino detectors to the sources of superhigh-energy particles on the RD stage is also determined by the intensity of the flux of high energy neutrinos.

In the considered case of superheavy particles the flux $F_\nu(E)$ of nonthermal neutrinos is determined by the relative concentration r the magnitude $\xi_\nu(E)$.

The phase space number density of nonthermal neutrinos is

$$\frac{dJ_\nu}{E^2 d\Omega} = \frac{d^3 F_\nu}{d^3 E} = \frac{d^3 n_\nu}{d^3 E} r n_\nu c \ . \qquad (927)$$

Here c is the speed of light and the spectrum is normalised on the modern number density of relic photons n_ν. The density

$$\frac{d^3 n_\nu}{d^3 E}$$

is related to the spectrum $\xi_\nu(E)$ of the neutrino cascade, induced by decays of superheavy particles,

$$\frac{d^3 n_\nu}{d^3 E} = \iint \xi_\nu \left(\frac{E}{(1 + z(t))} \right) \cdot \exp\left(-\frac{t}{\tau} \right) \cdot \frac{d^3 n_m}{d^3 p} \cdot dt \cdot d^3 p \ , \qquad (928)$$

where the dependence of ξ_ν on $E/(1+z)$ takes into account the red shift of neutrino energy, and the exponential factor in the integrand corresponds to the exponential law of decay. The momentum distribution of decaying particles is also taken into account in Eq. (928).

3.2. High-Energy Neutrino Background as the Probe for New Physics

The comparison of the differential, J_ν, and the integral, F_ν, spectra of non-thermal neutrino background with the corresponding characteristics of atmospheric neutrinos $J_\nu{}^a$ and $F_\nu{}^a$ makes it possible to estimate the sensitivity of neutrino telescopes to the existence of high-energy particle source in the early Universe.

The highest-energy neutrinos are of the most interest for such a comparison, since the calculations of Volkova and Zatzepin (1965) exhibit rapid fall of the atmospheric neutrino spectrum with the energy of neutrinos (see review of Chudakov, 1983).

To estimate the spectrum of neutrinos from decay of superheavy metastable particles, assume the instantaneous decay. In other words, we shall put the Dirac delta-function $\delta(t-\tau)$ instead of $\exp(-t/\tau)$ into Eq. (928).

We shall also neglect the momentum distribution of superheavy particles:

$$\frac{d^3 n_m}{d^3 p} = \delta(\mathbf{p}) \,. \tag{929}$$

In this case the neutrino distribution coincides with the spectrum

$$\frac{d^3 n_\nu}{d^3 E} = \xi_\nu\left(\frac{E}{(1+z_d)}\right) . \tag{930}$$

To find the function $\xi_\nu(E)$, the relation between the mass m of metastable particle and the value of $E_{\max}$ is essential.

In the case opposite to the condition (917),

$$m < E_{\max} \,, \tag{931}$$

neutrino interaction with the thermal background is negligible, and the function $\xi_\nu(E)$ is determined by the detailed properties of metastable particles and products of their decay.

In the case

$$m >> E_{\max} \tag{932}$$

the $(\nu\nu)$ interaction is essential.

Consider the latter case, assuming that the effects of this interaction dominate in the formation of nonthermal neutrino spectrum.

The matrix element of ($\nu\nu$) scattering due to neutral current interaction in the standard theory of electroweak interactions has the form

$$M = \frac{G_F}{4\sqrt{2}} \bar{\nu}_2 \gamma_\mu (1 + \gamma_5) \nu_1 \cdot \bar{\nu}_4 \gamma_\mu (1 + \gamma_5) \nu_3 \, , \qquad (933)$$

where G_F is the Fermi constant.

The differential rate of ($\nu\nu$) scattering is

$$\frac{d(\sigma v)}{2\pi p_\nu^2 dp_\nu} = \frac{G_F^2 (1 - \cos\theta)^2 E_1 \omega_1}{8\pi^2 p_\nu^2 p} \, , \qquad (934)$$

where E_1 and ω_1 are the initial neutrino energies, θ is the angle between their momenta,

$$E_\nu = p_\nu \qquad (935)$$

is the energy of final neutrino and

$$p^2 = E_1^2 + \omega_1^2 + 2E_1\omega_1 \cos\theta \, . \qquad (936)$$

With the use of the condition

$$\omega_1 >> E_1 \, , \qquad (937)$$

the differential rate of ($\nu\nu$) scattering obtains the form

$$\frac{d(\sigma v)}{dp_\nu} = \frac{G_F^2 (1 - \cos\theta)^2 E_1}{4\pi} \, . \qquad (938)$$

In the case of the scattering of identical neutrinos the right hand sides of Eqs. (934) and (938) are to be multiplied by the factor 1/2.

Consider, following Souslin et al. (1982) (see also Yuae and Buchler (1976)), the kinetic equation for the neutrino distribution function in a homogeneous medium.

The neutrino distribution function f_ν (E,t) is in this case the sum of the equilibrium distribution $f_\nu^{(0)}$ of the thermal neutrino background and of the high-energy distorsion of this distribution $f_\nu^{(1)}$, induced by decay of superheavy metastable particles.

The concentration of thermal neutrinos is evidently much greater as compared with the energetic neutrinos, i.e. the following inequality is valid:

$$n_\nu^{(0)} = \int f_\nu^{(0)} d^3 p >> \int f_\nu^{(1)} d^3 p \ . \tag{939}$$

Let us assume also the dominance of thermal neutrino energy density over the one of non-equilibrium particles, i.e. consider the case of a weak source of non-equilibrium neutrinos

$$\varepsilon_\nu^{(0)} = \int E f_\nu^{(0)} d^3 p >> \int E f_\nu^{(1)} d^3 p \ . \tag{940}$$

Then, one obtains the set of linear equation for the functions f_i with i=1,...,N neutrino species ($N = 3$)

$$\frac{df_{\nu i}^{(1)}}{dt} = \sum_{j,k} \int d^3 p_j \int d^3 p_k f_{\nu j}^{(1)} f_{\nu k}^{(0)} \frac{d^3 (\sigma v)^{jk}}{dp_i} - \\ - f_{\nu i}^{(1)} \sum_k \int d^3 p_k f_{\nu k}^{(0)} (\sigma v)^{ik} + S_i(p_i) \ . \tag{941}$$

Here

$$\frac{d^3(\sigma v)^{jk}}{dp_i}$$

is the differential rate of ($\nu\nu$) scattering and $S_i(p_i)$ is the function of the source.

If the neutrinos of all species have the same unperturbed distribution function, the quantity

$$f = \sum_i f_{\nu i}^{(1)}$$

satisfies the equation, having the same form as the equation for $f_\nu^{(1)}$ in the case of only one species of neutrino, given by

$$\frac{df_\nu^{(1)}(E,t)}{dt} = 2\int d^3p \cdot f_\nu^{(0)}(\omega)\cdot\omega\int_E^\infty dx f_\nu^{(1)}(x)\frac{G_F^2(1-\cos\theta)^2}{8\pi} -$$

$$- f_\nu^{(1)}E\int d^3p\cdot f_\nu^{(0)}(\omega)\cdot\frac{G_F^2(1-\cos\theta)^2}{8\pi} + S_\nu(E,t)\ . \quad (942)$$

Therefore we can consider the case of only one neutrino type.

Using the normalisation conditions

$$\int_0^\infty E\cdot f_\nu^{(1)}(E)dE = \frac{mn_m}{a_\nu} \quad (943)$$

and

$$\int_0^m E\cdot S(E)dE = mn_m\int_0^1 x\cdot S(x)dx = \frac{mn_m}{a_\nu}\ , \quad (944)$$

the functions $f_\nu^{(1)}$ and S_ν can be conveniently normalised as follows:

$$f_\nu^{(1)} = \frac{1}{m} f\left(\frac{E}{m}\right) n_m \ , \tag{945}$$

and

$$S_\nu(E,t) = n_m \frac{d\Gamma}{\Gamma dE} \delta(t-\tau) = S(E)\delta(t-\tau) \ , \tag{946}$$

where

$$S(E) = \frac{1}{m} S\left(\frac{E}{m}\right) n_m \ . \tag{947}$$

Here a_ν^{-1} has the meaning of the fraction of the rest energy of decaying particles, corresponding to the total energy of the initial non-equilibrium neutrino flux.

One can also introduce the dimensionless variable

$$x = \frac{E^{\mathrm{mod}}(1+z_d)}{m} \ , \tag{948}$$

where z_d is the red shift, corresponding to the period of decay and E^{mod} is the modern energy of non-equilibrium neutrino.

Then the equation for

$$f(x) \equiv f_\nu^{(1)}(x) \tag{949}$$

has the form

$$\frac{df(x,t)}{dt} = 2B\int_x^\infty dy f(y,t) - Bf(x,t) + S_\nu(x,t) \ , \quad (950)$$

where

$$B = \frac{G_F^2 \varepsilon_\nu m}{6\pi} \ , \quad (951)$$

and ε_ν is the energy density of the thermal neutrino background.

The solution of Eq.(950) can be easily obtained with the use of the Laplace transformation method. It has the form

$$f(x,t) =$$

$$= \exp(-B\tau x)\left\{S(x) + 2B\tau\int_x^1 S(y)dy + (B\tau)^2\int_x^1 yS(y)dy\right\} \ . \quad (952)$$

The quantity

$$B\tau \sim n_\nu \sigma_{\nu\nu} v t \quad (953)$$

has the meaning of the optical depth in the thermal neutrino background for energetic neutrinos with the energy

$$E \sim m \quad (954)$$

One finds from Eq. (952) that the larger is the opacity for initial energetic neutrinos from the source, the better is the approximation, in which the term, proportional to

$$\int_x^1 yS(y)dy$$

dominates in the distribution function. This term corresponds to the "multiplication" of high-energy neutrinos owing to neutrino scattering on the thermal neutrino background.

Putting

$$a_\nu \sim 1 , \tag{955}$$

one obtains

$$f(E) = \exp\left(-\frac{E}{E_0}\right)\frac{mn_m}{E_0^2} , \tag{956}$$

where

$$E_0 = \left(\frac{G_F^2 \varepsilon_\nu \tau}{6\pi}\right)^{-1} \approx E_{\max} . \tag{957}$$

Finally, the modern spectrum of high-energy neutrinos, predicted in this case, has the form

$$J_\nu(E) = \frac{rmn_\gamma}{(1+z_d)\left(E_{\max}^{\mathrm{mod}}\right)^2} \cdot \exp\left(-\frac{E}{E_{\max}^{\mathrm{mod}}}\right) \cdot \frac{c}{8\pi} . \tag{958}$$

Since there is no significant excessive background over the predicted flux of atmospheric neutrinos in neutrino telescopes, one can put an upper limit on the flux of nonthermal cosmological high-energy neutrinos

$$F_\nu(E > 1\ \mathrm{GeV}) < F_\nu^a(E > 1\ \mathrm{GeV}) . \tag{959}$$

Using Eqs. (958)–(959), one finds the restriction on the values rm and τ for superheavy metastable particles with the mass

$$m >> 10^6 \mathrm{GeV} \tag{960}$$

decaying at

$$t > 10^6 \mathrm{s}\ . \tag{961}$$

The restriction is shown in Fig.18.

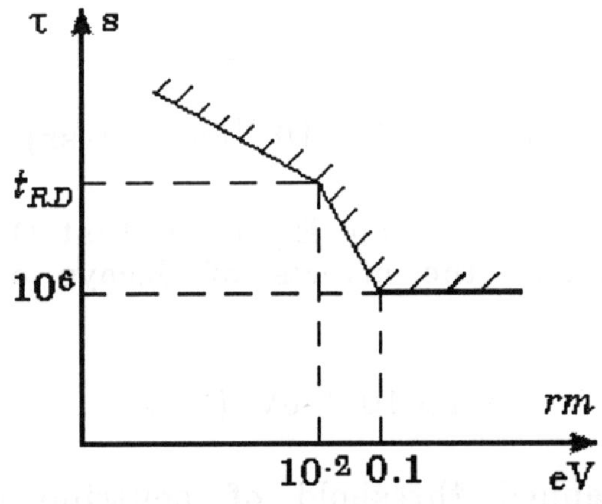

Fig.18. Constraints on superheavy metastable particles from the upper limit on super-high energy neutrino background.

In spite of the ambiguity of the form of the spectrum in the case (931), one can assume that there is no strong suppression of the high-energy tail up to the energy

$$E \le \frac{m}{2}\ . \tag{962}$$

Then the restriction, presented on the Fig.18, is also valid in this case.

3.3. DUMAND as the Tool for Cosmoparticle Physics

The planned Deep Underwater Muon and Neutrino Detectors (DUMAND) will open new possibilities for the tests of the existence of high-energy particle sources in the early Universe.

The planned minimal sensitivity of these detectors at

$$E \sim 1 \text{ TeV} \tag{963}$$

is

$$F_\nu(E > 1\text{TeV}) = 10^{-11}\left(\text{cm}^2 \cdot \text{s} \cdot \text{sr}\right)^{-1} . \tag{964}$$

One finds in this case from Eq. (958) that DUMAND will be able to detect the effects of decays of primordial particles with

$$rm > 10^{-10} \text{ eV} \cdot (1 + z_d) . \tag{965}$$

The planned threshold of neutrino detection in DUMAND is about 200 GeV. One obtains from Eqs. (924) and (925), that the condition

$$E_{\text{max}}^{\text{mod}} > 200 \text{ GeV} \tag{966}$$

is valid, if

$$\tau > 10^8 \text{ s} . \tag{967}$$

The values of *rm* and τ, accessible to DUMAND are presented in Fig.19 together with the restrictions on these values following from the data from neutrino telescopes, on the light element abundances (Chapter 6) and on the observed spectrum of the relic radiation (Chapter 5). One finds from Fig.19, that DUMAND can play the role of an unique detector for decays of particles with the masses, exceeding

$$m > 200 \text{ GeV} \cdot (1 + z_d) \; , \tag{968}$$

in the period $\tau > 10^8$ s.

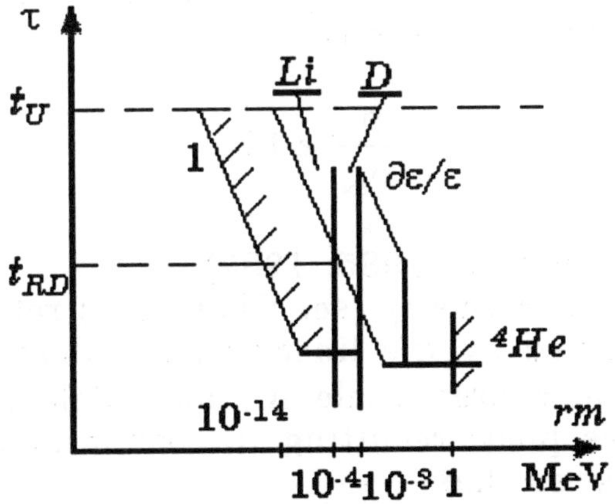

Fig.19. Constraints on *rm* and τ from 1- DUMAND in comparison with the constraints from neutrino telescopes, light element abundance and spectrum of relic radiation.

In studies of a "white spots" on the plane (*rm*, τ), it may be useful to probe such cosmological effects of super-high-energy physics, which can influence neither the light element abundance, nor the thermal background spectrum.

The existence of metastable particles with the Planck mass

$$m = m_{Pl} , \tag{969}$$

the so-called electrically charged maximons, can be a non-trivial consequence of quantum gravity (Markov, 1984). Decays of these particles can be accompanied by formation of superhigh-energy neutrino cascades. This case corresponds to the condition (932), so the spectrum of neutrinos, induced by maximon decays, is given by Eq. (958).

Note, that the process of cosmic string evolution may be an interesting example of the source of superhigh energy neutrino fluxes. The DUMAND experiment can probe in this case the relative contribution of cosmic strings into the total density, exceeding

$$\frac{\rho_s}{\rho_{tot}} \geq 10^{-11} \tag{970}$$

(Khlopov, Chechetkin, 1985; 1987).

The particle theory (see Chapter 2) relates this contribution, defined by the linear energy density of cosmic strings, to the energy scale Λ, at which the symmetry breaking takes place resulting in cosmic strings as topological defects, so that

$$\frac{\rho_s}{\rho_{tot}} = \Omega_s \sim \frac{\Lambda^2}{m_{Pl}^2} \ . \tag{971}$$

The existence of cosmic strings is related either to $U(1)$ symmetry breaking, or to the breaking of the symmetry leaving unbroken the subgroup of the symmetry, containing the strict discrete symmetry (see Chapter 2).

Therefore, DUMAND turns out to be sensitive to the pattern of symmetry breaking at scales, exceeding

$$\Lambda > 10^{14} \text{ GeV} . \tag{972}$$

Hypothetical particles, PBH and other objects, predicted by the modern particle theory and considered in the preceding chapters can be considered as various forms of the dark matter in the Universe. Analysis of possible influence of these forms of the dark matter on the physical processes in the Universe, makes it possible to use the observational data for definite conclusions on possible measure of such an influence, thus imposing restrictions on possible properties of hypothetical objects, and, consequently, on the parameters of initial models.

Such analysis leads to negative result, to restrictions on the properties of the dark matter in the Universe. On the other hand, there are the areas of the modern cosmology, in which one should assume the existence of the dark matter to give self-consistent explanation to the whole set of the observational data. As we shall see below, it is just the case in the confrontation of the theory of the large-scale structure formation in the Universe with the observations.

CHAPTER 8

THE NEW PHYSICS IN THE LARGE SCALE STRUCTURE FORMATION.

1. Problems of the large-scale structure formation

1.1. The Problem of Initial Fluctuations

The structure of inhomogenities in the expanding quasihomogeneous FRW Universe should be the result of a long stage of growth of small initial density perturbations. In the cosmological model such perturbations should be preassigned initially. They can not be ascribed to statistical fluctuations of the particle number density.

Indeed, there are

$$N \sim 10^{70} \tag{973}$$

baryons in the observed galaxy superclusters with the mass

$$M > 10^{13} M_{\odot} \,, \tag{974}$$

where

$$M_{\odot} = 2 \cdot 10^{33}\,\mathrm{g} \tag{975}$$

is the Solar mass.

The relative amplitude of statistical density fluctuations in these scales is

$$\left.\frac{\delta\rho}{\rho}\right|_{st} \sim \frac{\delta N}{N} \sim \frac{1}{\sqrt{N}} \sim 10^{-35} \ . \tag{976}$$

On the other hand, density fluctuations grow in the expanding Universe according to the law

$$\frac{\delta\rho}{\rho} \propto (1+z)^{-1} \ . \tag{977}$$

To form the inhomogeneities with

$$\frac{\delta\rho}{\rho} \sim 1 \tag{978}$$

at the present moment, there should have been density fluctuations with the amplitude

$$\delta = \frac{\delta\rho}{\rho} \sim 10^{-3} \tag{979}$$

in the period of recombination, corresponding to the red shift

$$z = z_{rec} \sim 10^{3} \ . \tag{980}$$

There are two different possibilities for the nature of initial perturbations in the cosmological models, assuming that baryonic matter and radiation are the only constituents of the Universe. (Here we put aside the so-called vortex theory of galaxy formation (Ozernoy, Chernin, 1966)).

According to the first possibility, the density perturbations are generated by metric perturbations, including both matter and radiation, so that the specific entropy is not changed. Such perturbations are called adiabatic, and the theory puttung their evolution into the bedrock of galaxy formation is called the adiabatic (A) theory of galaxy formation.

The other possibility is the perturbation of baryon-to-photon ratio at the homogeneous total (baryon and photon) density on the RD stage. The specific entropy of the matter is perturbed in this case, so the perturbations are called entropy perturbations, and the theory of their evolution leading to galaxy formation is called the entropy (E) theory of galaxy formation.

The case of the so-called isothermal density perturbations, in which the baryon charge inhomogenities are given on the homogeneous background of radiation, corresponds to the superposition of adiabatic and entropy perturbations.

Both adiabatic and entropy theories assume that the formation of inhomogenities takes place effectively at the stage of baryon matter dominance after recombination, when the radiation pressure does not influence the development of gravitational instability in the baryonic matter.

But A- and E-theories differ in the order of the formation of different objects.

They also differ in the predictions for the large-scale distribution of the matter. Beginning from the late seventies – the early eighties, a number of observational evidences was found in favour of the existence of evolved cellular structure in the Universe (De Vaucouleuers, 1959, 1976; Einasto et al., 1980; Tully, 1982).

The existence of this structure was predicted in the framework of the A-theory of galaxy formation (see review of Shandarin et al., 1983), and is a specific feature of this theory.

In contrast with the entropy (E) or vortex (Ozernoy, Chernin, 1966) theories, according to the A-theory the large-scale inhomogenities (so-called pancakes) (Zeldovich,

1970), should be formed first. Successive fragmentation of pancakes leads to the formation of galaxies and clusters of galaxies within the pancakes.

The E-theory in its most developed variants leads naturally to hierarchical clustering of inhomogeneities (Jones, 1976; Gott, 1977), so that small-scale inhomogeneities (at the scales of globular clusters) should be formed first. Their successive clustering results in the formation of inhomogeneities at larger scales (galaxies, clusters of galaxies).

The A-theory of galaxy formation turned out to be virtually the theory of the large-scale structure formation in the Universe. The existence of the pancakes (superclusters of galaxies) and giant (of the order of tens Mpc) "voids" (De Vaucouleuers, 1959, 1976; Einasto et al., 1980; Tully, 1982; Kirshner et al.,1981; Davis et al., 1981), exhibiting no apparent presence of galaxies within them, seemed to give final proof to the A-theory.

However, the theory of the large-scale structure formation met a number of dificulties in its confrontation with some observational facts. One of these difficulties is related to the ambiguous data on the mean density of the Universe, estimated by different methods. These questions were discussed in Chapter 3.

Data on the density of the luminous matter in galaxies and on the light element abundances favoured small mean matter density of the Universe

$$\Omega < 0.1 \div 0.2 \ . \tag{981}$$

On the other hand, data on the dynamics of the matter in galaxies, clusters of galaxies and galaxy superclusters favoured the higher total density

$$\Omega > 0.2 \ . \tag{982}$$

The strongest lower bound on the cosmological density followed from the analysis of the dynamics of large-scale structure formation and of the influence of initial density perturbations on the anisotropy of the relic radiation.

1.2. *The Problem of Self-Consistent Structure Formation and BBBR Anisotropy*

Soon after the discovery of relic radiation was, it was pointed out (Silk, 1968; Sunyaev, Zeldovich, 1970), that the temperature of this radiation can not be strictly constant along the whole sky.

Density and velocity perturbations, having evolved into the modern structure of the Universe, should have left the "footprints" in the relic radiation. The temperature of relic radiation should vary and the amplitude of the fluctuations of the temperature is unambiguously related to the amplitude of the initial density and velocity fluctuations in the Universe. There are only upper bounds on the amplitude of these fluctuations at the angular scale corresponding to clusters of galaxies

$$\frac{\delta T}{T} < 2 \cdot 10^{-5} \tag{983}$$

and the statements on the existence of the positive effect at the level

$$\frac{\delta T}{T} \leq 10^{-5} \tag{984}$$

at the angular scales

$$\theta > 1^{o} \tag{985}$$

(Smoot et al., 1992; Smoot, 1997; Strukov et al., 1996), fixing the amplitude of initial density and velocity perturbations at such a low level, that the observed structure could not be formed in the Universe at low values of the mean cosmological density. These data favour the value given by the Eq. (982)

$$\Omega > 0.2$$

or, in fact, even

$$\Omega \sim 1 \,. \tag{986}$$

These contradictions are naturally resolved qualitatively in the cosmological models with neutrinos with the mass

$$m_\nu \sim 30\text{eV} \tag{987}$$

(Marx, Szalay, 1976; Zeldovich, Sunyaev, 1980; Doroskevich et al., 1980a,b; Bisnovatyi-Kogan, Novikov, 1980; Hut, Klinkhammer, 1980; Schramm, Steigman, 1980; Sato, Sato, 1980).

Indeed, the observations of the luminous matter as well as the light element abundance provide only the estimation of the baryon component of the total density, whereas dynamical estimations and the arguments concerning the fluctuations of the temperature of relic radiation are related to the total density, including both baryonic and neutrino components.

Hence, the cosmology of neutrino rest mass promised a natural unique solution of a number of old astrophysical problems.

But the significance of cosmological models with massive neutrinos went far beyond the simple qualitative solution of these problems.

Based on the analysis of cosmological impact of experimental evidences for electron neutrino rest mass

$$14 < m_\nu < 46 \mathrm{eV} \tag{988}$$

(Lubimov et al.,1980), these models lead in their development to irreversible change of the ideas concerning the period of structure formation. They have clarified the internal inconsistency of models with baryon matter dominance in this period, making the existence of nonbaryonic dark matter to be an inevitable element of the theory of cosmological large-scale structure formation.

Combined analysis of the data on the angular distribution of the thermal electromagnetic background and on the large-scale distribution of the matter in the Universe makes it possible, in principle, to obtain the information on the physical nature of the dark matter (Doroshkevich, Khlopov, 1984).

Quantitative analysis of the development of cosmological structure in the models with massive neutrinos found some difficulties in the confrontation of detailed predictions of these models with the observational data on the large-scale distribution of the matter.

The quantitative discrepancy between the estimates of the dark matter density from the condition of formation of the structure at the observed isotropy of the relic radiation and the estimates of the dark matter density within the formed structure (Klypin, Shandarin, 1983; Frank et al., 1983; White et al., 1983) was demonstrated. This discrepancy can not be removed by simple change of massive neutrinos by some other hypothetical stable weakly (or superweakly) interacting particles.

One of the possible solutions rejects the idea, that the dark matter in the period of large-scale structure formation is the same, as in the modern Universe. In other words, the unstable dark matter is invoked.

Analysis of the evolution of the cosmological large-scale structure makes it possible to determine in this case the mass and lifetime of dark matter particles.

In the case of unstable massive neutrinos, it is possible to determine the parameters of hypothetical neutrino interactions, inducing the decay.

However, the problem on the physical nature of the dark matter has not found yet its final solution. The choice of definite candidate for dark matter particles determines the physics of dissipation of their density fluctuations, determining the short-wave cut in the spectrum of these fluctuations.

In the simplest case of the thermal background of weakly interacting particles (e.g. massive neutrinos) the scale of the short-wave dissipation R_c is determined by the mass m of these particles,

$$R_c \sim m^{-1} . \tag{989}$$

The scale R_c defines the characteristic size of the inhomogeneities, which are formed first.

Subject to this scale, the type of dark matter is specified as hot, warm or cold.

The scale R_c corresponding to superclusters (or clusters) of galaxies is referred to hot dark matter (HDM), e.g. to the primordial thermal gas of weakly interacting particles with the mass

$$m \sim 30 \div 100 \text{ eV} . \tag{990}$$

In the case of warm particles, e.g. of the primordial thermal gas of particles with the mass

$$m \sim 1 \text{ keV} \tag{991}$$

R_c corresponds to the scale of galaxies.

Inhomogeneities formed by the cold dark matter (CDM) have the scales R_c much smaller, than the scale of galaxies. CDM candidates are heavy weakly interacting particles with the mass

$$m >> 1 \text{ keV} \tag{992}$$

such as neutralino in supersymmetric models or heavy stable neutral leptons with the mass

$$m > (1 \div 10) \text{ GeV} . \tag{993}$$

Invisible axions have very small mass

$$m \sim 10^{-5} \div 10^{-3} \text{ eV} . \tag{994}$$

However, the dissipation scale of their density fluctuations (see Chapter 3) is very small, as in the case of heavy particles with the mass

$$m >> 1 \text{ GeV} . \tag{995}$$

That is why axions are also considered as CDM candidates.

The relation between the lifetime of dark matter particles τ and the age of the Universe t_U divides the dark matter on stable one,

$$\tau >> t_U , \tag{996}$$

and unstable one,

$$\tau < t_U . \tag{997}$$

Weakly interacting dark matter particles, mentioned above, represent collisionless gas in which dissipation proceeds at rather long timescale. The mirror and the shadow matter has internal interactions leading to dissipation rates comparable or even higher, than the rates for the baryonic matter. Hence, the dissipational forms of the dark matter are possible, resulting in the possible existence of compact dark matter objects. This possibility makes the interpretation of the observational data even more complicated.

1.3. The Problem of Physical Self-Consistency in the Theory of Large-Scale Structure Formation

The other problem of the theory of cosmological structure formation is the spectrum of initial density fluctuations.

The most popular is the so-called "flat" spectrum of fluctuations (Harrison, 1967; Zeldovich, 1970). Its advantages seem evident, since it assumes no preferred scale. The amplitude of fluctuatins is the same for all scales in the period, when the scale of the fluctuation is equal to the size of the horizon.

Flat spectrum of density fluctuations followed from the simplest variants of inflational scenario (Hawking, 1982; Brandenberger et al., 1982; Guth and Pi, 1982; Starobinsky, 1982). It is also formed in the case of density fluctuations, induced by the structure of long cosmic strings (Zeldovich, 1980).

However, the development of realistic models of multicomponent inflation, accounting for several types of scalar fields (Linde, Kofman, 1986; Sakharov, Khlopov, 1993) and the development of the theory of cosmic strings accounting for the contribution of closed loops of strings (Vilenkin, 1985) and their decay on smaller loops (Smith, Vilenkin, 1987), lead to more complicated forms of spectra of initial fluctuations.

Moreover, the variety of particle physics effects in the scenario of the very early Universe (see Chapters 3 and 4) opens new ways in the physical interpretaion of the phenomenology of adiabatic and entropy fluctuations and provides a nontrivial pattern of physical phenomena combined with the old ideas of A- and E-theories.

The initial density fluctuations, induced by inflaton and considered in the Chapters 3 and 4, are adiabatic and seem to give foundamental support to the A-theory. However, the evolution of these fluctuations in the CDM models results in the hierarchical clustering, typical for E-theory.

The realistic theory of cosmological structure formation seems to be multicomponent. It should reflect all main features of realistic models of the particle theory. This point is favoured both by the physical grounds of cosmological models and by the evident incapability of a simple one-parameter cosmological dark matter scenario to reproduce the whole set of the observational data.

Taking into account all these possibilities, following from the modern particle theory, one should take in mind that not only adiabatic, but also entropy fluctuations can find physical grounds in some scenarios of the very early Universe, based on the predictions of particle models.

Entropy fluctuations naturally follow from the inhomogeneous baryosynthesis, which can be related, for instance, to the spatial dependence of CP-violating phase, determining the value of the baryon asymmetry of the Universe (see Chapter 6). It can take place in the model of invisible axion (Ioshimura, 1983), or in the multicomponent model of spontaneous CP violation (Kuzmin et al., 1981). It may also take place in the framework of supersymmetric models, ascribing the baryon asymmetry of the Universe to the existence of the primordial condensate of scalar quarks (squarks) (Affleck, Dine, 1985; Linde, 1985).

In the latter case, the supersymmetric potential energy density is independent of the amplitude of the squark field, so that the amplitude of this field can be different in different spatial regions. As a result, the spatial dependence of the baryon number density arises.

So, during the last years, cosmology has expanded its ideas on both the possible physical nature of the matter, forming the cosmological structure of inhomogeneities, and the possible physical origin of initial density perturbations. Accounting for the physical motivations following from different parts of the particle theory one finds in the theory of the cosmological large-scale structure formation the nontrivial reflection of the physical mechanisms of inflation, baryosynthesis and mixture of various forms of the dark matter. Here the complicated knot of the three central problems of the modern cosmology is tied up. The cosmoparticle physics should develop all its skill to untie this knot and find physically self-consistent picture of the origin of galaxies.

Let us consider now some dark matter candidates in more details.

2. The Neutrino Rest Mass and the Large-Scale Structure of the Universe.

Experimental evidences in favour of the existence of electron neutrino rest mass of the order of 30 eV (Lubimov et al., 1980) inspired exstensive discussion of possible cosmological impact of massive neutrinos (Doroshkevich et al., 1980a,b; Zeldovich, Khlopov, 1981; Schramm, Steigman, 1980; Sato, Takahara, 1980; Bisnovatyi-Kogan, Novikov, 1980, Bisnovatyi-Kogan et al., 1980; Bond et al., 1980; Klinkhammer, Norman, 1980). The possibility to ascribe the existence of dark matter to massive neutrinos, first mentioned by Szalay and Marx (1976), is of special interest.

2.1. Massive Neutrinos in the Big Bang Universe

Following Doroshkevich et al. (1980a,b) and Doroshkevich, Khlopov (1981) consider the cosmological evolution of massive neutrinos.

As we have already mentioned in the Chapter 3, at

$$T \gg m_e \tag{998}$$

neutrino were in the thermodynamical equilibrium with the other particles — photons, electrons and positrons.

In this period, the mass of neutrino can be neglected,

$$T >> m_e >> m_\nu \ , \tag{999}$$

and therefore

$$E_\nu = p_\nu \ , \tag{1000}$$

where E_ν is the energy and p_ν is the momentum of neutrino.

According to the Fermi statistics, the occupation number of ultrarelativistic neutrino in the phase space is

$$n = \left(\exp\left(\frac{p_\nu}{T} \right) + 1 \right)^{-1} \tag{1001}$$

with the neutrino chemical potential

$$\mu = 0 \ . \tag{1002}$$

In this case

$$n_\nu = n_{\bar{\nu}} \ . \tag{1003}$$

At

$$T \leq 1 \text{ MeV} \tag{1004}$$

(see Chapter 3) the gas of neutrinos is collisionless.

The momentum of collisionless neutrinos decreases as

$$p \propto (1+z) \tag{1005}$$

and their phase space density (the occupation number) is conserved in the course of cosmological expansion.

According to (1001), it is equivalent to the decrease of the neutrino temperature T proportional to

$$T \propto (1+z) \ . \tag{1006}$$

At

$$T \cong 0.1 m_e \tag{1007}$$

electron-positron pair annihilation takes place, increasing the temperature of the thermal background radiation relative to the temperature of neutrino (see Chapter 3)

$$T_\nu = \left(\frac{4}{11}\right)^{1/3} T_\gamma = 1.83 \cdot (1+z) \text{ K} \tag{1008}$$

at

$$T_\gamma < 0.1 m_e \ , \tag{1009}$$

where

$$T_\gamma = 2.7 \cdot (1+z) \text{ K} \ . \tag{1010}$$

The distribution function (1001) gives the neutrino concentration

$$N_\nu = \int n \frac{d^3 p}{(2\pi)^3} = 75(1+z)^3 \text{ cm}^{-3} \ . \tag{1011}$$

Note, that the distribution function (1001) is not equilibrium one at

$$T_\nu << m_\nu \ , \tag{1012}$$

since the equality $E_\nu = p_\nu$ is not valid in the nonrelativistic case, and the relation

$$p_\nu = \left(E_\nu^2 - m_\nu^2\right)^{1/2} \tag{1013}$$

gives the phase space density in the form

$$n = \left[\exp\left(\frac{\left(E_\nu^2 - m_\nu^2\right)^{1/2}}{T_\nu}\right) + 1\right]^{-1} \approx$$

$$\approx \left[\exp\left(\frac{\left(2m_\nu E_k\right)^{1/2}}{T_\nu}\right) + 1\right]^{-1} , \tag{1014}$$

where the neutrino kinetic energy

$$E_k = \frac{p_\nu^2}{2m_\nu} \approx E_\nu - m_\nu << m_\nu \ , \tag{1015}$$

and

$$E_\nu + m_\nu \approx 2m_\nu \tag{1016}$$

at

$$T_\nu < m_\nu \ . \tag{1017}$$

Using Eq. (1001), one obtains after the averaging over the phase space (Doroshkevich et al., 1980a) the mean kinetic energy

$$\langle E_k \rangle = \frac{\langle p_\nu^2 \rangle}{2m_\nu} = 6.47\frac{T_\nu^2}{m_\nu} \tag{1018}$$

and the mean velocity

$$\langle v^{-2} \rangle = 0.385\frac{m_\nu^2}{T_\nu^2} \tag{1019}$$

$$\sqrt{\langle v^2 \rangle} = 3.6\frac{T_\nu}{m_\nu} = 664 \cdot (1+z)\ \frac{\text{km}}{\text{s}} \tag{1020}$$

at $m_\nu = 30$ eV . This mean velocity corresponds to the temperature

$$T_{\text{eqv}} = \frac{m\langle v^2 \rangle}{2} = 5.5 \cdot 10^{-5}(1+z)\ \text{K} \tag{1021}$$

of the Maxwell gas of neutrinos with the mass (987).

Neutrinos turn out to be nonrelativistic at the red shift

$$z = 4.5 \cdot 10^4 \left(\frac{m_\nu}{30\ \text{eV}}\right). \tag{1022}$$

In this period the law of the cosmological expansion changes. Nonrelativistic neutrinos soon begin to dominate in the total cosmological density and their dominance determines the law of cosmological expansion.

The approximate analytical expression for the equation of state of the Universe in the period of transition from the RD to the MD stage can be found on the basis of the neutrino distribution function (Doroshkevich, Khlopov, 1981).

Namely, one can find an approximation for the time dependences of the energy density ε_ν and pressure p of neutrino gas with the use of its distributon function (1001). It has the form

$$\varepsilon_\nu = \varepsilon_{eq} x_\nu^4 \int_0^\infty \frac{x^3\left(1+\frac{9x_\nu^2}{x^2}\right)^{1/2}}{\exp(x)+1} \approx \frac{\varepsilon_{eq}}{\sqrt{2}} x_\nu^4\left(1+x_\nu^{-2}\right)^{1/2} \quad (1023)$$

$$p = \varepsilon_{eq} x_\nu^4 \int_0^\infty \frac{x^3\left(1+\frac{9x_\nu^2}{x^2}\right)^{-1/2}}{\exp(x)+1} \approx \frac{\varepsilon_{eq}}{\sqrt{2}} x_\nu^4\left(1+x_\nu^{-2}\right)^{-1/2}, \quad (1024)$$

where

$$x_\nu = \frac{(1+z)}{(1+z_\nu)} \quad (1025)$$

and ε_{eq} is the neutrino energy density at

$$z = z_\nu = 5.5 \cdot 10^4 \left(\frac{m_\nu}{30\mathrm{eV}}\right)\left(\frac{H_0}{50\,\frac{\mathrm{km}}{\mathrm{s\cdot Mpc}}}\right)^2. \quad (1026)$$

Here H_0 is the modern value of the Hubble constant.

Eqs. (1023)–(1024) have correct asymptotics both in the case

$$z \gg z_\nu \tag{1027}$$

corresponding to the equation of state

$$p = \tfrac{1}{3}\varepsilon \ ,$$

and in the case

$$z \ll z_\nu \tag{1028}$$

when the equation of state is

$$p \ll \varepsilon \tag{1029}$$

that corresponds to the cosmological dust-like stage $p = 0$.

Using the approximate equation of state of the gas of massive neutrino, given by Eqs. (1023)–(1024), one can find the cosmological expansion law in the period, when massive neutrinos turn out to be nonrelativistic and start to dominate in the Universe.

The transition from the relativistic expansion law

$$a(t) \propto t^{1/2}$$

to the dust-like stage expansion law

$$a(t) \propto t^{2/3}$$

turns out to be rather long. It goes slower, than in the case of the transition from the radiation dominance stage to the baryonic matter dominance. The transition turns out to be even more longer, if there are several types of relativistic neutrinos besides the thermal electromagnetic background in the Universe, when the massive neutrino species starts to dominate in the cosmological density.

The question on the mean free streaming path of massive neutrino in the expanding Universe is of importance for the theory of evolution of density perturbatons.

At

$$z > z_\nu \tag{1030}$$

the particles are relativistic and their chaotic velocity is close to the velocity of light, so that the free streaming path is close to the size of horizon, which increases with the decrease of z.

The particles are nonrelativistic at

$$z < z_\nu \tag{1031}$$

and their free streaming path is much smaller, than the size of the horizon.

So the maximal free streaming path of neutrino in the expanding Universe is determined by the period

$$z \approx z_\nu \tag{1032}$$

when the particle turns out to be nonrelativistic. The value of this maximal free streaming path defines the most important parameter of the theory of gravitational instability in the gas of massive neutrinos: the scale R_ν of the cosmological structure formed by massive neutrinos.

2.2. Gravitational Instability of Massive Neutrino

Due to the isotropy and homogeneity of the unperturbed matter distribution in the linear theory of gravitational instability, it is convenient to use Fourier expansion of perturbations

$$\frac{\delta\rho}{\rho} = \frac{1}{(2\pi)^{3/2}} \int \delta(\mathbf{k}, t) \cdot \exp(i\mathbf{kr}) d^3k \ . \tag{1033}$$

Time evolution of perturbations is given by the transition function $C(k,t,t_{\rm in})$, relating perturbation at the time t with the initial perturbation at the time $t_{\rm in}$. One has the relation

$$\delta(\mathbf{k}, t) = \delta(\mathbf{k}, t_{\rm in}) \cdot C(k, t, t_{\rm in}) \ . \tag{1034}$$

The initial density perturbations $\delta(\mathbf{k},t_{in})$ are generated on the inflational stage (see Chapter 3) or by some other mechanism, related to the physics of the very early Universe.

The phases are usually assumed to be uncorrelated and the initial (stochastic) fluctuations are completely determined by the spectrum

$$b^2(k) = \left\langle \delta^2(k, t_{\rm in}) \right\rangle \ . \tag{1035}$$

The widely used "flat" spectrum (Zeldovich, 1972; Harrison, 1967), predicted by the simplest variants of inflational models, is given by

$$b^2(k) \propto k \ . \tag{1036}$$

Examine now the problem of the evolution of density perturbations of massive neutrinos. Irrespective to the physical mechanism of their origin, assume the existence of metric, density and velocity perturbations in the early Universe and consider their evolution.

Ultrarelativistic particles and radiation dominate on the RD stage of the evolution of Big Bang Universe. On this stage both the evolution of the Universe as a whole, and the development of density perturbations are analysed in the framework of hydrodynamical model of the

Universe with the equation of state of the relativistic liquid $p = \varepsilon/3$.

This model was studied yet even by Lifshitz (1946). The main results of his analysis are as follows (see review of Shandarin et al. (1983)).

1) Pressure does not influence the evolution of perturbations in the large scales. There is only one mode growing with time and there is another mode decreasing with time in these scales.

2) Gravity plays no role at small scales, and density perturbations transform into sound waves. The amplitude of these waves decreases slowly due to cosmological expansion in accordance with the theory of adiabatic invariants.

3) The border between large and small scales is the Jeans scale, corresponding to the distance covered by the sound wave in the expanding Universe at the time t

$$R_J = \frac{2ct}{\sqrt{3}} \ . \tag{1037}$$

In the period, when the temperature of radiation is much larger than the mass of neutrino, neutrinos are relativistic and this picture remains unchaged. Of course, the collisionless gas of neutrino can not be treated as the fluid with isotropic pressure, but it turns out (Khlebnikov, 1982), that the difference is insignificant for the evolution of long-wave perturbations.

So, in the early period, when neutrinos were ultrarelativistic, the Jeans mass M_J in neutrino perturbations was of the order of the mass M_h , contained within the cosmological horizon,

$$M_J \sim M_h = m_{Pl} \frac{t}{t_{Pl}} \ . \tag{1038}$$

The Jeans mass linearly increases with time, and is the boundary between the region I (see Fig.20), corresponding to the incresing neutrino perturbations, and the region II, in which no sound waves arise at small scales and neutrino perturbations dissipate due to the free streaming. The boundary between the large and small scales is still close to M_h.

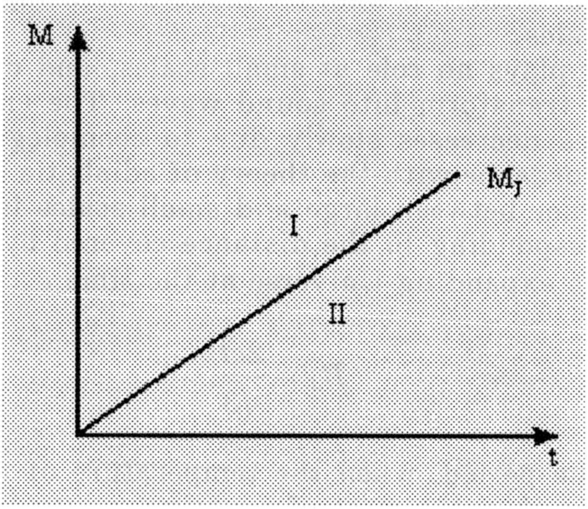

Fig.20. Evolution of density perturbations of relativistic neutrino.

Dissipation of neutrino density perturbations (region II in Fig.20) is induced by the free streaming of weakly interacting neutrinos from the regions of higher density to the regions of lower density. Detailed analysis of this problem is possible only on the basis of numerical simulations (see Bond, Szalay, 1982), but the large k asymptotics (small wavelengths)

$$kR_\nu \gg 1 \qquad (1039)$$

can be found analytically (Doroshkevich et al., 1980a)

Neutrino momentum distribution is given by the Fermi-Dirac function (1001) with zero chemical potential. Density perturbations are related to perturbations of the distribution function. Adiabatic perturbations correspond to inhomogeneity of the density and mean energy of neutrinos. Neutrino motion turns out to smooth these inhomogeneities in the scales, corresponding to neutrino free streaming, so that the perturbations at corresponding wavelengths dissipate.

The problem on the dissipation of waves in collisionless gas was considered in detailes for the Big Bang Universe in plasma and in neutral gas (Zeldovich, Novikov, 1975).

The difference between these cases and the case of massive neutrino in the Big Bang theory is due to the difference of the neutrino spectrum from the Maxwell distribution, resulting in the different law of dissipation.

The perturbations of neutrino gas at the wavelengths, much smaller, than R_ν, dissipate mainly in the period t_1, z_1, when neutrinos turn out to be nonrelativistic (see below).

The condition

$$\lambda < R_\nu \tag{1040}$$

in this period means that gravity can be neglected. Hence, the problem is trivialy reduced to the particle motion with constant velocity.

If the phase space density n is given by the function $n(\mathbf{p},\mathbf{x})$ at the time t_1, one obtains

$$n = n\left(\mathbf{p}, \mathbf{x} - \mathbf{v} \cdot (t - t_1)\right) \tag{1041}$$

at arbitrary moment t.

If the initial phase space density is given by

$$n = n_0(\mathbf{p}) + \delta n_0(\mathbf{p}) \cdot \exp(i\mathbf{kx}) \tag{1042}$$

at $t = 0$, one obtains for the perturbation of the phase space density

$$\delta n = \delta n_0(\mathbf{p}) \cdot \exp\left(i\mathbf{k}(\mathbf{x} - \mathbf{v}t)\right) \tag{1043}$$

and the density perturbation is given by

$$\delta\rho(\mathbf{x}, t) = \int \delta n(\mathbf{p}, \mathbf{x}, t) d^3 p = \int \delta n_0(\mathbf{p}) \cdot \exp(-i\mathbf{kx}) d^3 k \ . \tag{1044}$$

The velocity $\mathbf{v}$ should be treated as the function of momentum $\mathbf{p}$,

$$\mathbf{v} = \frac{c\mathbf{p}}{E} = c^2 \mathbf{p}\left(\left(m_\nu c^2\right)^2 + p^2 c^2\right)^{-1/2} . \tag{1045}$$

In the case of the expanding Universe one should evidently replace the product $\mathbf{kv}t$ by the integral

$$J = \int \mathbf{k} \frac{d\mathbf{x}}{dt} dt \ , \tag{1046}$$

where $\mathbf{k}$ and $\mathbf{x}$ are expressed in terms of the proper coordinates

$$d\mathbf{l} = a \cdot d\mathbf{x} \ , \tag{1047}$$

where a is the cosmological scale factor, so that

$$\frac{d\mathbf{x}}{dt} = \frac{\mathbf{v}}{a} \ . \tag{1048}$$

Moreover, the momentum of free (trial) particle decreases inversely proportional to the scale factor a :

$$\mathbf{p} = \mathbf{p}_0 \frac{a_0}{a} \ . \tag{1049}$$

Therefore the integral J finally has the form

$$J = \text{const} \cdot \int k_\nu \frac{c^2 p_0 a_0}{a^2 \left(p_0 c^2 \dfrac{a_0^2}{a^2} + m_\nu^2 c^4 \right)^{1/2}} dt \ . \tag{1050}$$

At the relativistic stage the integral can be reduced to

$$J = \text{const} \cdot \int k_\nu \frac{c}{a} dt \ , \tag{1051}$$

and it converges at

$$t \to 0$$

because

$$a = a(t) \propto t^{1/2} \ .$$

At the nonrelativistic stage one obtains

$$J = \text{const} \cdot \int k_\nu \frac{p_0 a_0}{m_\nu a^2} dt \ . \tag{1052}$$

Since

$$a = a(t) \propto t^{2/3}$$

at this stage, the integral converges at

$$t \to \infty \, .$$

Both estimations show that the integral should reach the maximum just in the period of transition from one stage to the other.

The order of magnitude estimation of the maximal value of J gives

$$J = k_\nu \frac{ct_1}{a(t_1)} \frac{p_0}{\langle p_0 \rangle} = \frac{k_\nu}{k_{\nu\,\max}} \frac{p_0}{\langle p_0 \rangle} \, , \tag{1053}$$

where $k_{\nu\max}$ is the maximal wave vector corresponding to the minimal wavelength R_ν, introduced above, p_0 is the momentum of a given particle at some early ultrarelativistic stage $t_0 << t_1$, $<p_0>$ is the neutrino momentum averaged over the spectrum at the same stage, the ratio $p_0/<p_0>$ being constant with time. We have already used $<p_0>$ in calculations of t_1 and k_ν.

The law of dissipation of perturbations depends significantly on the form of the function $\delta n_0(p)$.

If the perturbations in the nonrelativistic gas have the Maxwellian form

$$\delta n \propto \exp\left(-\frac{v^2}{\theta}\right) , \tag{1054}$$

one obtains for the fixed time interval

$$\int \exp\left(-\frac{v^2}{\theta} - ikvt\right) dv \approx \exp\left(-\frac{k^2 t^2 \theta}{4}\right) . \tag{1055}$$

But, if only particles at rest are perturbed, so that

$$\delta n(\mathbf{p}) = \mathrm{const} \cdot \delta(\mathbf{p}) \; , \tag{1056}$$

such perturbaton does not dissipate at all.

Using the same arguments in the case of invisible axion, one easily finds the explanation for the cold dark matter spectrum of primordial axions. Being the coherent oscillation of the axion field, the axion density fluctuations correspond to the particle distribution of the form

$$\delta n \propto \delta(\mathbf{v}) f(\mathbf{x}) \; , \tag{1057}$$

for which the short wavelength dissipation is absent.

In order to deduce the law of dissipation, one should clear up the nature of initial perturbation, generated on the "acausal" stage, when the wavelength was larger, than the size of horizon. This question is related to the mechanism of generaton of density fluctuations, being in the case of inflational models related to the parameters of scalar fields (see Chapter 3).

Remind that in the case of inflatonal models adiabatic density perturbations are generated due to the small difference of expansion rates in different spatial points, leading to the difference of temperature in these points and to the perturbations of the form

$$\delta n \propto \frac{\partial n}{\partial T} \exp(i\mathbf{kx}) \; . \tag{1058}$$

To prove this statement let us consider, following Doroshkevich et al. (1980a), the one dimensional calculation of perturbation in relativistic gas, induced by gravity.

Consider the initial distribution of phase density

$$n = n(p) = n\left(\frac{cp}{T}\right) . \tag{1059}$$

The effect of metric perturbation can be considered as an effect of the gravitational field. The change in the momentum of relativistic particle is in the given metric or in the given field proportional to momentum. Therefore

$$\frac{\partial n}{\partial t} = f(x) p \frac{\partial n\left(\frac{cp}{T}\right)}{\partial p} , \tag{1060}$$

where

$$f(x) \propto \exp(ikx) . \tag{1061}$$

The following relation is of importance

$$p \frac{\partial n\left(\frac{cp}{T}\right)}{\partial p} = -T \frac{\partial n\left(\frac{cp}{T}\right)}{\partial T} . \tag{1062}$$

If

$$\delta n \propto \frac{\partial n}{\partial T} ,$$

one has the desired formula

$$n = n_0 + \delta n = n\big(p, T + \delta T \cdot \exp(i\mathbf{kx})\big) . \tag{1063}$$

The same is approximately valid in the case of three spatial dimensions.

To estimate the effect of dissipation, consider

$$\delta\rho = \exp(ikx)\int \frac{d\left(\exp\left(\frac{cp}{T}\right)+1\right)^{-1}}{dT}\exp\left(\frac{i\mathbf{kp}R_\nu}{\langle p\rangle}\right)d^3p\,. \quad (1064)$$

One has the expansion

$$\left(\exp\left(\frac{cp}{T}\right)+1\right)^{-1} = \exp\left(-\frac{cp}{T}\right)-\exp\left(-2\frac{cp}{T}\right)+\ldots\,, \quad (1065)$$

where the first term dominates.

Integrating (1064) with the assumption

$$R_\nu k >> 1 \quad (1066)$$

we obtain

$$\left(\frac{\delta\rho}{\rho}\right)_\nu\Bigg|_{t>>t_\nu} = \left(\frac{\delta\rho}{\rho}\right)_\nu\Bigg|_{\frac{kct}{a}=1}\cdot\left(\frac{k_{\max}}{k}\right)^4\,. \quad (1067)$$

To find the prediction for the short wavelength amplitude at large

$$t >> t_\nu \quad (1068)$$

one should take into account that the shorter is the wavelength, the later starts to grow the fluctuation on the given wavelength on the nonrelativistic stage at t_1,

$$t > t_1 > t_\nu \quad (1069)$$

and the smaller is the growth of this fluctuation to the time t. It gives an additional suppression factor

$$(kR_\nu)^{-2} = \left(\frac{\lambda}{\lambda_J}\right)^2 \tag{1070}$$

in the amplitude of the fluctuation.

On the other hand, only fluctuations with the wavelength

$$\lambda > ct \tag{1071}$$

increase at the relativistic stage $t < t_\nu$ giving another suppression factor

$$\left(\frac{\delta\rho}{\rho}\right)_\nu \propto (kR_\nu)^{-2} \ . \tag{1072}$$

Finally, the approximate law of suppression of short wavelength fluctuations for the wave numbers $R_\nu k >> 1$ is given by

$$\left(\frac{\delta\rho}{\rho}\right)_\nu = \frac{\delta\rho}{\rho}(R_\nu)\cdot(kR_\nu)^{-8} \tag{1073}$$

so that the approximate formula for the distorsion of the initial amplitude of density fluctuation due to dissipation can be interpolated as

$$\left(\frac{\delta\rho}{\rho}\right)_f = \left(\frac{\delta\rho}{\rho}\right)_{in} \cdot \left(1 + k^2 R_\nu^2\right)^{-4} \ . \tag{1074}$$

The evolution of density fluctuations changes its character drastically in the period, when neutrinos become nonrelativistic.

It takes place at

$$t \sim t_\nu \tag{1075}$$

at the temperature

$$T \sim T_\nu \sim 10^5 \text{ K} , \tag{1076}$$

if the neutrino mass is $m_\nu \sim 30$ eV.

Note, that the evolution of the Universe as a whole is determined only by the total neutrino density, so that the variant 1) when the mass of neutrino of one type is equal to, say,

$$m_\nu \sim 30 \text{ eV} \tag{1077}$$

and the masses of the other types of neutrinos are small, and the variant 2) when the masses of all the types of neutrinos are equal

$$m_1 \sim m_2 \sim m_3 \sim 10 \text{ eV} , \tag{1078}$$

are equivalent.

The both variants are definitely different from the viewpoint of the evolution of density perturbaions. Here the heaviest neutrino plays the dominant role. It becomes nonrelativistic earlier, than neutrinos of other types, thus causing the change in the evolution of fluctuations. The contribution of nonrelativistic particles into the cosmological density soon starts to dominate, since the density of these particles decreases in the course of cosmological expansion more slowly, than the density of relativistic particles.

The evolution of nonrelativistic particles is described well by the hydrodynamical model with the equation of state $p = 0$.

The change of the equation of state leads, on the one hand, to the change of the cosmological expansion rate to

$$a(t) \propto t^{2/3}$$

and, on the other hand, to the change in the law of evolution of density perturbations.

2.3. *Structure Formation in the Neutrino-Dominated Universe*

In the period before $t \sim t_\nu$ the evolution of perturbations in neutrino, photon and baryon components was similar, so that to the moment (1075) the amplitude of these perturbations was the same in all components.

However, after $t > t_\nu$ the evolution of perturbations in these three components becomes different.

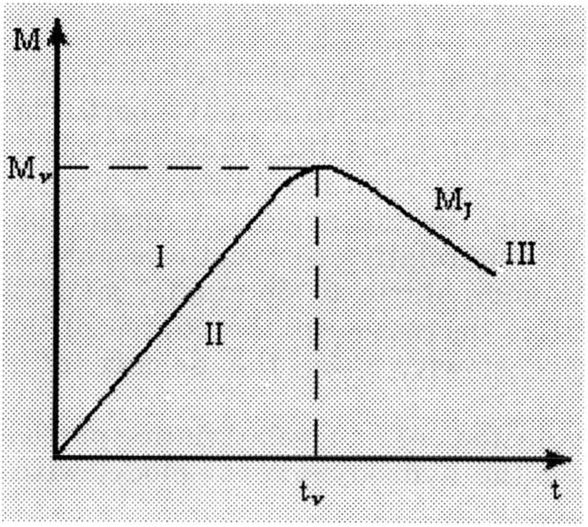

Fig. 21. Evolution of perturbations of non-relativistic neutrino.

At large scales, neutrino perturbations continue to grow, at small scales the perturbations continue to dissipate, but the boundary between the two regimes (Fig.21) moves to smaller scales. All perturbations at scales exceeding the Jeans scale at $t \sim t_\nu$ continue to grow

$$\frac{\delta\rho}{\rho} \propto (1+z)^{-1} . \tag{1079}$$

The Jeans scale, corresponding to the period $t \sim t_\nu$ plays an important role in the problems of the large-scale structure formation, determining many aspects of the predicted features of the Universe (Szalay, Marx, 1976).

This scale can be expressed in terms of the mass of the perturbed region, defined by the mass of neutrino and the Planck mass m_{Pl}, being the dimensional combination of fundamental constants (Markov, 1964; Szalay, Marx, 1976; Bisnovatyi-Kogan, Novikov, 1980)

$$M = \rho_{tot} R_\nu^3 = m_{Pl}\left(\frac{m_{Pl}}{m_\nu}\right)^2 . \tag{1080}$$

More precisely, the corresponding length scale is given by the value R_ν

$$R_\nu = 12.5(1+z)^{-1}\left(\frac{30\ \text{eV}}{m_\nu}\right) \text{Mpc} \tag{1081}$$

(Doroshkevich, Khlopov, 1981).

At $t > t_\nu$ perturbations in the photon and baryon components grow together with neutrino perturbations (region III on Fig.21) at the scales, exceeding the temporary Jeans scale

$$R_J = \frac{2ct}{\sqrt{3}} \tag{1082}$$

i.e. at

$$M > M_J \, . \tag{1083}$$

At smaller scales, the radiation pressure prevents the growth of perturbatons and, as before, perturbations in photon and baryon components convert into sound waves with the constant amplitude (Bisnovatyi-Kogan et al., 1980; Doroshkevich, Khlopov, 1981).

Only after the recombination of hydrogen at

$$T < T_{rec} \cong 4.5 \cdot 10^3 \text{ K} \tag{1084}$$

baryons cease to interact with the radiation, and the perturbations in the baryon component grow at all scales, exceeding

$$M_b \approx 10^5 M_\odot \, . \tag{1085}$$

The magnitude M_b corresponds to the Jeans mass calculated for the proper pressure of baryons.

However, the density perturbations in baryon and photon components have to the moment of recombination $t = t_{rec}$ at scales

$$R_\nu \le R \le R_J(t_{rec}) \tag{1086}$$

the amplitude much smaller, than the amplitude of neutrino perturbations at same scales.

At larger scales

$$R > R_J(t_{rec}) \tag{1087}$$

perturbations in all three components have grown simultaneously and remain equal.

The difference in the amplitudes of neutrino and baryon-and-photon perturbations provides the smallness of fluctuations of the temperature of the relic radiation in the neutrino-dominated Universe.

Released from the influence of radiation pressure, the amplitude of baryon perturbations reaches rapidly the amplitude of neutrino perturbations, so that the unique growing mode of density fluctuations is formed (see Fig.22).

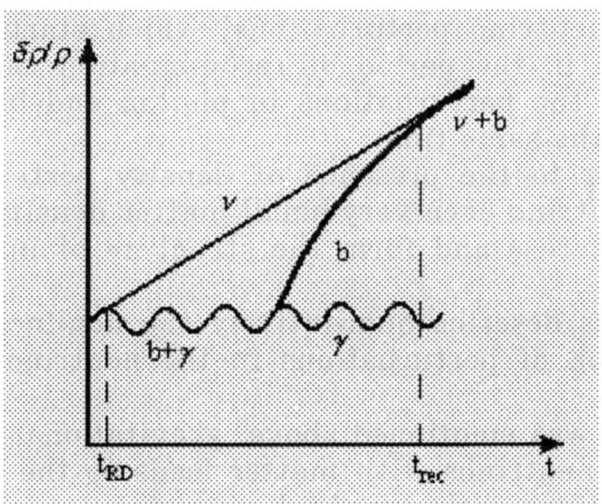

Fig.22. The growth of baryon density perturbations in the neutrino-dominated Unverse.

The amplitudes of perturbations of baryons and neutrinos do not coincide for some time after recombination. But, as it was shown by Doroshkevich et al. (1980a), the differences are virtually washed out in few cosmological times after recombination.

After recombination of hydrogen, the radiation ceases to interact with the matter and practically freely penetrate the neutral gas, so that the small-scale fluctuations of the temperature of the thermal background radiation gives the snapshot of density and velocity perturbations in photon and baryon components in the period of recombinaton (Silk, 1968; Sunyaev, Zeldovich, 1970; Doroshkevich, Khlopov, 1981). Thus the smallness of the amplitude of fluctuations of the relic radiation demonstrates the small amplitude of perturbations in baryon and photon components.

In the cosmological models without nonbaryonic dark matter (e.g. with massless neutrinos), the same perturbations determine the formation of the large-scale structure of the Universe. However, in the neutrino-dominated Universe, the formation of this structure is related to neutrino perturbations, which started to grow earlier, and in the period of recombination can be much larger than the perturbations in photon and baryon components at the same time.

Owing to the presence of density perturbations of massive neutrino even very small matter density perturbations can increae and form the structure of inhomogeneities.

Let us describe qualitatively the nonlinear stage of the evolution of perturbations in the neutrino-dominated Universe.

As it was mentioned above, gravitational interaction of massive neutrinos with matter leads to the unification of perturbations with very high precision.

According to the nonlinear theory of gravitational instability, the evolution of these perturbations leads to formation of "pancakes", i.e. of the flattened regions of high density with the size, determined by the dissipation scale of neutrino density perturbations. The phase correlation of neutrino and matter perturbations leads, in turn, to simultaneous contraction of the baryonic matter. But the further fate of these two components is different.

The gas component is stopped to contract and is heated in the shock wave, surrounding the contracted matter. It forms the usual gas "pancake", studied by

Sunyaev, Zeldovich (1972), Doroshkevich and Shandarin (1973), Doroshkevich et al. (1980b).

Neutrino form the collisionless "pancake" (Doroshkevich et al., 1980b,c) with the width, being 2-3 times larger than the width of the gas "pancake". Therefore the ratio of mean densities of contracted matter and neutrinos is by 2-3 times larger, than in the average over the space.

The calculations of photoionization of intergalactic gas by the radiation of "pancakes" (Doroshkevich et al., 1978; Doroshkevich, Khlopov, 1981) have shown, that in the absence of nonbaryonic dark matter the neutral hydrogen (HI) number density can not be lower, than

$$n_{\mathrm{HI}} \sim 3 \cdot 10^{-11} \ \mathrm{cm}^{-3} \tag{1088}$$

at $\Omega = 0.1$, whereas the studies of spectra of distant quasars (Gunn, Peterson, 1965) evidence, that

$$n_{\mathrm{HI}} < 10^{-11} \ \mathrm{cm}^{-3} \tag{1089}$$

or even

$$n_{\mathrm{HI}} < 10^{-12} \ \mathrm{cm}^{-3} \ . \tag{1090}$$

Estimations, made in the model of the neutrino-dominated Universe by Doroshkevich et al. (1980b) and Doroshkevich, Khlopov (1981), show, that significant increase of the amount of the thermal energy in the contracted gas and of the luminosity of this gas in the hydrogen ionizing range provide the explanation of low concentration of neutral hydrogen between the clusters of galaxies.

The analysis of cosmological impact of neutrino rest mass have found better agreement between the theory of large-scale structure formation and observations.

Based on this theory, the relation between the predicted parameters of the structure and anysotropy of the thermal background, on one hand, and the mass of neutrino, on the other hand, was established in a transparent way.

The successive development of the theory of large-scale structure formation, based on the theory of singularities of Lagrange transformations (Arnold et al., 1982) the so-called "theory of catastrophes" (see review Arnold, 1982), on the methods of statistical analysis of the observed properties of the structure correlation analysis (Zeldovich et al., 1982), on the cluster-analysis (Doroshkevich et al., 1983b; Fry, 1983), and on the methods of the percolation theory (Shandarin, 1983), provided detailed quantitative grounds for the proper choice of the candidate for dark matter particles.

The development of the models of the neutrino-dominated Universe has uncovered the need for non-baryonic dark matter. The successive development of the theory of large scale structure formation have found that massive stable neutrinos with the mass $m_\nu \sim 30$ eV do not satisfy the set of requirements for the proper candidate for dark matter particles.

So the structure of the Universe began to play the role of specific detector of the dark matter particles.

Let us consider this particular aspect of the large scale structure formation theory, i.e. the use of the set of data on the structure formation for determination of possible properties of the dark matter particles.

CHAPTER 9

PROBES FOR THE DARK MATTER PARTICLES

1. The Structure of the Universe as the Detector for the Dark Matter Particles.

1.1. The Structure Formation Constraints on Unstable Particles

To present the connection between the observational data and the parameters of dark matter particles in a straightforward way, let us turn to the description of gravitational instability, used in Chapter 4, where primordial black hole formation was considered.

This description is useful in order to clearly demonstrate the dependence of the parameters of cosmological structure and the anisotropy of the temperature of thermal background on the parameters of dark matter particles.

To illustrate this dependence, remind the approach of Chapter 4 to the problem of metric perturbations in the very early Universe.

The flat spectrum of metric perturbations corresponds to the constant amplitude at scales, exceeding the cosmological horizon. The density fluctuations at these scales grow as

$$\frac{\delta\rho}{\rho} \propto t \tag{1091}$$

The initial spectrum of density perturbations was taken in Chapter 8 for some period in the very early Universe (e.g., after the end of inflation) and its time evolution was determined by the transition function.

Here, for the sake of simplicity, we shall use the constancy of metric perturbations behind the horizon and assume that the spectrum of metric perturbations at the scales under consideration is flat, being determined by a small dimensionless quantity

$$\delta \ll 1 \,. \tag{1092}$$

The flatness of the spectrum means, that the value $\delta\rho/\rho$ at the scale of the cosmological horizon coincides with δ at any moment of time.

Such description does not account for a number of factors, induced by the change (or even changes) of the equation of state.

However, these numerical factors are insignificant for the purpose of illustration of main features of relations between the parameters of dark matter particles and the observational data.

The evolution of perturbations at scales, smaller than the horizon, is determined by the properties of considered particles and by the law of cosmological expansion.

Let us restrict ourselves by the class of weakly interacting particles.

In the context of the problems of gravitational instability, the primordial black holes are also referred to this class.

The primordial magnetic monopoles (in the absence of primordial magnetic fields) also can be referred to this class. Owing to their large mass and low density of plasma on the RD stage after electron-positron annihilation, the drag force in plasma is negligible. Putting aside magnetic charge fluctuations, monopoles can be considered in this case as weakly interacting particles.

So, consider the evolution of density fluctuations of particles with the mass m, the frozen relative concentration $r = n_m/n_\gamma$, and the lifetime τ .

Consider first sufficiently long-living particles with

$$\tau > t_U \ , \tag{1093}$$

where t_U is the age of the Universe.

Assume in the accordance with the simplest inflational models (see Chapter 3) that $\Omega = 1$ and that the considered particles dominate in the modern cosmological density. This condition fixes the quantity

$$(rm)_0 = 6.6 \text{ eV}\left(\frac{H_0}{50\,\frac{\text{km}}{\text{s}\cdot\text{Mpc}}}\right)^2\left(\frac{2.7\text{ K}}{T_0}\right)^3 , \tag{1094}$$

where H_0 and T_0 are the modern Hubble constant and the temperature of the thermal background radiation, respectively.

From the condition, that the total and critical densities are equal, one finds that any form of stable particles should satisfy the condition

$$rm < (rm)_0 \ . \tag{1095}$$

This condition may be weaker by a factor 2–3 in the case when we use the upper limit on the total density, following from the lower estimations of the age of the Universe.

In any case, the particles, for which

$$rm \gg (rm)_0 \tag{1096}$$

is predicted, should be unstable and their lifetime should be less, than the age of the Universe,

$$\tau < t_U \ . \tag{1097}$$

At

$$t < \tau \tag{1098}$$

and

$$\tau > \frac{m_{Pl}}{(rm)^2} \tag{1099}$$

such particles dominate in the Universe.

After the decay unstable of particles, relativistic products of their decay dominate, experiencing the red shift, so that at $t = t_U$ the effective value of rm , being the measure of the contribution of unstable particles into the total density, does not exceed

$$(rm)_{eff} = rm\left(\frac{\tau}{t_U}\right)^{1/2} . \tag{1100}$$

Thus, the upper limits on the total cosmological density do not exclude the existence unstable particles with $rm \gg (rm)_0$ and the lifetime

$$\tau < t_U\left(\frac{(rm)_0}{rm}\right)^2 . \tag{1101}$$

Let us consider now the refinement of these limits from the analysis of the influence of stable and unstable weakly interacting particles on the evolution of density perturbations.

At the temperature $T >> m$ the particles of the mass m are relativistic and their density perturbations within the cosmological horizon dissipate due to free streaming.

In the period, when $T \sim m$, particles turn out to be non-relativistic. The size of horizon in this period is

$$R_m \sim \frac{m_{Pl}}{T^2} \sim \frac{m_{Pl}}{m^2} \ . \tag{1102}$$

In the modern Universe this scale is equal to

$$R_m^{\mathrm{mod}} = R_m(t = t_U) = R_m \cdot (1 + z_m) = R_m \cdot \frac{m}{T_0} = \frac{m_{Pl}}{mT_0} \ , \tag{1103}$$

where T_0 is the modern temperature of the thermal electromagnetic background and z_m is the red shift, corresponding to the period, when $T \sim m$.

The scale R_m defines the minimal scale of inhomogeneities, formed by the considered particles.

Owing to weak (logarithmic or small power) growth of inhomogeneities of non-relativistic weakly interacting particles in the period, when the temperature falls from the value $T \sim m$ down to the value

$$T \sim rm << m \tag{1104}$$

the density fluctuations at this scale grow a little, so that it is just this scale that determines the scale of the structure of inhomogeneities, formed by the weakly interacting particles with the mass m.

So, as it was established in Chapter 8 for neutrinos, there is an unambiguous relation between the mass of weakly interacting particles and the scale of the cosmological structure, formed by these particles. One has

$$R_m \propto \frac{1}{m} \, . \tag{1105}$$

Based on the A-theory, the scale of the observed large scale structure of the Universe can be related to the parameters of the spectrum of initial perturbations, and the scale of perturbations can be found (Shandarin et al., 1983).

The correlation function of relative density perturbations can be taken in the form

$$\xi(r_{12}) = \frac{\left\langle \left(\rho(\mathbf{r}_1) - \langle\rho\rangle\right) \cdot \left(\rho(\mathbf{r}_2) - \langle\rho\rangle\right) \right\rangle}{\langle\rho\rangle^2} = \sigma^2 \left(1 - \frac{r_{12}^2}{2r_c^2} + \ldots\right) , \tag{1106}$$

where $<\rho>$ is the mean cosmological density, σ is the dispersion of perturbations and r_c is the correlation radius. From the observational data the estimation for the correlation radius (Doroshkevich, 1983a,b) gives

$$r_c \sim 10 \div 30 \text{ Mpc} \, . \tag{1107}$$

The equality of the values r_c and R_m^{mod}

$$R_m^{\text{mod}}(m) = r_c \tag{1108}$$

gives the interval

$$m = 40 \div 120 \text{ eV} \tag{1109}$$

for the possible value of the mass m (Doroshkevich, Khlopov, 1984)

We see, that the scale of the observed large-scale structure of the Universe permits to determine the mass of weakly interacting particles, forming the structure.

Note, that all particles with the fixed value of rm have also characteristic scale in the spectrum of their perturbations, corresponding to the horizon at the moment t_m , when the particles begin to dominate in Universe. This scale is

$$R'^{\mathrm{mod}} = \frac{m_{Pl}}{(rm)T_0} \quad . \tag{1110}$$

For any form of stable non-relativistic particles, dominating in the modern Universe at $\Omega = 1$ that scale corresponds to the allowed value of r_c.

As will be discussed later, the presence of this characteristic scale can be considered as the evidence, that weakly interacting particles of any kind can form something like the observed structure (Melott et al., 1983). This possibility can not be excluded on the basis of the existing observational data. Moreover, it is just this possibility, that is considered as the standard Cold Dark Matter model of the large-scale structure formation.

Our simplified approach makes it possible to relate the value of δ with the predicted anisotropy of relic radiation.

The value of δ determines the amplitude of density and velocity perturbations in plasma, inducing the fluctuations of the temperature of the relic radiation in the period of recombination, so that a relation is valid:

$$\frac{\delta T}{T} = \gamma\delta < \left(\frac{\delta T}{T}\right)^{obs}_{\max} , \tag{1111}$$

where the numerical factor $\gamma \sim 0.1 \div 1$ is defined by the details of the processes in which the temperature fluctuations are formed in the thermal radiation, as well as by the conditions under which these processes take place.

On the other hand, δ determines the time scale of the structure formation. The growth of perturbations, starts from the moment of time t_m and follows the law

$$\frac{\delta\rho}{\rho} \propto t^{2/3} \ .$$

The existence of the observed structure of the Universe corresponds to the growth of perturbations up to

$$\frac{\delta\rho}{\rho} \sim 1$$

and demands that

$$t_m \cdot \delta^{-3/2} < t_U \ . \qquad (1112)$$

Consider now the constraints on the possible parameters of hypothetical massive weakly interacting metastable particles with the lifetime

$$\tau < t_U \ , \qquad (1113)$$

following from the existence of the large scale structure of the Universe.

Assume, that the particles decay only on weakly interacting particles, so that the constraints found in the preceding paragraph of this chapter are not appropriate.

If the value of rm for these particles exceeds significantly the value $(rm)_0$, given by Eq. (1094), the existence of relativistic products of decay of the considered particles should lead to strong deceleration of the growth of perturbations in the period, determined by the values rm and τ.

Based on the fact that, on the one hand, the large scale structure is formed and, on the other hand, the thermal background is isotropic to the observed extent, one can impose constraints on rm and τ in order to exclude the effect of suppression of growth of perturbations, contradicting the combination of the both facts.

The two cases are possible.

In the first case, unstable particles decay before the non-relativistic particles with

$$rm = (rm)_0 \tag{1114}$$

start to dominate in the Universe, i.e. at

$$t_m > \tau \,. \tag{1115}$$

It means, that the considered unstable particles do not survive to the period of large scale structure formation and there should exist much more long-living particles, forming the structure.

Let us suppose for definiteness that the dark matter particles, forming the structure, are stable with $rm = (rm)_0$.

Then, if the rm-value of unstable particles exceeds

$$rm = (rm)_0 \cdot \left(\frac{t_m}{\tau}\right)^{1/2} \tag{1116}$$

the stable particles with $rm = (rm)_0$ start to dominate not in the period $t = t_m$ when their density exceeds the density of the thermal background radiation and primordial massless neutrinos, but later, when their energy density turns out to be larger, than the energy density of relativistic products of the decay of unstable particles, i.e. at the moment t_1, determined from the equation

$$(rm)_0 = rm \cdot \left(\frac{\tau}{t_1}\right)^{1/2} . \tag{1117}$$

One should change t_m in Eq. (1112) by t_1, given by

$$t_1 = \left(\frac{rm}{(rm)_0}\right)^2 \tau > t_m . \tag{1118}$$

The set of inequalities (1111)–(1112) leads to the constraint on the values rm and τ :

$$rm < (rm)_0 \left(\frac{t_U}{\tau}\right)^{1/2} \left[\frac{1}{\gamma}\left(\frac{\delta T}{T}\right)^{obs}_{\max}\right]^{3/4} . \tag{1119}$$

The constraint turns out to be by three orders of magnitude stronger, than the similar constraint, followed from the lower estimations of the age of the Universe.

Consider now the second case

$$\tau > t_m . \tag{1120}$$

In this case the period of the growth of perturbations is longer by the factor $rm/(rm)_0$, if $rm > (rm)_0$ .The growth of the time scale of the evolution of perturbations can be found from the following argument.

Relativistic products of decay of unstable particles should experience the red shift, making their energy density smaller by the factor $rm/(rm)_0$, in order to make their contribution into the total density compatible with the upper limits on the cosmological density and to provide the stage of dominance of stable particles with $rm = (rm)_0$, at which the effective growth of perturbations continues.

In this case the existence of the large-scale structure is crucial in obtaining the constraints on rm and τ .

Unstable particles have dominated in the Universe starting from the period

$$t'_m = \frac{m_{Pl}}{(rm)^2} < t_m = \frac{m_{Pl}}{(rm)_0^2} < \tau \tag{1121}$$

before their decay at $t < \tau$, thus providing the effective growth of perturbations at the scales, ranging from their minimal nondissipating scale

$$R_{\min}^{\text{mod}} = \frac{m_{Pl}}{mT_0} \tag{1122}$$

to the scales, corresponding to the horizon in the period, when the particles start to dominate in the Universe,

$$R_h^{\text{mod}} = \frac{m_{Pl}}{(rm)\cdot T_0} \; . \tag{1123}$$

Growth of perturbations in these scale can provide the galaxy formation, which might not be prevented by the intermediate RD stage of dominance of relativistic decay products.

However, the formation of the large-scale structure means, that the perturbations at the scale of the large-scale structure have grown up to

$$\frac{\delta\rho}{\rho} \sim 1 \, ,$$

and this growth can start within the cosmological horizon from the initial amplitude δ only after the perturbation "comes out of the horizon", i.e. after

$$t > \frac{r_c}{c \cdot \left(1 + z(t)\right)} \,, \tag{1124}$$

where r_c is the correlation radius of the structure, c is the speed of light and $z(t)$ is the red shift, corresponding to the period, when the scale of the perturbation turns out to be equal to the scale of the horizon.

The growth of these large-scale perturbations is suppressed on the intermediate RD stage of dominance in the Universe of relativistic products of unstable particles decays. For sufficiently long RD stage, it prevents the formation of the large-scale structure in view of the upper limits on the amplitude δ, following from the observed isotropy of the thermal electromagnetic background.

So one should restrict the possible duration of the intermediate RD stage, resulting in the following constraint on rm and τ :

$$rm < \frac{t_U}{t_m} \delta^{3/2} (rm)_0 = \frac{t_U}{t_m} (rm)_0 \left(\frac{1}{\gamma} \left(\frac{\delta T}{T} \right)^{obs}_{\max} \right)^{3/2} \,, \tag{1125}$$

$$t_m < \tau < t_U \,. \tag{1126}$$

Constraints (1115), (1119) and (1125)–(1126) are shown in Fig.23. These constraints do not specify the detailed properties of hypothetical particles. Only possible contribution into the cosmological density of the considered particles or their decay products is used.

In this sense these constraints are as universal, as the constraints on the weakly interacting particles following from ^{4}He abundance, age of the Universe or upper limits on PBH concentration in the Universe.

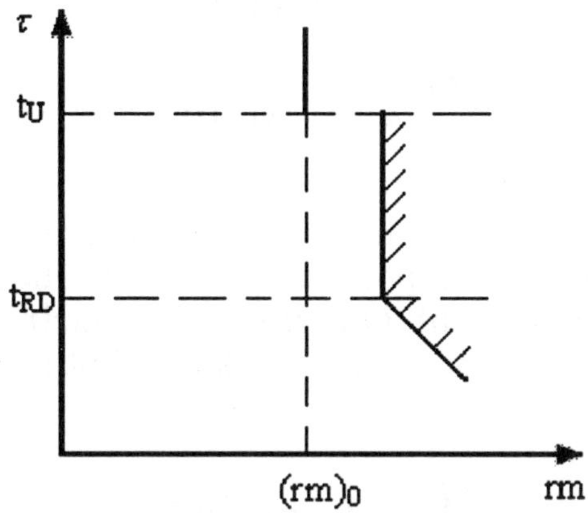

Fig.23. Constraints on unstable particles from the condition of large-scale structure formation.

1.2. The Troubles of Simple Dark Matter Models

The problems of the model of neutrino-dominated Universe were realised after the detailed quantitative analysis of predictions of the simple model of the dominance of stable neutrinos with the mass $m_\nu \sim 30$ eV . This model predicted clear cellular structure (Klypin, Shandarin, 1983; Frank et al., 1983; White et al., 1983; Centrella, Melott, 1983) well confirmed by the observations both qualitatively (see review by Oort, 1983) and by a number of quantitative parameters. They were:

i) estimation of the contrast of the structure (Kuhn, Uson, 1982);

ii) the results of the cluster-analysis and of the methods of the percolation theory (Zeldovich et al., 1982; Doroshkevich et al., 1983; Shandarin, 1983),

iii) comparison of the scales obtained in the theory with observations (Doroshkevich, 1983a,b) and, finally,

iv) some specific features of correlation relations (Fry, 1983).

The predicted values of the large-scale fluctuations of the temperature of relic radiation

$$\frac{\delta T}{T} \approx (2 \div 3) \cdot 10^{-5} \tag{1127}$$

were still compatible with the observational upper limits at small angles, corresponding to perturbations on the scale of superclusters of galaxies (Shandarin et al., 1983; Cecarelli et al., 1983; Starobinsky, 1983; Fixen et al., 1983)

$$\frac{\delta T}{T} < (4 \div 6) \cdot 10^{-5} \ . \tag{1128}$$

However, quantitative comparison of numerical models with the observations disclosed the following problems of the model of stable massive neutrinos (Klypin, Shandarin, 1983; Frank et al., 1983; White et al., 1983).

First, the evolved cellular structure at $\Omega = 1$ decays so rapidly, that it turns out to be impossible to make the formation of first "pancakes" (the observed distant quasars!) at the red shifts

$$z \sim 4 \div 5 \tag{1129}$$

compatible with the existence of the evolved cellular structure at $z \sim 0$.

Second, similar conclusions followed from the studies of the time evolution of correlation function, deduced from the results of numerical simulations.

Third, one could not explain on the basis of this model the observed parameters of galaxy clusters — their masses, central densities and the fraction of the matter, contained in them.

All these troubles are related to the too dense and too massive pancakes, predicted in the framework of neutrino-dominated Universe. Such pancakes decay rapidly and evolve in a system of giant clusters of galaxies.

The simplest solution of these problems is the transition to smaller cosmological densities, say, to

$$\Omega \sim 0.3 \ . \tag{1130}$$

But this solution implies smaller period of growth of perturbations. In order to form the structure, the greater initial amplitudes of density perturbations δ are needed, which is in contradiction with the observational upper limits on the fluctuations of the temperature of relic radiation.

Moreover, the solution with $\Omega \sim 0.3$ should correspond to very specific type of inflational models.

Melott et al. (1983) and Bond et al. (1982) have turned to dark matter models, assuming the dominance in the Universe of heavy weakly interacting particles with the mass

$$m >> 30 \text{ eV} \tag{1131}$$

or coherent axion field oscillations, having the same type of the spectrum of density fluctuations.

In such models spectrum of density perturbations is characterised by two scales.

The smaller scale corresponds to horizon in the period, when the particles became nonrelativistic, and the larger scale is related to the horizon in the much later period, when the considered particles began to dominate (Shandarin et al., 1983; Bond et al., 1982).

In that two-scale model the "pancakes" correspond to the smaller scale, but there is some correlation in their distribution at the distances, corresponding to the larger scale. The latter, as we have already mentioned , in the cosmological model with the critical density, i.e. with $\Omega = 1$, is rather close to the scale of the observed large-scale structure. Such models may provide the formation of galaxies (and quasars) before the large-scale structure is formed, removing some problems of neutrino-dominated Universe.

However, in the simplest versions of these Cold Dark Matter (CDM) scenarios it was impossible to reconstruct clear structure at $z \sim 0$.

The second trouble was that galaxies in voids were inevitably predicted.

And, finally, it was hardly possible to make compatible the observed and predicted dark matter densities in galaxies and galaxy clusters, if the "pancake" formation took place at $z \sim 4 \div 5$.

To remove these problems, more refined versions of CDM scenario were evolved, invoking the so-called "biasing" hypothesis. It assumed that galaxies are formed in the cases, when the amplitude of fluctuations exceeds the mean dispersion by a certain "biasing" factor. This hypothesis induces biasing in the mass and light distribution, but the physical grounds for it are not clear.

To conclude the discussion of the problems of simple dark matter scenarios, the problem of small-scale distribution of the dark matter should be considered.

The existence of inhomogeneities at the scales of galaxies is the evident advantage of the cold dark matter models. Massive dark halos of galaxies, proved by the observed galactic rotation curves, and finding recent confirmation in the effects of gravitational lensing, have natural explanation in the framework of CDM models.

On the other hand, the existence of such halos, especially in the case of dwarf galaxies, as well as the very origin of galaxies at large red shifts can not be easily explained in the Hot Dark Matter (HDM) models.

Indeed, the suppression of small-scale fluctuations, being so important for the large-scale structure formation in the HDM models, in the neutrino-dominated Universe, in particular, causes problems for formation of small-scale dark matter objects within neutrino "pancakes".

Moreover, there is effective suppression of pure baryon density fluctuations at the scales, smaller, than the supercluster scale in this model. We shall discuss below the mechanism of this suppression in view of its important application for the case of multicomponent dark matter.

So, assume, that there are some sources of baryon charge density fluctuations, so that entropy fluctuations exist at small scale.

Consider now the evolution of such fluctuations in the neutrino dominated Universe.

Linearised equations for the evolution of small baryon density fluctuations may be easily found (Doroshkevich et al, 1980a) in the Newtonian approximation.

Before recombination in the Universe, containing massive neutrino and the mixture of baryons, electrons and radiation the equations have the form

$$\ddot{\delta}_\nu + \frac{4}{3}\frac{\dot{\delta}_\nu}{t} - \frac{2}{3}\frac{\delta_\nu}{t^2} = 0 \tag{1132}$$

$$\ddot{\delta}_m + \frac{4}{3}\frac{\dot{\delta}_m}{t} + \frac{c^2 k_1^2}{3}\left(\frac{t_1}{t}\right)^{4/3} = \frac{2}{3}\frac{\delta_\nu}{t^2} \ , \tag{1133}$$

where

$$\delta_\nu \equiv \left(\frac{\delta\rho}{\rho}\right)_\nu \quad \text{and} \quad \delta_m \equiv \left(\frac{\delta\rho}{\rho}\right)_b \ ,$$

the dots mean the time derivatives, c is the speed of light and k_1 is the wave vector of the perturbations at the period t_1, when massive neutrino started to be non-relativistic.

The equations were deduced under the assumptions:

i) nonrelativistic neutrino dominate in the Universe in the period

$$z_r < z < z_1 \ , \tag{1134}$$

where the red shift $z_r \sim 1500$ corresponds to the period of recombination,

ii) the cosmological model is close to the model of the "flat" Universe, i.e. to the model with $\Omega = 1$,

iii) the scales, smaller than the horizon, were considered and the gravitational interaction of baryons and radiation was neglected.

It follows from Eqs. (1132)–(1133), that at $z_r < z < z_1$ neutrino perturbations grow as

$$\delta_\nu = A\left(\frac{t}{t_1}\right)^{2/3} \tag{1135}$$

and baryon density perturbations evolve as

$$\delta_m = A + t^{-1/3}\left(C_1 \sin\left(\sqrt{3}k_1\left(\frac{t}{t_1}\right)^{1/3}\right) + C_2 \cos\left(\sqrt{3}k_1\left(\frac{t}{t_1}\right)^{1/3}\right)\right). \tag{1136}$$

The constants C_1 and C_2 can be easily found from the initial condition at $t = t_1$.

It is seen, that the oscillating component dissipates slowly and due to the presence of the growing mode of neutrino the constant component

$$\delta_m = \frac{2A}{k_1^2 t_1^2} \tag{1137}$$

arises, growing with the wavelength and reaching δ_ν near the Jeans scale R_J.

After the recombination, the baryon matter and the radiation cease to interact with each other, and the third term in the left-hand side of Eq.(1133) can be neglected. Putting the initial conditions at the moment of recombination t_r

$$\delta_m(t_r) = 0 \tag{1138}$$

and

$$\dot{\delta}_m(t_r) = 0 \ , \tag{1139}$$

one finds, that for neutrino perturbations growing according Eq. (1135), the baryon density fluctuations evolve as

$$\delta_m = \delta_\nu \left[1 + 2\frac{t_r}{t} - 3\left(\frac{t_r}{t}\right)^{2/3} \right] . \tag{1140}$$

If

$$t = t_r(1 + \Delta) \ , \tag{1141}$$

then

$$\delta_m = \frac{8}{3}\delta_\nu \Delta^2 \ , \tag{1142}$$

and

$$\delta_m = \frac{1}{2}\delta_\nu \tag{1143}$$

at the red shifts

$$z \sim 10 \tag{1144}$$

only.

At

$$z > z_1 \tag{1145}$$

entropy fluctuations are the baryon charge perturbations on the homogeneous background of radiation and neutrino.

At

$$z < z_1 \tag{1146}$$

baryon density fluctuations induce neutrino density fluctuations at scales, smaller, than the horizon. The growth of neutrino perturbations is given by the equation

$$\ddot{\delta}_\nu + \frac{4}{3}\frac{\dot{\delta}_\nu}{t} - \frac{2}{3}\frac{\delta_\nu}{t^2} = \frac{2}{3}\frac{\Omega_m}{\Omega_\nu}\frac{\delta_m}{t^2} , \tag{1147}$$

which has the solution

$$\delta_\nu = \frac{\Omega_m}{\Omega_\nu}\delta_m\left(-1 + \frac{3}{5}\left(\frac{t}{t_1}\right)^{2/3} + \frac{2}{5}\frac{t_1}{t}\right) . \tag{1148}$$

For

$$\frac{\Omega_\nu}{\Omega_m} = 150 ,$$

the amplitudes of perturbations δ_m and δ_ν can grow only 8 times to the modern period, corresponding to $z = 0$.

Neutrino perturbations at the scales

$$R < R_\nu \tag{1149}$$

do not grow due to strong dissipation. After recombination, the growth of δ_m is completely determined by the value of δ_ν . For

$$\delta_\nu = 0 \tag{1150}$$

the equation

$$\ddot{\delta}_m + \frac{4}{3}\frac{\dot{\delta}_m}{t} - \frac{2}{3}\frac{\Omega_m}{\Omega_\nu}\frac{\delta_m}{t^2} = 0 \tag{1151}$$

has the solutions, corresponding to the decreasing mode

$$\delta_m \propto t^{-1/3} \tag{1152}$$

and to the weakly growing mode

$$\delta_m \propto t^n , \tag{1153}$$

where the index n is

$$n = \frac{1}{6}\left(\sqrt{1 + 24\frac{\Omega_m}{\Omega_\nu}} - 1\right) \approx 2\frac{\Omega_m}{\Omega_\nu} \approx \frac{1}{75} \tag{1154}$$

for

$$\frac{\Omega_m}{\Omega_\nu} = \frac{1}{150} .$$

1.3. Arguments for the Existence of Unstable Dark Matter Particles

All troubles of the model of stable neutrino dominance in the Universe are related in essence to the contradiction between the condition of the structure formation, implying

$$\Omega \approx 1 \,, \tag{1155}$$

and the condition of the conservation of the already formed structure, corresponding to smaller density of the matter within evolving structure

$$\Omega < 0.5 \,. \tag{1156}$$

One can interpret this contradiction as the evidence, that during relatively short period in the beginning of nonlinear evolution, the structure loses the most part of the dark matter, contained in it.

The formulation of the problem in this manner shows the definite way to solve it. It makes clear that the solution may be related to the instability of the most part of the dark matter, dominated in the Universe in the period of the structure formation.

In fact, the same contradiction may be formulated in the framework of simplest inflationary models as the contradiction between the predicted critical density $\Omega = 1$ and the indications, that the density within the observed inhomogeneities corresponds to $\Omega > 0.5$ (Gelmini et al., 1984; Steigman, Turner, 1984; Turner et al., 1984).

Hence, this problem is formulated as the problem of the two types of dark matter. There should exist homogeneously distributed dark matter, dominating in the total density, and the nondominating dark matter within the inhomogeneities.

In the framework of the theory of the large-scale structure, the existence of two modern forms of dark matter (homogeneous and inhomogeneous) reflects the existence of the two stages of dark matter evolution. At the linear stage of the growth of perturbations, the whole dark matter was distributed inhomogeneously, but in the beginning of the nonlinear evolution of inhomogeneities the most part of the dark matter should be outside them.

In the CDM scenario, all dark matter is in fact inhomogeneous, but its distribution differs from the distribution of the visible matter due to biasing. The dark matter associated with galaxies represents only small part of the total dark matter.

However, the most part of the modern dark matter may be really homogeneous and in this case (Doroshkevich, Khlopov, 1984a,b) both the solution of the problem of the two types of the modern dark matter, arising in the inflational models, and the solution for the problem of the structure formation and conservation may be found on the basis of models with unstable dark matter particles. Then the condition of the structure formation (see Sec.1.1) imposes lower limit on the lifetime of these particles and the condition of slowing down the nonlinear evolution of the structure imposes the upper limit on this lifetime.

One can use the theory of the large-scale structure formation to determine the desired properties of unstable particles that form the structure.

Indeed, the scale of the observed structure (its correlation length) determines the mass of the dark matter particles. Then the conditions of

i) structure formation;

ii) slowing down of the nonlinear evolution of the structure

fix the lifetime of unstable dark matter particles.

In the framework of the Unstable Dark Matter (UDM) scenario the problem of the confrontation of inflational models with the observations can be naturally solved. Here the decay products of unstable particles form the modern homogeneous dark matter, and stable dark matter particles (or simply baryons (Gelmini et al., 1984)), contributing as the nondominant fraction into the total density, represent the dark matter within the modern inhomogeneities.

Let us demonstrate now the principal possibility to determine from the set of astronomical data the type of unstable dark matter particles. This is possible on the base of old Big Bang model, if such primordial particles are fermions.

If the particles were in the equilibrium with the hot plasma in the early Universe, they decouple the earlier, the weaker is their interaction. The earlier particles decouple from the other particles in plasma, the more species of the other particles were in the equilibrium and the smaller is the relative concentration of the decoupled particles, r , (see Chapter 3).

If the primordial concentration of particles is not determined by their decoupling, but is the result of some non-equilibrium process, the frozen concentration of fermions can be only smaller than the equilibrium concentration of relativistic particles taken at the moment of their decoupling.

In the case of bosons, their generation in the coherent state is possible, as it takes place for invisible axion or inflaton.

In the case of fermions, the maximal value of rm at the fixed m corresponds to the one, given by the decoupling of relativistic particles from the equilibrium hot plasma in the early Universe.

This statement follows naturally from the fact, that the non-equilibrium process leads to the increase of the specific entropy and, consequently, to the increase of the thermal photon number density.

Moreover, since any non-equilibrium process, taking place after decoupling of the considered particles, leads also to the growth of the specific entropy (and the photon number density), the maximal frozen concentration of fermions is determined in the framework of the Big Bang model by the number of species of particles, being in the equilibrium in the period of decoupling (see Chapter 3).

On the other hand, in the models of unstable particles their value of rm may exceed $(rm)_0$, corresponding to $\Omega = 1$ for stable particles.

The greater is rm , the earlier particles start to dominate in the Universe and the longer is the period of growth of their small initial density perturbations. For larger rm the structure may be formed at smaller amplitude of fluctuations, δ , as compared with the case of the stable particles.

The relation

$$\frac{\delta T}{T} = \gamma\delta$$

puts into correspondence the amplitude δ and the amplitude of fluctuations of the temperature of the thermal electromagnetic background ($\delta T/T$). Using the observational data on the black body radiation anisotropy one obtains from this relation the estimation of the value of δ and, consequently, the lower limit on the value of rm.

Putting together this lower limit and the estimation of the mass m from the correlation length of the large scale structure, one obtains the lower limit on r, which may be compared with the maximal possible value of r , estimated for decoupling of particles in the thermal equilibrium. Thus, in principle, the value of r can be fixed, giving some indirect information on the coupling strength of the dark matter particles.

The condition of sufficient growth of large scale perturbations, deduced in Sec. 1.1, causes serious trouble for the models of unstable neutrino of the type of Gelmini et al. (1984) and Davis et al. (1981).

In the model of Gelmini et al. (1984) the mass of unstable neutrino with the relative concentration $r = 3/11$ was taken to be equal to

$$m_\nu = 550 \text{ eV} \qquad (1157)$$

giving the value

$$rm = 150 \text{ eV} . \tag{1158}$$

From the arguments of Sec 1.1. one immediately finds, that the formation of the large scale structure is virtually impossible in this model.

The simple arguments, given in this Section, may be modified with the account for more nontrivial cosmological scenarios and cosmological consequences of the particle theory.

The extension of simple inflational model to more realistic multicomponent scenarios, accounting for the set of Higgs fields, maintaining the hierarchy of symmetry breaking, can lead to rather peculiar spectrum of density perturbations (Linde, 1984b; Linde, Kofman, 1986), if the phase transition corresponding to the symmetry breaking takes place on the inflational stage. These spectra, defined by the parameters of scalar field potentials, strongly deviate from the "scale-free" spectrum.

Rather complicated "hidden sector" of particle theory arises in the case of mirror or shadow matter, having non weak internal interactions and superweak interaction with the ordinary particles. The existence of sufficiently strong internal interactions modifies the relationship between the dissipation scale of perturbations and the parameters of dark matter particles.

For example, in the case of mirror matter (see Chapter 10) the existence of mirror photons and mirror electromagnetism leads to the similar dissipation scale as in the ordinary baryon matter. In the same manner, as in the ordinary plasma, this scale is determined by the recombination of mirror electrons and nuclei into the mirror atoms and not by the masses of the mirror particles.

Invisible axions (see Chapter 2) give another example of nontrivial relation between the dissipation scale and the mass of dark matter particles. In spite of very small mass of axion the dissipation scale of axion density perturbations turns to be very small. The reason is that in the case of coherent axion field oscillations (see Chapter 3) this scale is determined by the product of the axion mass and the amplitude of axion field oscillations.

So the restrictions, given above, generally constrain some definite combination of the parameters of dark matter physics.

Note, that the data on the observed nonthermal electromagnetic background in different ranges (see Chapter 2 and e.g. Zeldovich, Novikov, 1975; Dolgov, Zeldovich, 1980) impose rather stringent restrictions on the possible modern energy density of electromagnetic radiation from the dark matter decay. One has the upper limit

$$\rho_\gamma < 10^{-9}\rho_{cr} \tag{1159}$$

in the X ray and gamma ray range,

$$\rho_\gamma < 10^{-6}\rho_{cr} \tag{1160}$$

in the infrared and optical range, and

$$\rho_\gamma < 10^{-11}\rho_{cr} \tag{1161}$$

in the radio wave range.
The constraint on the branching ratio of photon decays of dark matter particles follows from these data. Such a restriction was discussed in connection with the possible neutrino decay

$$\nu_H \to \nu_L + \gamma$$

in the modern Universe (Zeldovich, Khlopov, 1981; Petcov, 1977; Shabalin, 1980; Aliev, Vysotsky, 1981; Berezhiani et al., 1986).

Crude estimation of upper bounds on the possible branching ratio of photon decay modes can be obtained from the upper limits, given by Eqs. (1159)–(1161), by multiplying them by the factor, accounting for the decrease of relative contribution of radiation into the total density due to the dust-like expansion after the dark matter decay. These upper bounds provide additional information on the possible physical nature of dark matter particles and their interactions.

In the class of unstable weakly interacting particles, the problem of formation and evolution of the structure can find solution, implying the existence of weakly interacting particles with the mass

$$30 < m < 100 \text{ eV} \tag{1162}$$

and the lifetime

$$10^8 < \tau < 10^9 \text{ years} \tag{1163}$$

with the dominant decay into weakly interacting particles.

Among the known particles massive neutrino can be the candidates, provided that the new neutrino interaction exist, inducing the neutrino decays with the characteristics given by Eqs. (1162) and (1163). The simplest variant of the model of the structure formation by unstable dark matter — the cosmological model of unstable neutrino was elaborated in (Doroshkevich, Khlopov, 1984 a,b; Doroshkevich et al., 1985, 1987, 1988, 1989).

Unstable dark matter finds deep physical grounds in the gauge theory of broken symmetry of quark and lepton families, what we consider further in Chapter 11.

2. Dark Matter in the Galaxy

2.1. The Condensation of Dark Matter in Galaxies

At the stage of galaxy formation, massive neutrinos, neutralinos, axions and most of the other forms of dark matter particles represent collisionless gas. They can interact with matter only by gravitation. Therefore, no energy dissipation due to radiation takes place in the gas of collisionless dark matter particles, as it is in the case of the ordinary matter.

Nevertheless, as it was shown by Zeldovich et al. (1980), the motion of dark matter particles in the non-static gravitational field of ordinary matter, which contracts due to the energy dissipation via radiation, provides an effective mechanism of energy dissipation for neutrinos as well. Consequently the contracting ordinary matter induces the condensation of the dark matter that leads to significant increase in the dark matter density in the Galaxy as compared with its mean cosmological density.

Let us consider the mechanism, suggested in (Zeldovich et al., 1980) and developed by Fargion et al. (1995) in more details. For the sake of simplicity, we refer to the gas of collisionless dark matter particles as to the gas of massive neutrinos.

When ordinary matter ("baryons") contracts due to the energy dissipation via radiation, weakly interacting dark matter particles ("neutrinos") move in a potential, which varies with time.

Since the energy of a particle moving in a time-dependent potential is generally not conserved, neutrinos can reduce their energy and, consequently, increase their density. Let us treat the simplest case of particle motion along radial orbits in the central part of the contracting baryon system, where the density is independent of radius. If $\rho_\nu(t)$ and $\rho_b(t)$ are, respectively, the central densities of neutrinos and baryons, the motion of neutrinos is determined by the equation

$$\frac{d^2r}{dt^2} = -\omega(t)^2 r \,, \tag{1164}$$

where

$$\omega(t)^2 = \frac{4\pi G}{3}\left(\rho_\nu(t) + \rho_b(t)\right) . \tag{1165}$$

Suppose that the baryon density slowly increases. For slowly varying ω , the amplitude of oscillations is given by the adiabatic invariant

$$\frac{E}{\omega} = \omega R^2 = \text{const} , \tag{1166}$$

where E is the energy and R is the amplitude of the oscillations.

With the increase of ω , the oscillation amplitude of neutrinos decreases and their density increases respectively, according to the equation

$$\frac{\rho_\nu(t)}{\rho_\nu(0)} = \left(\frac{R(0)}{R(t)}\right)^3 = \left(\frac{\omega(t)}{\omega(0)}\right)^{3/2} = \left(\frac{\rho_\nu(t) + \rho_b(t)}{\rho_\nu(0) + \rho_b(0)}\right)^{3/4} . \tag{1167}$$

Due to radiation energy losses, the baryon density can grow so that the condition

$$\rho_b(t) >> \rho_\nu(t) \tag{1168}$$

finally holds. Then we obtain from Eq.(1167) that the density of neutrinos increases with time as

$$\frac{\rho_\nu(t)}{\rho_\nu(0)} = \left(\frac{\rho_b(t)}{\rho_\nu(0) + \rho_b(0)}\right)^{3/4} . \tag{1169}$$

The analytical treatment, given above, was completed in Zeldovich et al. (1980) by numerical models, proving the law of condensation given by Eq.(1169).

The considered process of condensation of a collisionless gas may take place in any collapsing system of ordinary matter, provided that at all stages contraction is dominantly supported by self–gravitation.

It is generally assumed that this condition is not satisfied at the initial stages of formation of objects smaller than globular clusters, at which the development of the thermal instability and the effects of the outer pressure of the hot gas are dominant (see Chapter 10).

Thus the considered mechanism should be effective in the course of galaxy formation but does not seem to work in the process of formation of globular clusters and smaller astronomical objects (stars, in particular).

If massive neutrinos (or any other form of collisionless dark matter particles) dominate in the cosmological density, such a mechanism provides an explanation for the formation of massive halos of galaxies by these particles. It was used e.g. in (Doroshkevich et al., 1980b) for the scenario of the neutrino-dominated Universe to explain why massive neutrinos remain at the periphery of galaxies and do not contribute much to the density in the central parts of galaxies.

The assumption that the hypothetical particles dominate in the galactic halo allows one, without any recourse to an explicit dynamical model of halo formation, to obtain estimation for the density distribution of particles. Using the fit for mass distribution in halo, one can fit the particle distribution in the Galaxy and evaluate possible effects of their weak annihilation, as it was done by Silk and Srednicky (1983) for supersymmetric particles.

However the universality of the mechanism of condensation (1169) permits its application to the case of a small contribution of the hypothetical particles to the total density, thus providing a reasonable estimation for the expected distribution of the particles in the Galaxy and their possible effects, even if they do not play any significant role in the dynamics of halo formation. It strongly increases the sensitivity of astrophysical data to the existence of such particles. That leads to stringent constraints on their allowed parameters.

2.2. WIMP annihilation in the Galaxy

There are many theories of cosmic rays origin, connecting it with the activity of galaxy nuclei, supernova explosions, processes in magnetospheres of pulsars and different acceleration mechanisms in galactic magnetic fields (Ginzburg, 1990; Hill et al., 1987).

Cosmoarcheology opens new possible mechanisms for cosmic ray and gamma background production, related to cosmological consequences of particle theory, leading to sources of energetic particles in the Universe.

It offers new insight into relationship between physics of cosmic rays and super high energy physics and cosmology, provides exotic sources of cosmic ray and gamma backgrounds and opens new aspects in experimental physics of cosmic rays and gamma astronomy. For instance, cosmic gamma ray background provides information on the existence of antimatter domains (see Stecker, 1985 and Refs. therein) or evaporating PBHs (Sedelnikov, Khlopov, 1996) in the Universe.

Extensive study of cosmoarheological chains, linking the data on cosmic rays and gamma backgrounds to possible tests of the models of inflation, baryosynthesis and dark matter, as well as of various cosmological consequences of GUT, SUSY particle models and of theoretical mechanisms of CP violation, mass of neutrino etc., is assumed by AstroDamus project (Khlopov et al., 1996).

One of the effects of the new physics in cosmic rays is related to the effect of weak annihilation of weakly interacting massive particles (WIMP) in the Galaxy. Following Zeldovich et al. (1980) and Fargion et al. (1985) let us consider this effect on the example of heavy stable neutrino.

The condensation of heavy neutrinos in the Galaxy leads to the increase in the rate of the neutrino annihilation resulting in a copious production of cosmic rays. The most stringent limit on the mass of heavy neutrinos can be obtained by considering the electronic component of cosmic rays, which is most sensitive to the new sources of cosmic rays.

The flux of relativistic electrons from neutrino annihilation in the Galaxy is given by

$$J = T_e \frac{dn_\nu}{dt} \frac{c\delta}{8\pi} \ \mathrm{cm}^{-2}\mathrm{s}^{-1}\mathrm{ster}^{-1} \ , \tag{1170}$$

where

$$\frac{dn_\nu}{dt} = n_\nu^2 \sigma_\nu v \tag{1171}$$

is the rate of neutrino annihilation in the Galaxy per unit volume, σ_ν is the cross section of the neutrino annihilation in the Galaxy,

$$T_e \approx 10^7 \ \text{years} \tag{1172}$$

is the life-time of cosmic rays in the Galaxy, c is the speed of light, δ is the number of relativistic electrons which are produced in one act of the neutrino annihilation.

In order to obtain the constraint on the heavy neutrino mass, we consider first the annihilation of heavy neutrinos into e^+e^- pairs in the Galaxy. Since heavy neutrinos in the Galaxy are non-relativistic, the ultrarelativistic electrons in the annihilation reaction

$$\nu\bar{\nu} \to e^- e^+ \tag{1173}$$

are produced practically monochromatic with the energy $E \cong m$ so that even for $m = 1$ TeV the energy spread is $\Delta E << 1$ GeV.

Electrons can also be produced in the secondary processes from the decays of μ-, τ-leptons and quarks, but these processes contribute only to the soft part of the cosmic ray spectrum and therefore we shall neglect them.

The annihilation cross section for the process (1173) in the Galaxy can be written as

$$\sigma_\nu v \approx 2.9 s M_W^2 D_Z 10^{-36}\ \text{cm}^2 \cdot c\ , \tag{1174}$$

where D_Z is the Z-boson propagator given by

$$D_Z s \propto \frac{s}{\left(s - M_Z^2\right)^2 + \Gamma^2 M_Z^2}\ , \tag{1175}$$

the relativistic invariant $s = (p_\nu + p_{\bar{\nu}})^2 \cong 4m^2$, and Γ is the Z-boson width.

Therefore for the flux of cosmic electrons we have

$$J_e \approx 10^{13} n^2 s M_W^2 D_Z\ \text{cm}^{-2}\text{s}^{-1}\text{ster}^{-1}\ . \tag{1176}$$

The experimental energy spectrum of cosmic electrons (Lang, 1974) integrated over the energy resolution ΔE of the detector is

$$J^{\text{exp}} \approx 1.16 \cdot 10^{-2} \left(\frac{E}{1\ \text{GeV}}\right)^{-2.6} \left(\frac{\Delta E}{1\ \text{GeV}}\right) \text{cm}^{-2}\text{s}^{-1}\text{ster}^{-1}\ , \tag{1177}$$

where $3 < E < 300$ GeV and $\Delta E << E$.

Comparing Eqs. (1176) and (1177) one finds that the existence of heavy neutrinos is forbidden in the mass range

$$60 < m < 115 \text{ GeV} . \qquad (1178)$$

The absence of the constraint in the mass range 44–60 GeV is due to the smallness of the relic neutrino concentration (the annihilation cross section of neutrinos in the early Universe is very large in the vicinity of the Z-boson peak). The constraint in the region of large neutrino masses is a consequence of the rapid decrease of the flux (1176)

$$J_e \propto m^{-8} \qquad (1179)$$

as compared with the experimental flux (1177)

$$J_{\mathrm{exp}} \propto m^{-2.6} .$$

2.3. Direct and Indirect Searches of WIMPs in the Galaxy

Let us combine the indirect information discussed above with the experimental data on the direct underground search for WIMP–nuclei elastic scattering in order to obtain bounds on the mass of stable neutrino of the fourth generation, following the recent results of Fargion et al. (1997) .

For the sake of definiteness, consider the standard electroweak model, including one additional family of fermions. The heavy neutrino ν and heavy charged lepton L form a standard $SU(2)_L$ doublet. In order to ensure the stability of the heavy neutrino, we assume that its mass does not exceed the mass of respective charged lepton M_L, $m < M_L$, and that the heavy neutrino is a Dirac particle.

It is known that modern laboratory experimental results do not contradict the existence of heavy Dirac neutrinos with mass

$$m > \frac{M_Z}{2} \,, \tag{1180}$$

where M_Z is the mass of the Z-boson.

In the early Universe at high temperatures, such heavy neutrinos should be in thermal equilibrium with other particle species. As the temperature in the Universe drops, heavy neutrinos become non-relativistic at $T \sim m$ and their abundance falls off rapidly according to exponential law. In the further expansion of the Universe, as the temperature decreases below the freeze-out value T_f, the weak interaction processes become too slow to keep neutrinos in equilibrium with other particles. As a consequence, the number density of heavy neutrinos is given by their frozen out concentration being equal now to

$$n = \frac{6 \cdot 10^3}{\sqrt{g_*}} \left(\frac{m_p}{m_{Pl}}\right)\left(\frac{m_p}{m}\right)\left(\frac{\rho_{cr}}{10^{-29} h^2 \, \frac{g}{\text{cm}^3}}\right)\left[\int_0^{x_f} dx m_p^2 (\sigma v)\right]^{-1} \text{cm}^{-3}, \tag{1181}$$

where

$$\rho_{cr} = 1.879 \cdot 10^{-29} h^2 \, \frac{\text{g}}{\text{cm}^3} \tag{1182}$$

is the critical density of the Universe; h is the dimensionless Hubble constant H normalised as

$$h = \frac{H}{100 \, \frac{\text{km}}{\text{s} \cdot \text{Mpc}}} \tag{1183}$$

m_{Pl} is the Planck mass; m_p is the proton mass;

$$x = \frac{T}{m}; \quad x_f = \frac{T_f}{m}; \tag{1184}$$

σ is the annihilation cross section; v is the relative velocity of the neutrino-antineutrino pair in its center-of-mass frame; g_* is the effective number of relativistic degrees of freedom at $T = T_f$. The contribution of bosons and fermions to g_* is equal 1 and 7/8, respectively. The freeze-out temperature T_f can be computed by iterations from

$$x_f^{-1} = \ln\left(\frac{0.0955 m_{Pl} \sigma m \sqrt{x_f}}{\sqrt{g_*}}\right) \tag{1185}$$

and in the considered case is

$$T_f \approx \frac{m}{30} \ . \tag{1186}$$

In general, the following processes could lead to the annihilation of heavy neutrinos in the Universe

$$\nu\bar{\nu} \to f\bar{f}, W^+W^-, ZZ, ZH, HH \ .$$

However (see Fargion et al., 1997 and Refs. wherein) the dominant processes are

$$\nu\bar{\nu} \to f\bar{f} \tag{1187}$$

below the threshold for W^+W^- production, and

$$\nu\bar{\nu} \to W^+W^- \tag{1188}$$

above this threshold.

Fig.24 shows the dependence of cosmological density $\rho = 2mn$ of heavy neutrinos (and antineutrinos) as a function of neutrino mass. In the region $m \sim m_Z/2$ the neutrino density is extremely small as a result of the huge value of the annihilation cross section at the Z-boson pole. When the neutrino mass increase, the cross section for neutrino annihilation drops and this leads to an increase of the neutrino density, which reaches its maximum value at $m \sim 100$ GeV.

At $m > m_W$ the annihilation channel into W^+W^- opens and gradually becomes the dominant one, and since its cross section grows like m^2, the present neutrino density drops again.

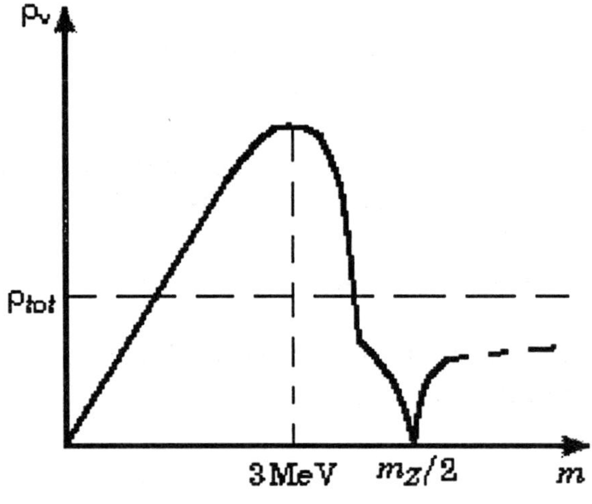

Fig.24. Dependence of the neutrino density in the Universe on the mass of neutrino

As is seen in Fig.24, neutrino density is small in comparison with critical density. However neutrino condensation, discussed in Sec 2.1, leads to significant increase in the neutrino density in the Galaxy as compared to mean cosmological density. One obtains for the central density of neutrino

$$n_{0G} \approx n\left(\frac{\rho_{0G}}{\rho_b}\right)^{3/4} , \tag{1189}$$

where

$$\rho_{0G} \approx 10^{-20}\ \frac{\text{g}}{\text{cm}^3} \tag{1190}$$

is the central density of the matter in the Galaxy and ρ_b is the density of baryon matter in the Universe. To evaluate neutrino density from below the upper limit on the baryon matter density should be taken.

It is often assumed that the density of dark matter halo in the Galaxy decreases with the distance from the center according to the law

$$\rho(r) = \frac{\rho_0}{1+\left(\frac{r}{a}\right)^2} , \tag{1191}$$

where

$$2 < a < 10\ \text{kpc} \tag{1192}$$

is the core radius of the halo.

Assume that neutrino density also follow the distribution (1191). Taking the minimal value for the core radius a = 2 kpc and substituting into Eq.(1191) the distance

$$r_{Sun} \approx 8.5\ \text{kpc} \tag{1193}$$

from the Sun to the centre of the Galaxy, we find that the density in the solar neighbourhood is reduced by the factor ≈ 19 in comparison with the central neutrino density given by Eq. (1189). Therefore, taking into account (1189)–(1193), we obtain that the density of heavy neutrinos in the solar neighbourhood have to satisfy the condition

$$n_{Sun} > 3.3 \cdot 10^6 n \ . \tag{1194}$$

Let us apply the results obtained above to the experimental data on the direct search for WIMPs by WIMP-nucleus elastic scattering.

Neutrinos scatter from nuclei by Z-boson exchange and therefore the axial coupling, which only produces small spin-dependent effects, can be neglected. In the nonrelativistic limit, the neutrino cross section due to coherent scattering on nuclei is given by

$$\sigma_{el} \approx \frac{m^2 M^2}{2\pi(m+M)^2} G_F^2 Y^2 K^2 \ , \tag{1195}$$

where

$$K = N - \left(1 - \sin^2 \theta_W\right) Z; \tag{1196}$$

N, Z are numbers of neutrons and protons, respectively; M is the mass of target nucleus, G_F is the Fermi constant; Y is the average hypercharge (Y~1). The nuclear degrees of freedom can be taking into account by a nuclear form factor.

One finds from the results of underground experiments with Ge detector (Caldwell, 1995) that the existence of very heavy neutrinos is forbidden in the mass region

$$60 < m < 290 \text{ GeV} \ . \tag{1197}$$

Note that the constraint (1197) represents the minimal region excluded by Ge underground experiments because this result has been obtained with the use of the conservative set of values for astrophysical and cosmological parameters. But even such conservative treatment leads to more restrictive limits on the mass of heavy neutrinos than Eq. (1178) obtained from the analysis of the spectrum of electrons in cosmic rays.

If Higgs meson exists, the bound (1197) does not exclude the possibility for heavy neutrinos to have mass in the region

$$\left|m_H - m\right| < \Gamma_H \ , \tag{1198}$$

where Γ_H is the width of Higgs meson, because in this case the s-channel annihilation

$$\nu\bar{\nu} \to H \to \text{anything} \tag{1199}$$

could reduce significantly the neutrino density in the Galaxy.

Recently preliminary results on the search of underground WIMPs, using the annual modulation signature with large mass NaI(Tl) detectors, were published (Bernabei et al., 1997). The overall analysis have shown that there is an indication on the single crystal response (however, as it was mentioned by Bernabei et al. (1997), only very large exposure would possibly allow to reach a firm conclusion).

Bernabei et al. (1997) note that the region of the signal is well embedded in estimations for neutralino signal based on the minimal supersymmetric standard model. However this case can correspond also to the elastic scattering by nucleus of NaI of relic neutrinos with the mass between 45 GeV and 50 GeV. This region is consistent with the present laboratory bound

$$m > 45 \text{ GeV} . \tag{1200}$$

The independent confirmation of this event and the test for its possible physical nature can be obtained in experiments with cosmic rays, by AMS spectrometer, which is in preparation for the Alfa space station.

The detection of monochromatic positrons with energy above 45 GeV would be a clear signature of the annihilation of Dirac neutrinos in the galactic halo because the direct annihilation of Majorana fermions into electron-positron pair in the Galaxy is severely suppressed. On the other hand, the detection of the effect in the continuum spectrum of positrons could be an indication on the annihilation of neutralinos.

The search for effects of cosmic heavy neutrinos may be supplemented by the search for such neutrinos at accelerators (Fargion et al., 1996). The reaction

$$e^+e^- \to \nu\bar{\nu}\gamma \tag{1201}$$

gives the possibility of analysing the mass region

$$m \sim \frac{m_Z}{2} , \tag{1202}$$

which is difficult for an astrophysical investigation because of small value of neutrino density but can be important in relation with an event possibly observed in DAMA experiment (Bernabei et al., 1997).

If heavy neutrinos exist, there could be also an interesting hadronless signature for Higgs meson Bremsstrahlung production at accelerators

$$e^+e^- \to ZH \to Z\nu\bar{\nu} \to l^+l^-\nu\bar{\nu} , \tag{1203}$$

and this mode could be the dominant one.

Thus only complex investigations including underground experiments, accelerator searches, and astrophysical investigations can clarify the physical nature of dark matter.

Note in conclusion, that the condensation in the Galaxy is also important for direct searches for cosmic light neutrinos and axions.

Light neutrinos with the mass $m \sim 30$ eV and the velocity dispersion in the Galaxy ~ 300 km/s have the wavelength of the order of

$$\lambda \sim \frac{\hbar}{mv} \sim 6 \cdot 10^{-4} \text{ cm} . \tag{1204}$$

Owing to weak vector current interaction, the galactic neutrinos can scatter coherently on the bodies with the size $a < \lambda$.

The momentum transfer in this scattering causes small acceleration, which may be in principle detectable (Scwartsman et al., 1982).

The existence of axion photon $\alpha\gamma\gamma$ interaction leads to the possibility of axion–photon conversion in radio–frequency cavity (Sikivie, 1983). This principle is used in axion halo– and helio–scopes, used for experimental search for axion fluxes in the galactic halo and from the Sun, respectively.

Shadow and mirror matter represent the form of dark matter candidates which are hardly accessible to direct experimental means. Astronomy turns to be the unique source of information on the existence and possible properties of this form of the dark matter, which we discuss in the next Chapter.

CHAPTER 10

MIRROR WORLD IN THE UNIVERSE

1. Mirror Particles

1.1. The Equivalence of Left and Right

Putting aside the "Through the Looking Glass" mirror world of Lewis Carroll (1871), the fundamental reasons for the existence of mirror partners for the known elementary particles were first put forward in the same famous Nobel-prize-won paper of T.D.Lee and C.N.Yang (1956) , in which they questioned the parity conservation in the weak interactions.

In the quantum field theory, the P-parity transformation, i.e. the inversion of the space coordinates

$$\mathrm{P}: \quad (t, \mathbf{x}) \rightarrow (t, -\mathbf{x}) \tag{1205}$$

results in the transformation of the field operator

$$\mathrm{P}: \quad \Psi(t, \mathbf{x}) \rightarrow \Psi^{P}(t, -\mathbf{x}) \; . \tag{1206}$$

Before the work of Lee and Yang (1956) it was assumed that parity is conserved in all fundamental interactions of elementary particles. It means, that the inversion of coordinate axes results in the transformation of the field of definite particle into the other field describing particles that also exist in the Nature.

For example, the left-handed state of a field with nonzero spin transform into the right-handed state of the same field, so that

$$\Psi_L^P(t,-\mathbf{x}) = \Psi_R(t,\mathbf{x}) \tag{1207}$$

and conversely, the right-handed state of the field transforms into the left-handed state.

The field operator, corresponding to a particle with a zero spin, can be invariant under P-transformation, so that as the result of P-inversion of spatial coordinates

$$\Psi^P(t,-\mathbf{x}) = +\Psi(t,\mathbf{x}) \ . \tag{1208}$$

Such field correspond to a scalar particle.

The field operator of a spinless particle can change its sign after P-transformation,

$$\Psi^P(t,-\mathbf{x}) = -\Psi(t,\mathbf{x}) \tag{1209}$$

and that corresponds to a pseudoscalar particle.

Parity conservation means that the Lagrangian is invariant under P-inversion, i.e. it is a scalar.

P-inversion may be viewed as the reflection in a mirror. Until 1956, it was assumed, that the mirror reflection of any fundamental particle process is either the same process, or some other process, also existing in the Nature.

Parity violation in the weak interactions put forward the processes, for which this fundamental rule is not valid.

So, neutrino, created in beta-decay has only one helicity. P-transformation gives the neutrino state, which is not produced in weak interactions.

In the beta-decay of polarised nuclei electrons are emitted preferentially along the direction of nuclear spin. Mirror reflection of this process results in the process with the opposite preferential direction of electron momentum relative to the polarisation of the nucleus, that is not observed in the Nature.

In the conclusive Section of their work Lee and Yang (1956) considered the evident theoretical problem, related to parity violation. P-violation leads to the non-equivalence of the right- and left-handed coordinate systems.

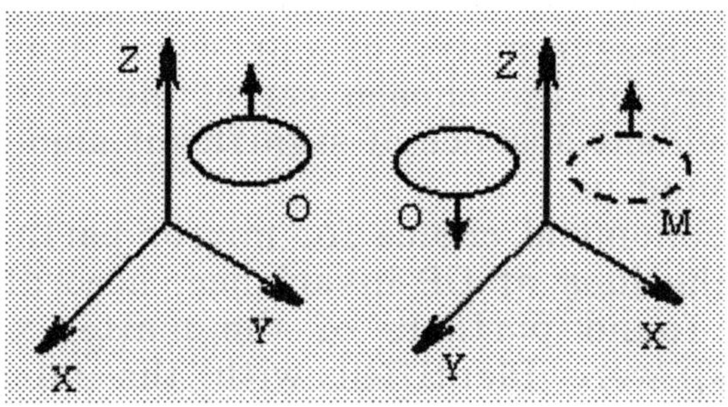

Fig.25. Mirror reflection of P-non-conserving process leads to the process, which does not exist in the Nature.

Indeed, P-transformation of the coordinate system, in which P-violating process is described, corresponds to the transition from the left-handed to the right-handed co-ordinate system, or to the mirror reflection of the process. As a result, due to P-nonconservation such a transformation leads to the process, which does not exist in the Nature (see Fig.25). So P-violating process fixes the preferential coordinate system.

On the other hand, the existence of preferential coordinate system means, that the empty space time has some preferential orientation. The only difference between the left- and right- handed coordinate systems is the direction of the rotation along the short arc from the axis x to the axis y around the fixed z axis. It seems doubtful, that God has some reasons to choose some preferential direction of such rotation for the empty space.

To restore the equivalence of left- and right- handed coordinate systems, Lee and Yang (1956) assumed, that mirror partners should exist for all known particles. In this case P-inversion should be accompanied by the mutual interchange of the ordinary (O) particles and their mirror (M) partners.

If P-violation in the mirror world has the opposite sign relative to the ordinary particles, combined reflection and mirror partner interchange does not cause any change in the description of the original process and its mirror image. Then P-asymmetry of our world turns to reflect the asymmetry of the chosen set of elementary particles due to the evident absence of significant amount of mirror particles around us.

The economical choice for mirror partners was soon offered by the idea of CP-invariance, according to which the already found partners of the matter particles, i.e. antiparticles, were identified with the mirror partners (Lee, 1957, Landau, 1957).

Antiparticles seem to exhibit the main property of the mirror particles — their strict symmetry to the respective ordinary partners. The absolute values of all their fundamental parameters (masses, charges, lifetimes, branching ratios) were assumed to be strictly equal to the analogous parameters of the corresponding particles.

Indeed, CP-transformation puts the observed left-handed state of neutrino into correspondence with the state of the right-handed antineutrino,

$$\mathrm{CP}: \ \nu_L \rightarrow \bar{\nu}_R \tag{1210}$$

and since right-handed antineutrinos are also produced in weak interactions the symmetry between left and right is restored.

In the same manner, the β^- decay of polarised nucleus is put by CP-parity transformation in correspondence to the β^+ decay of antinucleus. The theory of P-violating weak interactions predicts the opposite direction of the preferential emission of positrons as compared to the emission of electrons.

Hence the combination of mirror reflection with the interchange of particles and antiparticles seemed to support the equivalence of left- and right- handed coordinate systems.

However, the discovery of CP violation made the identification of the mirror partners with antiparticles erroneous, and the question on the true set of mirror particles and their expected properties was put forward again (Kobzarev et al., 1966).

It turned out, that in the case, when mirror particles are not identified with antiparticles, they can not share the same interactions, the ordinary particles participate.

Kobzarev et al. (1966) mentioned that the strict symmetry between ordinary and mirror electrons leads to serious problem in the atomic physics, if ordinary and mirror worlds have identical electromagnetic interactions. One expects doubling of the atomic states in this case owing to the additional degrees of freedom, related to mirror electron states.

Similar problem appears in the hadron physics, if the strong interaction of ordinary and mirror particles are identical. Due to mirror particles, the doubling of some hadron states, e.g. the states of neutral pions, should take place.

The discovery of intermediate W- and Z-bosons of weak interactions and the measurement of their widths excludes common weak interaction for ordinary and mirror particles. So not only ordinary matter particles but also gauge bosons of their interaction should have their mirror partners.

1.2. Fractons and Alice Strings

The simplest way to include mirror particles into the particle model is to add to the $SU(2)\mathrm{x}U(1)\mathrm{x}SU(3)_c$

gauge symmetry of the standard model the same symmetry ascribed to mirror particles.

One should assume that mirror particles correspond to trivial representation of the ordinary gauge group and, vice versa, the states of ordinary particles transform as scalars under the mirror gauge group transformations.

It means that particles of definite mirrority (i.e. ordinary or mirror) are sterile relative to the interactions of the other mirrority.

So, the mirror electron interacts with the mirror photon with the same strength as the ordinary electron interacts with the ordinary photon. However, the mirror electron does not interact with the ordinary photon, and the ordinary electron does not interact with the mirror photon.

One may expect that the states referred to nontrivial representations of the gauge groups of both mirrority should correspond to hypothetical particles with very peculiar properties (Khlopov, 1980).

Indeed, the state of X-lepton, transforming relative to the group $[SU(2)\mathrm{x}U(1)\mathrm{x}SU(3)_c]_O\mathrm{x}[SU(2)\mathrm{x}U(1)\mathrm{x}SU(3)_c]_M$ as

$$X\left(2,\ \frac{1}{3},\ 1;\ 1,\ 1,\ 3\right) \tag{1211}$$

shares the properties of ordinary lepton and mirror quark, and is a doublet relative to electroweak ordinary gauge symmetry and a triplet relative to mirror colour transformations.

Owing to mirror colour confinement, such states should be bound together with some other mirror coloured states into composite particles, giving states that look like anomalous leptons.

The bound states of X-leptons and mirror partners of ordinary quarks q_M such as baryon-like state (Xq_Mq_M)

behaves in ordinary electromagnetic processes as a colourless particle with fractional charge, called fracton.

Fractons also arise as bound states of X-quarks, which correspond to the representation of the gauge group $[SU(2)\mathrm{x}U(1)\mathrm{x}SU(3)_c]_O\mathrm{x}[SU(2)\mathrm{x}U(1)\mathrm{x}SU(3)_c]_M$

$$X\ \left(1,\ 1,\ 3;\ 2,\ \frac{1}{3},\ 1\right). \tag{1212}$$

X-quarks have fractional mirror electric charge and are neutral with respect to ordinary electromagnetic interaction.

X-quarks bound with the ordinary quarks in the baryon–like or meson–like hadrons form fractionally charged composite colourless particles such as (Xuu) baryon with the electric charge $+4/3$.

In the early Universe, X-quarks (or X-leptons) freeze out with their antiparticles and can form fractons in the period of QCD confinement. The specifics of fractons as compared to the free quarks (Zeldovich et al., 1967) is that the primordial fracton concentration can strongly decrease in the modern astrophysical objects, say, in Earth, due to the process of X-quark recombination

$$(Xuu)+(\overline{X}u)\rightarrow(X\overline{X})+\text{ordinary hadrons} \tag{1213}$$

with the successive decay of X-quarkonium

$$(X\overline{X})\rightarrow\text{ordinary hadrons} \tag{1214}$$

or

$$(X\overline{X})\rightarrow\text{mirror particles}\ . \tag{1215}$$

X-quark recombination (1213)–(1215) induced by the "Coulomb" attraction of uncompensated mirror electric charges of X-quark and antiquark reduces fracton abundance and makes their existence compatible with the negative results of the searches for fractionally charged particles (Khlopov, 1980).

In the framework of grand unification, the gauge symmetry $[SU(2)\text{x}U(1)\text{x}SU(3)_c]_O\text{x}[SU(2)\text{x}U(1)\text{x}SU(3)_c]_M$ of ordinary and mirror particles is embedded within the unifying GUT symmetry group, G_{OM}.

Breaking of G_{OM} results in the separation of ordinary and mirror particle sectors under the condition of strict discrete symmetry between them. Topology of G_{OM} breaking is reduced under this condition to the case of $U(1)$ symmetry breaking and provides the existence of topological string solution.

One may illustrate this strict topological result (Tyupkin, Shwartz, 1982) by the following simple argument.

Breaking of G_{OM} divides the Higgs multiplet H of the unifying group on the subsets of Higgs fields H_O and H_M corresponding to unbroken subgroups of separated symmetries of ordinary and mirror particles.

The discrete symmetry between the multiplets H_O and H_M makes these two sets of fields equivalent to imaginary and real part of one complex field that reduces the analysis to the case of broken $U(1)$ symmetry. Then the topological argument is valid proving the existence of string solution in the considered case of broken non-abelian symmetry.

Shwartz (1981) found that this string solution exhibits the property to change the mirrority of object in the result of its motion along the closed path around the string. Shwartz mentioned that Lewis Carroll's Alice could pass through the looking glass along the closed path around such string and called the solution Alice string.

In the early Universe the phase transition to the phase of broken G_{OM} symmetry with separated sectors of ordinary and mirror particles should result in the appearance of a network of cosmic strings.

The evolution of the string network within the cosmological horizon leads to straightening of long strings and decay of small string loops. As a result of this evolution the string network contribution into the total density is given at the time t by

$$\rho_s \sim \frac{\mu}{t^2} , \qquad (1216)$$

where μ is the mass per unit string length related to the scale Λ of G_{OM} symmetry breaking as

$$\mu \sim \Lambda^2 . \qquad (1217)$$

One finds from Eqs. (1216) and (1217) that the relative contribution of strings into the total density is given by

$$\Omega_s = \frac{\rho_s}{\rho_{cr}} \sim G\mu \sim \left(\frac{\Lambda}{m_{Pl}}\right)^2 . \qquad (1218)$$

Owing to the negative pressure corresponding to the false vacuum inside the string space–time properties of cosmic strings differ from the ones of one–dimensional structures of the ordinary matter. General relativity predicts zero gravitational potential outside the string but there should exist an angular cut in the flat space–time along the string. The angular size of the cut is of the order of

$$\delta \sim G\mu \sim \left(\frac{\Lambda}{m_{Pl}}\right)^2 . \tag{1219}$$

Due to the cut of space–time along the string, its intersection with the line of sight on astronomical object causes the effect of gravitational lens. Instead of the object one observes its two images whose angular coordinates are separated by the value (1219).

String motion produces the wake in the passed matter. The matter perturbed by the string moves in the wake towards the plane of string motion with the velocity v_w of the order of

$$\frac{v_w}{c} \sim (G\mu)^{1/2} \sim \frac{\Lambda}{m_{Pl}} . \tag{1220}$$

It is interesting to note, that the strict results of general relativity calculations of these phenomena turn out to coincide formally with the Newtonian treatment of a one–dimensional matter string with the mass per unit length μ moving with the speed of light (Gasilov et al., 1987).

The Alice string is the typical cosmic string and should share all the properties mentioned above. However the change of relative mirrority of objects around the Alice string adds new aspects in their observational effects.

The object defined as ordinary along the path to the left of the Alice string is defined as mirror object along the path to the right of this string.

Therefore the Alice string crossing the line of sight changes the mirrority of object relative to the observer. Ordinary stars become mirror and invisible. Mirror stars turn out to be ordinary and visible.

The objects of mixed mirrority radiate both ordinary and mirror photons. The Alice string crossing the line of sight interchanges the mirrority of radiation. Mirror photon radiation to the left of the string turns out to be the process of ordinary photon emission to the right of the string and vice versa.

For non-correlated processes of ordinary and mirror radiation, the Alice string crossing the line of sight induces the interchange of mirrority, which can be observed as anomalously rapid variation of luminosity. Blinnikov and Khlopov (1982) noted that such effect may explain the 100 s X-ray luminosity variation of QSO 1525+227 observed in 1981 by Einstein X–ray observatory.

Gravitational lensing by the Alice string induces the two images of different mirrority. In the second image one observes the mirror radiation of the first image.

If the radiation is non-correlated, the usual identification of string-induced gravitational lensing by the identical patterns of the spectra of the two images can not be valid in the case of the Alice string (Sazhin and Khlopov, 1989).

The account for the different mirrority of images may revive the string lensing interpretation of broad pair Q1146+111 B,C and offer such interpretation for "twin galaxy" field 0249-186 or double quasar GL Q2345+007 (Oknyansky, 1996).

The successive development of mirror particle model should account for the possible breaking of the strict symmetry between ordinary and mirror particles.

The shadow world appears in this case. Strict correspondence between the original and mirror image transforms in this case into the peculiar distortion of the size and form of the object in its shadow.

Cosmoparticle physics should develop the system of tests both for the mirror and shadow matter. The cosmological evolution of mirror or shadow matter is the important input for such analysis.

2. Mirror Particles in the Early Universe

2.1. Inflation and Constraints on the Domain Structure

In the framework of inflational cosmology, the ratio of the ordinary and mirror matter densities in the Universe is determined by the relative probability of ordinary and mirror particle production in the course of reheating after inflation.

In the absence of mutual strong, weak and electromagnetic interactions, one can not exclude that some interaction at the energy scale

$$F >> 10^2 \text{ GeV} \tag{1221}$$

provides transitions between ordinary and mirror particles. Following Khlopov et al. (1989, 1991) one can estimate the cross section for such transition, given in the units (230), by

$$\sigma(T) = \begin{cases} \dfrac{\gamma}{T^2} & \text{for } T > F \\ \dfrac{\gamma T^2}{F^4} & \text{for } T < F \,, \end{cases} \tag{1222}$$

where $\gamma < 1$.

This interaction can provide the thermal equilibrium between the ordinary and mirror particles, if the rate of its reactions

$$n\sigma v \sim T^3 \sigma \tag{1223}$$

exceeds the rate of the cosmological expansion, given at the stage of relativistic expansion by

$$\Gamma \sim \frac{T^2}{m_{Pl}} \,. \tag{1224}$$

One obtains (Khlopov et al., 1989, 1991) that the condition

$$n\sigma v > \Gamma \tag{1225}$$

takes place for

$$F < \gamma m_{Pl} \tag{1226}$$

and for the reheating temperature

$$T_R > \left(\frac{F}{\gamma m_{Pl}}\right)^{1/3} F \tag{1227}$$

only. Provided that these conditions are valid independently on the mechanism of inflation after its end, the cosmological reheating leads to equilibrium symmetric distribution of ordinary and mirror matter.

If no such interaction (except for gravity) exist, or the mutual interaction is too weak to provide the above conditions, the equilibrium distribution between the ordinary and mirror matter still may be established, if the decay of the inflaton field is induced by the interaction mutual for ordinary and mirror matter, in particular, by gravity.

The latter is true in the case of inflation, induced by R^2 effects in the polarisation of gravitational vacuum, or in the case of inflaton, having symmetric interaction with the ordinary and mirror matter.

If the products of inflaton decay have definite mirrority, the symmetry between the ordinary and mirror particles demands the existence of the mirror partner for such inflaton.

Within the frame of the model of chaotic inflation (Linde, 1984), the initial amplitudes of the ordinary and mirror inflatons may be different, and that leads to the formation of the domain structure in the distribution of the ordinary and mirror matter (Dubrovich, Khlopov, 1989).

In the regions, where the amplitude of the ordinary inflaton is higher, than the one of the mirror inflaton, the ordinary particles should dominate after inflation and the admixture of the mirror particles should be exponentially small. On the contrary, dominance of the mirror inflaton results in the negligible density of the ordinary particles after inflation.

Since the inflated regions are generally much greater than the observed part of the modern Universe, the mirror asymmetrical inflation, driven by the ordinary inflaton, corresponds to the exponentially small density of the mirror matter within the modern cosmological horizon. The inverse case of the inflation, driven by the mirror inflaton and resulting in the exponentially small density of the ordinary matter, is evidently excluded for the presently observed part of the Universe.

If the inflaton has no definite mirrority and the equal amounts of the ordinary and mirror particles are produced after inflation, the domain structure may be formed due to the random local asymmetry of the amplitudes of ordinary and mirror scalar fields in various periods after the general inflation, say, in the phase transitions.

The scale of such domain structure is determined by the concrete parameters of the fields, involved in the process, and it may be much smaller than the scale of the modern cosmological horizon (Dubrovich, Khlopov, 1989; Khlopov et al., 1989, 1991).

In the case of such small-scale structure its scale must be either smaller, than the size of the horizon in the period of the Big Bang nucleosynthesis, or larger than the size of superclusters. These constraints follow from the analysis of the effects of mirror domain structure on the physical processes in the early Universe after Big Bang nucleosynthesis (Dubrovich, Khlopov, 1989).

Mirror domain looks like empty void for ordinary particles. At the radiation-dominated stage the radiation pressure pushes the ordinary matter into domain, that leads to violent processes of semirelativistic shock wave collisions.

Nuclear collisions in the shock wave fronts influence the light elements abundance, and the condition that there is no overproduction of D, ^{3}He, ^{6}Li, ^{7}Li and ^{7}Be imposes severe constraints on strong shock waves.

The energy release from even weak shock wave collisions distorts the Planckean form of the black body background radiation spectrum, what strengthens the constraints on the domain structure up to the scales of superclusters of galaxies.

The case of the allowed small-scale structure, corresponding to masses

$$M << M_\odot \tag{1228}$$

has no practical difference by its cosmological features from the case of initially homogeneous mixture of the ordinary and mirror matter.

The allowed large scale mirror domains, corresponding to the mass scales

$$M \geq 10^{16} M_\odot \tag{1229}$$

would look like giant voids in the distribution of the ordinary matter and in an especially highly peculiar case might have lead to the "island" model of the Universe (Dolgov, Kardashev, 1986; Dolgov et al., 1987).

The observed isotropy of the thermal electromagnetic background excludes the case, when the modern outer border of the mirror domain is beyond the cosmo-

cosmological horizon. It excludes the structure of such domains (Dubrovich, Khlopov, 1989) in the scales

$$l_h(t_{\rm rec}) \cdot \sqrt{1+z_{\rm rec}} < l < l_h(t_U) \ , \tag{1230}$$

where l_h $(t_{\rm rec})$ is the size of the horizon in the period of recombination at the red shift $z = z_{\rm rec}$ and l_h (t_U) is the size of the modern horizon.

Note, that the constraints on the domain structure are based on the averaged over the Universe effects in the assumption on the symmetric distribution of mirror and ordinary domains. In the asymmetric case, when ordinary matter dominates within the modern cosmological horizon, isolated mirror domains are not excluded, provided, that their effect, averaged over the observed part of the Universe is within the observational upper limits.

2.2. The Inhomogeneous Baryosynthesis and the Island Distribution of Mirror Baryons

The necessity in the special set of mirror partners, not coinciding with antiparticles, follows from the fact, that in the presence of CP violation in the world of the ordinary particles, the equivalence between the left- and right-handed coordinate systems should be restored. For the mirror partners, CP-violating effects are equal by the magnitude and have the opposite sign to the corresponding effects of ordinary particles. In particular, it leads to the opposite sign of CP-violating effects in the processes of baryosynthesis in the mirror world.

So, the generation of the baryon excess in ordinary particles formally corresponds to the generation of antibaryon excess in mirror particles, provided that there is a strict symmetry between the processes of baryosynthesis in the ordinary and mirror worlds.

However, since the sign of the baryon number for mirror particles is not observable we shall refer to the "baryon" excess in the cases of the both mirrority.

Since the evolution of ordinary and mirror matter is symmetric, local processes of baryon excess generation in the very early Universe lead to simultaneous production of the equal baryon excesses in the ordinary and mirror matter. In the absence of domain structure ordinary and mirror baryons are produced in the Universe with the equal local densities.

In the presence of domain structure the domain scales and the averaged densities of ordinary and mirror baryons within domains should be equal.

If the baryon excess generation is not related to CP-violating local baryon-violating processes in the ordinary and mirror matter, the interesting possibility of "entropy" density perturbations in the relative distribution of the ordinary and mirror baryon excesses arises, in principle, at any scale (Khlopov et al., 1989, 1991).

In particular, the mechanism of baryosynthesis in supersymmetric GUT models (Affleck, Dine, 1984; Linde, 1985), ascribes the cosmological baryon asymmetry to the existence of primordial condensate of supersymmetric scalar quarks and leptons. Even for the strict symmetry in the supersymmetric potentials of the ordinary and mirror particles, such mechanism provides inhomogeneous relative distribution of ordinary and mirror baryon excesses.

In the difference from the mirror domains considered in the previous Section the baryon excess inhomogeneities represent local suppression of the baryons of the other mirrority. The energy densities of relativistic mirror and ordinary particles are equal in this regions.

At radiation-dominated stage the existence of regions of strong domination of mirror baryona over ordinary baryons (or vice versa) does not cause any violent dynamic effect and is not excluded by the observational data.

In such regions, called baryon islands of the fixed mirrority (Khlopov et al., 1989, 1991), astronomical objects may be formed at any scale, up to the scale of the modern horizon.

2.3. The Big Bang Nucleosynthesis and the Mirror World.

Note, that in all cases, except for the very large scale mirror domains, relativistic mirror particles are present in the same amount as the ordinary relativistic particles in the period of the Big Bang nucleosynthesis. It means that after the first second of the cosmological expansion, when the freezing out of the neutron to proton ratio in the ordinary matter takes place, one should add the contribution of mirror photons, mirror electron-positron pairs, right-handed neutrinos and left-handed antineutrinos into the total density.

Such doubling of the relativistic species in the period of nucleosynthesis should lead to the growth of the primordial ordinary ^{4}He abundance up to (Blinnikov, Khlopov, 1980, 1982, 1983; Khlopov, Chechetkin, 1987; Carlson, Glashow, 1987)

$$Y_{\text{prim}} \geq 28\% \ . \tag{1231}$$

Taking into account the widely accepted upper limit on the primordial helium abundance, $Y_{\text{prim}} < 25$ % (see Chapters 3 and 5), we come to the conclusion (Carlson, Glashow, 1987) that homogeneous mixing of ordinary and mirror mater is excluded by the observation.

Taken this fact for granted, one should consider some asymmetrical solution for the mirror matter, or conclude that the symmetry between ordinary and mirror matter is broken, and the shadow matter should be considered as the most realistic case.

Even in this case the successive discussion of the symmetric mirror matter is important, since the quantitative definiteness of this model is very instructive in finding qualitative astronomical effects to be shared by more general asymmetrical cases.

However, one should take in mind, that the observed ^{4}He abundance (see Chapter 5 and Refs. wherein) is given by

$$Y_{obs} = (28 \pm 12)\%$$

that by itself does not contradict the prediction of the mirror world model. Recent observations of radio recombination lines in HII regions together with the analysis of ionisation structure of these regions show (Ershov et al., 1988; Tsivilev, 1991; Gulyaev et al., 1997) that primordial abundance is in the interval (23.8 – 28.4)% what also does not exclude the case of the symmetric mirror world.

One does not observe directly the primordial helium, and in the absence of direct methods has to use reasonable interpolations of the observational data to the pregalactic abundance. However reliable this procedure is, it is model dependent, and in the absence of model independent results, the question on the existence of homogeneously mixed ordinary and mirror matter can not get its final answer.

3. Formation of Astronomical Objects of the Mirror Matter

With all reservations made in the previous Section, we shall follow in our analysis the general scenario of the cosmological evolution of homogeneously mixed mirror and ordinary matter in the period of galaxy formation (Blinnikov, Khlopov, 1980, 1982, 1983; Khlopov et al., 1989, 1991).

Starting from its early versions, this scenario, assuming the equal cosmological densities of the mirror and ordinary matter, invokes the dark matter in the form of massive neutrino (Blinnikov, Khlopov, 1980, 1982, 1983) or of some other unspecified physical nature (Khlopov et al., 1991), dominating in the Universe in the period of the galaxy formation and maintaining the dominant part of the modern cosmological density. This assumption together with the account for inflation and baryosynthesis, considered in the previous Section, transfers the analysis into the framework of realistic inflationary cosmological model with baryosynthesis and dark matter.

The scenario assumes the dominance of ordinary and mirror radiation on the RD stage with equal density together with relativistic light neutrino with a small admixture of ordinary and mirror baryons with equal densities. The scenario can naturally account for the existence of a small (at the RD stage) admixture of equal densities of nonrelativistic ordinary and mirror particles, such as ordinary and mirror neutralino, ordinary and mirror axions etc.

Note again, that the equal densities of mirror and ordinary particles in the period of Big Bang nucleosynthesis are in the contradiction with the widely accepted conjecture on the primordial ^{4}He abundance.

So, taking this conjecture for granted, one should invoke some mechanism, suppressing the contribution of mirror matter in the first second of expansion.

However it practically does not influence the further discussion, which concerns the evolution of mirror baryons, whose contribution into the cosmological density in the period of Big bang nucleosynthesis is negligible in the case of the strict symmetry between ordinary and mirror matter as it is for ordinary baryons.

After the end of RD stage the non-relativistic dark matter particles start to dominate in the Universe and the gravitational instability evolves within the cosmological horizon.

The concrete choice of the model of the large scale structure formation, as well as of the form of the dark matter determining it, is not essential for the bulk of our further conclusions on the effects of mirror baryon matter.

Mirror baryons with the averaged density equal to the one of the ordinary baryons, maintain admixture of non-relativistic matter participating the general development of the gravitational instability. This development is determined by the form of the dark matter, dominating in the cosmological density.

The influence of the mirror matter on the observed properties of the large scale structure is possible only in the case of large-scale mirror domains, corresponding to the mass scales

$$M > 10^{16} M_{\odot} \tag{1232}$$

or large-scale island distribution of baryons.

In this case, the scale of mirror baryon inhomogeneities can be arbitrary, so that the formation of "pure" mirror objects is possible in all the scales. Having this possibility in mind, we consider below the effects of mirror objects in all possible astronomical scales.

Note, that the large-scale island baryon distribution may provide the physical mechanism for the "biasing" in the distribution of the luminous and dark matter. Mirror baryon islands must look in this case like voids, in which ordinary matter galaxies are absent.

However, the problem of rapid structure evolution seems to retain in this case, being inherent to all models of structure formation by the stable dark matter (Doroshkevich et al., 1989).

Following Blinnikov, Khlopov (1980, 1983) and Khlopov et al. (1989, 1991) consider the "pancake" scenario of structure formation.

The main feature of this scenario is retained in the unstable dark matter model, more realistically reproducing the observational data.

Moreover, as we shall see further, the main features related to formation of mirror baryon objects are adequate for any other dark matter model, e.g., the CDM scenario.

In the absence of island or domain structure of baryon distribution, mirror (M) and ordinary (O) baryons form together with the dark matter the common nonlinear flattened structure (the pancake). At the border of the O+M gas protocluster (pancake) the shock wave is formed.

The cooling of the gas compressed by the shock wave due to O- and M- electromagnetic radiation leads to the fragmentation of the pancake.

Fragmentation of the ordinary and mirror baryon matter within the pancake goes independently, since the O-gas pressure does not act on the M-matter and vice versa, whereas the gravitational potential of fragments is too low to cause the capture of fragments of the other mirrority.

This is the crucial point in the argument for separation of mirror and ordinary astronomical objects at the scales, at which the thermal instability is essential. Homogeneously mixed ordinary and mirror baryonic matter forms dominantly "pure" stellar objects of definite mirrority up to the scales of globular clusters.

Let us consider in more details this principal feature of the evolution of homogeneously mixed ordinary and mirror gases.

3.1. The Separation of Ordinary and Mirror Matter

Following Doroshkevich et al. (1978) and Blinnikov, Khlopov (1980, 1983) the mass of O- and M-gas fragments are determined by the scale of the evolving thermal instability, given by (Field, 1965; Doroshkevich, Zeldovich, 1981)

$$l_{th} \cong \left(\frac{\lambda T}{\varepsilon}\right)^{1/2}, \tag{1233}$$

where the electron thermal conductivity is equal to

$$\lambda = 10^{-6} T^{5/2} \frac{\text{erg}}{\text{cm} \cdot \text{s} \cdot \text{K}} , \qquad (1234)$$

T is the temperature, and the energy losses due to the bremsstrahlung radiation are equal

$$\varepsilon = 10^{-27} n^2 T^{1/2} \frac{\text{erg}}{\text{cm}^3 \cdot \text{s}} . \qquad (1235)$$

The mass of fragments is then given by

$$M_{\text{th}} \cong m_p n l_{\text{th}}^3 \cong 5 \cdot 10^9 M_\odot \left(\frac{T}{10^6 \text{K}} \right)^{13/2} \left(\frac{10^{-15} \frac{\text{erg}}{\text{cm}^3}}{P} \right)^2 , \qquad (1236)$$

where the pressure on the shock front $P = nkT$ is the universal function of the moment of the pancake formation, given by

$$P = 4 \cdot 10^{-15} (1 + z_0)^4 \frac{\text{erg}}{\text{cm}^3} \qquad (1237)$$

for $\Omega = 0.3$ (Doroshkevich et al., 1978).
Taking the red shift in the period of the pancake formation equal to

$$z_0 \sim 5 \div 2 , \qquad (1238)$$

one obtains the mass of O and M fragments of the order of

$$M_{\text{th}} \sim 10^2 \div 10^5 M_\odot , \qquad (1239)$$

respectively, since the temperature of the hot gas is always close to

$$T = 10^6 \text{ K} . \tag{1240}$$

Similar estimations of the mass of the ordinary gas fragments were obtained by Doroshkevich (1980) for the different mechanism of fragmentation.

The fragments are not gravitationally bound and exist as separated objects owing to the outer pressure of the hot gas with the temperature $T \sim 10^6$ K .

Since the pressure of the ordinary hot gas does not act on the mirror matter cloud and vice versa, there is no reasons for mixing of O- and M- gas clouds, if such mixing was not given in the initial conditions for the density fluctuations, from which the clouds evolve. The strict identity of the initial conditions for the fragmentation of the ordinary and mirror matter seems, however, highly artificial.

O-clouds and independently M-clouds can merge with the clouds of the same mirrority, grow their mass, bind gravitationally and contract, forming stellar clusters.

Then the joint merging of O- and M- clouds with the higher mass of the order of

$$M \sim 10^6 \div 10^9 \ M_{\odot} \tag{1241}$$

takes place forming large gravitationally bound systems. Gravitational capture of O-fragment by M-fragment and vice versa is in principle possible at this stage. Following Blinnikov and Khlopov (1983) let us estimate the probability of this capture.

Let the protocluster of the mass M and the typical radius R decay on N fragments with the mass $m = M/N$ and with the typical size r . If we even assume that the O- and M-clouds are at rest and are at the mean distance r

from each other, the time scale of their merging is of the order of several free-fall times, given by

$$t_{ff} \sim \left(\frac{r^3}{Gm} \right)^{1/2} \tag{1242}$$

and this value is of the order of

$$t \sim (G\rho)^{-1/2} \ , \tag{1243}$$

where $\rho \sim M/R^3$, what is close to the hydrodynamic time scale of the protocluster. So, during the merging time estimated above, the Maxwell distribution of the clouds on their relative velocities is established with the dispersion of the order of the virial velocity of the fragments in the cluster, given by

$$v_0^2 \sim \frac{GM}{R} \ . \tag{1244}$$

It means that the probability of the low relative velocity of clouds is small.

The total probability w that the M-fragment will be captured by one of the O-fragments is determined by their neighbourhood in the phase space, i.e. is given by the product of the probability w_s that the M-fragment is near the O-fragment (at the distance r) and the probability w_v that the relative velocity of M-fragment is smaller than the escape velocity from the O-fragment, that corresponds to relative velocity of the fragments

$$v < v_1 \cong \left(\frac{Gm}{R} \right)^{1/2} , \tag{1245}$$

Note that the probability of merging for two clouds of ordinary matter is given by their size and the rate of energy dissipation.

The total probability of gravitational capture of mirror matter fragment by the cloud of the ordinary matter is

$$w = w_s \cdot w_v \ , \tag{1246}$$

where

$$w_s = 1 - \left(1 - \frac{r^3}{R^3}\right)^N \tag{1247}$$

and

$$w_v = v_0^{-3} \int_0^{v_1} v^2 \exp\left(-\frac{v^2}{v_0^2}\right) dv \cong \left(\frac{v_1}{v_0}\right)^3 = \left(\frac{mR}{rM}\right)^{3/2} \cong \left(\frac{R}{Nr}\right)^{3/2} \tag{1248}$$

(Blinnikov, Khlopov, 1983).

At $N >> 1$ the maximum in the probability, given by Eq. (1246), is formally reached at the earliest stages of the contraction for

$$r \sim \frac{R}{N^{1/3}} \ . \tag{1249}$$

Then

$$w \cong \frac{1}{N} \ . \tag{1250}$$

Since $N \sim M/m$, one obtains for the mass of the protocluster

$$M \sim 10^{14} M_\odot \ , \tag{1251}$$

and for the masses of fragments

$$m \sim 10^6 \div 10^9 M_\odot \tag{1252}$$

the upper limit for the probability of the O- and M- cloud merging

$$w \leq \frac{m}{M} \sim 10^{-8} \div 10^{-5} \ . \tag{1253}$$

The probability of merging of clouds of the same mirrority, i.e. of the two O-matter or two M-matter fragments is determined by the cross section of their collision, given by

$$\sigma_c \cong 4\pi r^2 \left(1 + \frac{v_1^2}{v_0^2}\right) . \tag{1254}$$

So the collision time scale for one fragment is of the order of

$$t_c \sim \frac{R^3}{N \sigma_c v_0} \ . \tag{1255}$$

In the beginning of contraction, when r is large, this time scale is smaller than the time of the free crossing of the cluster, that corresponds to

$$t_c < \frac{R}{v_0} \ . \tag{1256}$$

Thus, only one of N fragments could have comparable amounts of O- and M-matter, which should separate within it in the course of the successive fragmentation on stars.

The admixture of M-matter in the rest of O-fragments (and vice versa, of the O-matter in M-fragments) is provided by the gas homogeneously filling the O+M pancake. So the relative fraction of such admixture is given by a small value

$$f \sim \frac{\langle \rho \rangle}{\rho_i} \tag{1257}$$

being the ratio of the mean density of gas within the pancake (or later — within the galaxy) and the density within the considered object.

At the stage of the thermal instability the pressure of the hot gas with the temperature $T \sim 10^6$ K is nearly equal to the pressure of the cold gas with the temperature $T \sim 10^4$ K so that the relative fraction of the admixture of the matter of the other mirrority is given by

$$f \sim \frac{\langle \rho \rangle}{\rho_i} \cong \frac{10^4}{10^6} = 10^{-2} \ . \tag{1258}$$

This fraction freely (taking apart the effects of accretion, to be considered below) passes the inhomogeneities and does not participate the processes of star formation etc. At later stages, the density contrast grows, making respectively smaller the fraction of the other mirrority admixture. So, in the Galaxy the mean density is equal to

$$\langle \rho \rangle \cong 10^{-24} \ \frac{\text{g}}{\text{cm}^3} \tag{1259}$$

and in the volume of the Solar system the free stream of M-matter provides (taking apart M-matter inside the Sun to be considered below) the admixture not exceeding

$$M_o \leq 10^{21}\ \text{g} \tag{1260}$$

and the fraction of the M-matter inside the Earth is estimated (Blinnikov, Khlopov, 1983) to be smaller, than

$$f \sim 10^{-24} \tag{1261}$$

what seems practically impossible to be detected.

Therefore, in what follows, by the O- and M-matter separation within single fragments with the mass

$$M < 10^6 \div 10^9\ M_\odot \tag{1262}$$

we mean, that the probability for the fraction of the admixture of the matter of the other mirrority to exceed 1% is not larger, than

$$w \leq 10^{-5}\ . \tag{1263}$$

The largest single fragments with the mass

$$M \sim 10^9\ M_\odot \tag{1264}$$

seem to form dwarf galaxies. The admixture of the ordinary matter in the dwarf mirror galaxies should not exceed 1% and they may be observed by the effect of gravitational lens.

On the other hand, in the case of the hot dark matter scenario both the admixture of M-matter and hot dark matter (massive neutrino) in the dwarf O-galaxies should be small and they should not practically contain dark matter.

In the case of cold dark matter scenario small-scale inhomogeneities of dark matter is expected, that can lead to the presence of dark matter as well as to the comparable amounts of O- and M-matter in dwarf galaxies.

At the scale

$$M < 10^6 \ M_\odot \tag{1265}$$

the estimation of O- and M-matter mixing, given above for the case of hot dark matter scenario, is completely valid also in the case of cold dark matter scenario.

O- and M-fragments, corresponding to these smaller scales, form together with the dark matter (both hot and cold) normal and giant galaxies. In the establishment of the equilibrium in the gas component, the processes of dissipation play important role.

Since the laws of dissipation are strictly symmetrical in O- and M-gases and the condensation of already formed stellar systems is determined by the unique gravitational interaction, one should expect, that O- and M-matter practically separated at the scale of single fragments have the same distribution at the scale of galaxies.

Practically complete separation of O- and M-matter at the scale (1265) leads to formation of stars and stellar systems, composed of the matter of the definite mirrority (taking apart small admixture due to accretion and specific conditions under which the mixing is enhanced – see below). With the same reservations, the formation of double M-O–stars or stars, composed of the comparable amount of mirror and ordinary matter is in the general case strongly suppressed.

Indeed, in the general conditions in the Galaxy, such suppression can be estimated following Blinnikov and Khlopov (1983). This estimation is strongly enhanced in some specific cases, such as in the molecular clouds or in globular clusters, what will be considered in Sec.4.2.

The process of the stellar capture in binaries needs either triple collisions or binary collisions with strong tidal dissipation. At the normal stellar number density in the Galaxy, given by

$$n_s \sim 0.1 \text{ pc}^{-3} \tag{1266}$$

these processes practically don't go. They can be considered in the dense stellar clusters or galactic nuclei only. Blinnikov and Khlopov (1983) have estimated following Fabian et al. (1975) and Press, Teukolsky (1977) the rate of binary formation due to tidal dissipation.

In the cluster of the radius R , containing N stars of the solar type, the rate of formation of the O-matter binaries is given by

$$\dot{N}_{bin} \cong 3\left(\frac{N}{10^5}\right)^{1.4}\left(\frac{R}{5 \text{ pc}}\right)^{-2.4}\left(10^9 \text{ years}\right)^{-1} . \tag{1267}$$

In the clusters of ordinary stars the rate of O-M binary is given by this expression multiplied by the factor, accounting for the small fraction of M-matter, and the estimation is typically very small.

For instance, if the mixture of M-matter is even of the order of $f \sim 1\%$ the rate of O-M binary formation is

$$\dot{N}_{bin}(\text{O - M}) \sim 0.3 \cdot \left(10^{10} \text{ years}\right)^{-1} \tag{1268}$$

(Blinnikov, Khlopov, 1983).

In galactic nuclei, where the comparable amounts of O- and M-matter are expected, the number of O-M pairs should be of the same order as the number of ordinary binaries, but separate stars are not observed there.

Due to assumed strict symmetry of O- and M-interactions, the M-star evolution should go in the same way it takes place for the O-star of the same mass. As a result of such evolution M-star forms M-white dwarfs, M-neutron stars, black holes (whose properties are evidently independent on the physical nature of the matter they are originated from), or it is completely disrupted.

On the other hand, the averaged densities of O- and M- matter (and of O- and M- gas, in particular) in galaxies should be equal. One also expects strict symmetry in the distribution on the different types of objects and in the mass and velocity distributions of the objects of each type.

In particular, the strict symmetry of distributions should lead to the existence of compact nonluminous objects in the spherical component of our Galaxy with the density of the order of the baryon density. Such objects (MACHOs) are observed in the halo of the Galaxy by the effect of gravitational micro-lensing, and one should seriously consider the interpretation of these observations on the basis of the existence of a mirror world.

With the comparable orbital momentum of the galactic O- and M-components, the unique disc with the comparable amount of O- and M-matter should be formed in the Galaxy. Formation of separate O- and M-discs seems impossible (Blinnikov, Khlopov, 1983). Their mutual attraction still at the stage of their formation should lead to merging into unique disc component. The averaged density of the dark matter (the M-matter) in the disc should be equal to the one of the visible matter.

The question on the existence of this "local dark matter" has rather long history (Oort, 1965; Gliese, 1977). The observational estimations (Hill et al., 1979; Jones, 1972; Balakirev, 1976; House, Kilkenny, 1980) that the density of the dark matter should not exceed 30% of the density of the visible matter in the vicinity of the Solar system were found by Blinnikov and Khlopov (1983) not to be sufficiently precise to exclude the density of the local dark matter of the order of the visible matter density. This possibility finds support in the series of papers (Bahcall, 1984; Bahcall et al., 1992), where the total local density in the vicinity of the Solar system is estimated as

$$\frac{\rho_{tot}}{\rho_v} \sim 2.3 \,, \tag{1269}$$

where ρ_v is the averaged density of the visible matter objects in the local part of the disc component of the Galaxy.

Note, that the formally symmetric mirror world implies

$$\frac{\rho_{tot}}{\rho_v} = 2 \tag{1270}$$

but the account for the ordinary stars with low luminosity, such as brown dwarfs, and for respective M-stars as well as for small contribution into the density of disc of the dark matter particles, forming the nonluminous halo of the Galaxy,

$$\frac{\rho_{dm}}{\rho_v} < 0.1 \tag{1271}$$

make the result of Bahcall, provided that they are confirmed (see critisism of Kuijken and Gilmore (1989, 1991)), a serious argument, favouring the existence of the mirror world.

One should stress, that though the local dark matter, if it exist, provides rather small contribution into the total density, being of the order of the baryon density

$$\Omega_{ldm} \sim \Omega_b \leq 0.1 \tag{1272}$$

and plays no important dynamical role in the cosmological evolution, its physical significance can not be overestimated.

The reason is that all weakly interacting particles, considered as the candidates to cosmological dark matter, represent collisionless gas, which condense in the visible part of Galaxy according the law (see Chapter 9)

$$\frac{\rho_{dm}(t)}{\rho_{dm}(0)} \propto \left(\frac{\rho_b(t)}{\rho_b(0)}\right)^{3/4}$$

so that no more than 10% of the visible density can be provided by the dark matter halo particles. Hence, the local dark matter with the density comparable with the baryon density should have nontrivial physical nature. It should have effective mechanisms of dissipation as the baryon matter does.

The simplest solution is to try to explain the existence of the local dark matter by some faint baryonic objects such as brown dwarfs. In the case such attempts are found unsuccessful, one should face on the problem of the existence of dissipational dark matter, such as the mirror or shadow matter. It invokes dark matter particles weakly interacting the ordinary matter but having much stronger and even long range interactions between themselves.

3.2. The Gas Accretion on the Astronomical Objects of the Other Mirrority

Sufficiently pure separation of the O- and M-matter in the astronomical objects with the mass

$$M < 10^6 \, M_\odot$$

makes the process of accretion of the O-gas on the M-objects (and vice versa, of the M-gas on the O-objects) the most important for the mixing of O- and M-matter in these objects.

Let us consider this process, following Blinnikov and Khlopov (1983).

According to the classical theory of accretion the gravitating body with the mass M and with the velocity v captures the matter from the distance

$$R_A \sim \frac{2GM}{v^2} \ . \tag{1273}$$

For the subsonic movement, the velocity v should be changed by the speed of sound v_s .

From the standard formula for the mass flow defined as

$$\dot{M} = \pi R^2 \rho v \tag{1274}$$

one obtains, that with the constant boundary conditions the incoming matter comprises for the time t the mass fraction, given by

$$\frac{\Delta M}{M} \cong 10^{-5} \left(\frac{M}{M_O} \right) \left(\frac{10 \ \frac{\text{km}}{\text{s}}}{v} \right)^3 \left(\frac{\rho}{10^{-24} \ \frac{\text{g}}{\text{cm}^3}} \right) \left(\frac{t}{10^{10} \ \text{years}} \right) . \tag{1275}$$

Thus, even for large globular clusters, having the mass of the order of

$$M \sim 10^6 M_\odot \ , \tag{1276}$$

and the velocity of their movement of the order of

$$v \sim 100 \ \frac{\text{km}}{\text{s}} \ , \tag{1277}$$

the fraction of the accreted matter is negligible.

The same is true for the dwarf galaxies, having the mass

$$M \sim 10^9 M_\odot \tag{1278}$$

and the velocity of motion

$$v \sim 10^2 \div 10^3 \ \frac{\text{km}}{\text{s}} \ . \tag{1279}$$

Due to very low density of gas in the galaxy clusters, being of the order of

$$\rho \sim 10^{-27} \ \frac{\text{g}}{\text{cm}^3} \ , \tag{1280}$$

one obtains from Eq. (1275) that

$$\frac{\Delta M}{M} \sim 10^{-5} \div 10^{-2} \ . \tag{1281}$$

In fact, Eq. (1275) is strictly valid only for the accretion on stars. In the case of stellar clusters and galaxies it provides only upper limit on the mass fraction of the accreted matter. The reason is, that the typical velocity v of these objects

$$v \geq v_1 \tag{1283}$$

exceeds the escape velocity v_1 , so that the capture radius

$$R_A \leq r \tag{1284}$$

turns out to be smaller than the radius r of the cluster. It makes the gravitational potential insufficiently effective for the capture.

If the radius of the accreting M-star is small, as compared to the capture radius

$$r \ll R_A \ , \tag{1285}$$

the fall of the ordinary matter down on such star results in the luminosity of the order of

$$L \sim \frac{G\dot{M}M}{r} \ . \tag{1286}$$

If the density of the accreting matter is of the order of (1280) this luminosity is very small even in the case of the accretion on the M-neutron star, i.e. such star may be observed only at relatively small distance of some tens pc.

For the period of the M-neutron star evolution, being of the order of

$$t \sim 10^{10} \text{ years} \ , \tag{1287}$$

and for the radius of the M-neutron star

$$r \sim 10^6 \text{ cm} \tag{1288}$$

with the mass

$$M \sim 1\, M_\odot \tag{1289}$$

the mass of the ordinary matter stored by it can be estimated as

$$\Delta M \sim 10^{26} - 10^{28} \text{ g} \tag{1290}$$

(Blinnikov, Khlopov, 1983).

After the accreted O-matter loses angular moment, it contracts. Following the conditions of equilibrium it is compressed in the gravitational field of the M-star up to the densities

$$\rho \sim 10^9 \div 10^{10} \ \frac{g}{cm^3} \tag{1291}$$

within the size of the M-neutron star (1288).

The magnetic field frozen in the accreted matter experiences rapid change when the matter reaches the central optically opaque central part of the O-matter core of the M-star. V. Shwartsman estimated the size of that core to be of the order of

$$r_{cO} \sim 10^5 \ \text{cm} \ , \tag{1292}$$

and the time scale of the rapid variation of the luminosity in the photoshere of the O-core, induced by the recommutation of the magnetic fields, should be

$$\Delta t \sim \frac{r_{cO}}{c} \sim 3 \cdot 10^{-6} \ \text{s} \ . \tag{1293}$$

As Shvartsman pointed out, it is possible to use the methods of the searches for effects of short variation by MANIA experiment (Shwartsman, 1977) and to distinguish the effect of M-neutron star from similar effects of single ordinary neutron stars and black holes.

The idea was based on the estimation, that the gravitational radius of a black hole with the mass equal to the few Solar masses, and the radius of the surface of the ordinary neutron star are by the order of magnitude larger, than the radius of the photoshere of the O-core in the M-neutron star. In the case of M-neutron star it provides the time variation which is so short, that it can not be explained by the similar effects from the single black hole or ordinary neutron star.

For larger radius of M-star, its O-luminousity due to O-matter accretion falls down respectively, and the accreted O-matter cools down and forms objects like planets or comets in the potential well of the M-star, provided that such objects are not disrupted by tidal forces. The indirect effects of such M-star may be observed only at very near distance from the Solar system, not exceeding 1000 astronomical units. Owing to very small probability for the O-M binary formation, this close neighbourhood of the M- star to the Sun may take place in the cross rather improbable passing of the M-star in the vicinity of the Sun.

The most stringent constraints on the existence of the close invisible companion of the Sun of any possible physical nature come from the sky distribution of pulsars with anomalously low deceleration rates. The acceleration of the Solar system in the direction of the invisible companion shifts this distribution to the blue end, thus inducing anisotropy in it. Using the observational data on the sky distribution of pulsars with the measured deceleration rates, Blinnikov and Khlopov (1983) deduced the constraint on any invisible companion object of the mass M at the distance d from the Sun

$$M \le 1 M_{\odot} \left(\frac{d}{10^3 \text{ a.u.}} \right)^2 . \tag{1294}$$

According to the Eq. (1275), with the average density of M-matter in the galactic disc of the order of

$$\rho \sim 10^{-24} \ \frac{\text{g}}{\text{cm}^3}$$

and with the velocity of Sun relative to disc component, being of the order of

$$v \sim 10 \div 20 \ \frac{\text{km}}{\text{s}}$$

the Sun during its lifetime can accrete approximately

$$\Delta M \sim 10^{27} \div 10^{28} \text{ g} \tag{1295}$$

of the mirror matter. Some part of this M-matter can have the angular moment and can form the planet rotating inside the Sun near its surface, where the gravitational force is maximal. Blinnikov and Khlopov (1980) estimated, that if the planet has the mass

$$M_o \sim 10^{26} \text{ g} \tag{1296}$$

and moves at the depth of about

$$h = 2 \cdot 10^4 \text{km} \tag{1297}$$

under the surface of the Sun, this planet should cause nonradial oscillations of the Sun with the observed period

$$T = 160 \text{ min} \tag{1298}$$

(Severnyi et al., 1976, 1979).

The existence of such planet composed of the mirror matter does not contradict the observations of the shifts in the Mercury orbit. The planet may be detected by precise gravimetric measurements.

The specific angular momentum, which is necessary for the formation of such planet, and which is by the order of the magnitude lower than the one of the Mercury, can appear due to M-gas density fluctuations of the order of 1% at the accretion radius

$$R_A \sim 10^{14} \text{ cm} \tag{1299}$$

M-planet might be formed outside the Sun and then imbed it along the spiral due to the tidal friction.

Consider following Blinnikov, Khlopov (1980, 1983) the effects induced by the motion of the M-planet inside the Sun.

According to the standard Solar model, at the depth

$$h = 2 \cdot 10^4 \text{km}$$

inside the Sun the temperature is

$$T \sim 10^5 \text{ K} \tag{1300}$$

so that the speed of sound is

$$v_s \cong 3 \cdot 10^6 \ \frac{\text{cm}}{\text{s}} \ . \tag{1301}$$

The velocity of the M-planet motion along the circular orbit is equal to

$$v = 4.4 \cdot 10^7 \ \frac{\text{cm}}{\text{s}} \tag{1302}$$

so that the accretion radius on it is

$$r_A \cong 10^4 \text{ cm} \ . \tag{1303}$$

If the radius of the M-planet is of the order of

$$r \geq 10^8 \text{ cm} \ , \tag{1304}$$

one has

$$\frac{r}{r_A} >> \frac{v}{v_s} , \tag{1305}$$

and a rather unusual case of accretion takes place (Blinnikov, Khlopov, 1980).

Still for

$$r > r_A \tag{1306}$$

the solar matter is not captured by the mirror planet, and if the condition given by Eq. (1305) also holds, the velocity perturbations of the incoming gas Δv is everywhere much less, than the speed of sound,

$$\Delta v << v_s . \tag{1307}$$

It means, that even if the accretion shock wave is formed, it is weak and the maximal heating of the solar matter is in this case very small, so that the corresponding energy release per unit time is of the order of

$$\dot{E} \sim \dot{M} \frac{(\Delta v)^3}{v_s} , \tag{1308}$$

where

$$\dot{M} \sim \pi r^2 \rho v \tag{1309}$$

is the mass of the solar matter, being perturbed (but not captured!) by the M-planet per unit time.

Thus, the M-planet should cause inside the Sun gravitational and weak acoustical effects only.

It is interesting to note, that the search for the mirror matter inside the Earth seems rather unreasonable. The probability for the existence of the primary M-matter object inside the Earth is negligible, and the accretion of the M-matter can not lead to any storage of the M-matter in the Earth, since the M-matter velocity near the Earth is determined by the gravitation of the Sun and can not be less than

$$v_{oE} = 2^{1/2} v_E = 42 \ \frac{\mathrm{km}}{\mathrm{s}} \ , \tag{1310}$$

where the velocity of the orbital motion of the Earth is equal to

$$v_E \cong 30 \ \frac{\mathrm{km}}{\mathrm{s}} \ . \tag{1311}$$

Thus the velocity of the M-matter accreted by the Solar system is in the vicinity of the Earth greater than the escape velocity from the Earth and no capture of the M-matter by the Earth can take place.

4. Observational Physics of the Mirror Matter

On the basis of the scheme of the evolution of the mirror matter described above, contemplated by possible effects of island baryon distribution, let us consider the observational effects of mirror matter predicted on different astronomical scales.

It is clear from the above discussion, that all possible observational effects of the mirror matter must be induced exclusively by its gravitational interaction with the ordinary matter. With the account for possible island distribution of mirror baryons the existence of any type of mirror object is possible.

From the most general viewpoint one can distinguish two types of effects. They are the case of pure gravitational interaction and the situations, in which gasodynamics effects are also induced by gravity.

In the first case, the effects of mirror matter induce peculiar velocities of the ordinary objects. Effects of this kind may be called "kynematic". It is clear that they are mostly pronounced in the cases, when the ordinary object is in the gravitational field, induced by the mirror configuration of much larger mass.

In the second case, one considers effects, arising due to gravitational action of different types of objects on the gas of the other mirrority.

Let us discuss some examples of the interactions of various objects of different mirrority, which may be accessible for observational discovery in more details.

4.1. Galaxies and Galaxy Clusters of Definite Mirrority

In the case of island baryon distribution on the scales of galaxies or clusters of galaxies these astronomical objects are of definite mirrority, i.e. are dominantly composed either of the ordinary or the mirror matter. Possible admixture of the ordinary matter in the mirror galaxies or mirror matter in the ordinary galaxies may be originated from either the presence of small additional admixture, determined by the local mirror asymmetry in the process of baryosynthesis (see Sec. 2.2) or from the accretion of intergalactic gas on such objects.

One can consider the following observational effects of existence of mirror galaxies and clusters of galaxies.

I. The capture of ordinary galaxies by the cluster of mirror galaxies can result in the appearance of an object with large peculiar velocity or of small groups of galaxies with velocities leading to anomalous virial paradox, i.e. with the velocity dispersion up to

$$\left\langle v^2\right\rangle^{1/2} \sim (1 \div 2)\cdot 10^3 \ \frac{\text{km}}{\text{s}} \ , \tag{1312}$$

what is typical for dense rich clusters of galaxies (Vorontzov-Veliaminov, 1972).

Peculiar velocity component of massive galaxy can be found, in principle, at the level of

$$v_{pec} \geq 10^3 \ \frac{\text{km}}{\text{s}} \tag{1313}$$

by the methods, suggested by Zeldovich and Sunyaev (1982) for the measurement of peculiar velocities of galaxy clusters, i.e. by the measurement of distortions of black body background radiation induced by its scattering on the electrons of the gaseous halo of the galaxy.

The probability for the capture of galaxy by rich cluster seems to be rather large.

Indeed, assume that the dissipation of energy, necessary for the capture, takes place at the distance between the centers of galaxies of the order of the diameter of galaxy d. Then the cluster, containing N galaxies and having the diameter D , will capture background galaxies with the probability

$$w = \pi \cdot d^2 \cdot n \cdot \frac{2}{3} D = 4N \left(\frac{d}{D} \right)^2 \cong 0.01 \div 1 \ . \tag{1314}$$

Here

$$n = \frac{N}{\frac{\pi}{6} D^3} \tag{1315}$$

is the number density of galaxies in the cluster. The numerical estimation is given for the rich cluster of

$$N = 10^3 \div 10^4 \tag{1316}$$

galaxies the ratio of galaxy and cluster diameters was taken equal to

$$\frac{d}{D} = 10^{-3} \div 10^{-2} \ . \qquad (1317)$$

II. In the described above process of the capture by the mirror galaxy cluster of the ordinary matter galaxies, the latter will inevitably lose significant amount of gas. As a result the poor cluster of galaxies, formed in the potential well of the mirror cluster, can have significant amount of Inter Stellar Gas (IGG), filling the region with the size typical for rich galaxy clusters.

The amount and, consequently, the density of IGG must be by the factor of

$$k = \alpha \frac{N_M}{N_O} \qquad (1318)$$

times smaller, than in the rich clusters. Here N_M and N_O are the numbers of galaxies in the rich cluster of M-galaxies and of the O-galaxies captured by it. The numerical factor $\alpha < 1$ takes into account that in the difference to rich clusters, capturing the galaxies of the same mirrority, the cooling flow (see for review Fabian et al., 1984; Gilfanov, Sunyaev, 1987; Gorbatsky, Kritzuk, 1987) will not be maintained and gas is not lost by the cluster.

Thus at

$$\frac{N_O}{N_M} \approx 10^{-2} \qquad (1319)$$

one may expect the suppression factor to be equal to

$$k^{-1} \approx 0.03 \qquad (1320)$$

and the emissivity of IGG to be the factor of

$$k^{-2} \approx 10^{-3} \div 10^{-2} \tag{1321}$$

smaller than in the case of rich galaxy clusters.

The next generation of X-ray telescopes will make it possible to search for QSO up to the red shifts equal to $z = 5 \div 10$ and for IGG with the red shifts $z = 1 \div 3$ in the rich galaxy clusters with the red shifts up to $z = 2 \div 4$.

Hence, in the case, considered here, IGG may be observed for the red shifts

$$z = 1 \div 3 \ . \tag{1322}$$

Observation of hot intergalactic gas without visible rich galaxy cluster can be strong argument, favouring the existence of the mirror (shadow) matter.

III. The gravitational interaction of ordinary and mirror galaxies results in the distortion of their form. In the ordinary galaxy the distortion must be observed in the absence of its visible source. The evolved numerical methods of the calculation of tidal interactions between the galaxies (see Toomre, Toomre, 1972; Korovyakovskaya, Korovyakovsky, 1982; Naguchi, 1987 and ref. therein) can provide solution for the inverse problem of determination of the parameters of the body, inducing the perturbations, by the form of the distorted galaxy.

Contrary to the perturbation, induced by a single black hole of the same mass, the perturbation by the mirror galaxy is not related to the observational effects of accretion on such black hole. Corresponding effects of accretion on the active nuclei of a mirror galaxy will be suppressed by the mass ratio of the galactic nucleus M_{nuc} to the mass of the whole galaxy M_{gal} being of the order of

$$\frac{M_{nuc}}{M_{gal}} \leq 10^{-2} \div 10^{-4} \tag{1323}$$

for active galactic nuclei.

IV. The gravitational perturbation, induced by the M-galaxy, can initiate the burst of star formation in the O-galaxy rich by gas or in the protogalactic O-gas cloud. As a result the O-galaxy will be observed as the irregular.

The phenomena III and IV can be also induced by non-luminous gravitationally bound clusters of dark matter, which may be formed due to the biasing in the distribution of baryons and dark matter.

However, the collisionless gas of dark matter particles can not form dense inhomogeneities, typical for the mirror matter, in which the mechanisms of dissipation exist. So one must look for the effects, induced by the mirror matter of the moderate concentration, differing from the effects of both black holes and diffused weakly interacting dark matter.

V. If the activity of galactic nuclei is determined by the existence of black holes with the masses of the order of

$$M_{BH} \sim 10^6 \div 10^{10} M_\odot \,, \tag{1324}$$

the symmetry between O- and M-matter provides the conjecture on the exsistence of such black holes in the nuclei of mirror galaxies. In this case, M-galaxies with active nuclei will be observed as the separate supermassive black holes and their observational properties will be determined by the amount of the ordinary matter in their neighbourhood.

As it was mentioned by Blinnikov and Khlopov (1983), the presence of M-matter in the mixed O-M galaxy makes it easy to explain the origin of closed binary supermassive black holes in galaxy nuclei.

VI. The massive M-galaxies can induce the effect of gravitational lens without optically visible source of this effect.

VII. Rapid motions of M-matter masses may be the source of gravitational waves without any other observational effects. One can discriminate them from the supermassive black holes, which also may be the source of gravitational radiation, by the effects of accretion on such black holes.

VIII. The presence of O-gas in the M-galaxy will result in the observed single gas clouds or small mass galaxy with the large internal velocity dispersion, i.e. with the large internal virial paradox.

As the possible candidates in such fomations intergalactic gaseous clouds can be considered. In a number of the neighbouring groups of galaxies cold gas is observed in the form of massive HI clouds with the mass

$$M > 10^6 M_\odot \tag{1325}$$

(Haynes, 1979) and with the sizes, typical for galaxies and being of the order of

$$r \sim 20 \div 25 \text{ kpc} . \tag{1326}$$

The most massive from the known HI clouds is discovered (see Khlopov et al., 1991 for Refs.) in the galaxy group G11 in the Lion constellation near the galaxy M96. This cloud has the size no less, than

$$r \geq 30 \div 100 \text{ kpc} . \tag{1327}$$

The mass of neutral HI hydrogen in it is estimated as

$$M > 10^9 M_\odot \tag{1328}$$

and the concentration

$$n = 4 \cdot 10^{-9} h_{50} \text{ cm}^{-3} , \tag{1329}$$

where the dimensionless Hubble constant is defined as

$$h_{50} = \frac{H}{50 \, \frac{\text{km}}{\text{s} \cdot \text{Mpc}}} \, . \tag{1330}$$

The surface brightness of this cloud is in the optical range smaller, than 30 stellar magnitudes per the square arc second.

Such clouds may cause effect in the absorption spectra of QSO.

4.2. The Mirror Matter at the Scale of Globular Clusters

Let us consider now the case of separation of M- and O-matter at the scale of globular clusters.

Globular clusters are among the oldest astronomical objects, being formed possibly before the galaxy from the inhomogeneities with the mass

$$M \leq 10^6 M_\odot \, . \tag{1331}$$

They seem to be the objects of definite mirrority even for the homogeneous initial mixing of the O- and M-matter (see Sec. 3).

I. The capture of O-stars by the M-globular cluster can lead to the formation of diffused cluster of ordinary stars, surviving for very long time and possessing strong virial paradox.

The better chances to form such objects by captures of background O-stars have M-globular clusters, moving near the galactic plane along the orbit with not large eccentricity. But, following Khlopov et al. (1989; 1991), the better chances for diffused cluster to be formed in the gravitational field of M-globular cluster are in the case, when the diffused cluster is formed from O-gas, captured by the mirror globular cluster in the period of separation of the matter of different mirrority. According to Blinnikov, Khlopov (1983) (see Sec.3) the fraction of gas of the other mirrority is in this case of the order of

$$f \approx 10^{-2} \ . \tag{1332}$$

The time scale of decay of normal diffused clusters (Wielen, 1971) is

$$t_{dc} = (1 \div 3) \cdot 10^{8} \text{ years} \ , \tag{1333}$$

whereas such cluster, being formed in the potential well of mirror globular cluster, can have the age

$$t_{dc} \approx 10^{10} \text{ years} \ . \tag{1334}$$

There are several old diffused clusters, observed in the Galaxy with so great lifetime.

For NGC 188 the age is estimated (Vandenberg, 1983; Jane, Demarque, 1983) as

$$t_{dc} \approx (5 \div 10) \cdot 10^{9} \text{ years} \ . \tag{1335}$$

For M67 the age estimated in (Nissen et al., 1987) is

$$t_{dc} \approx (5 \pm 0.5) \cdot 10^{9} \text{ years} \ . \tag{1336}$$

For NGC 752 this age is equal to (Twarog, 1983)

$$t_{dc} \approx 2 \cdot 10^{9} \text{ years} \ . \tag{1337}$$

For NGC 2243 and Melloffe 66 their age is estimated by Gratton (1982) to be equal to

$$t_{dc} \approx 6 \cdot 10^{9} \text{ years} \ . \tag{1338}$$

Stars, captured by mirror globular cluster, can have various ages contrary to the usual diffused clusters. This fact should be taken into account in the determination, whether the star belongs to the diffused cluster.

II. After the capture of ordinary star by mirror globular cluster or of mirror star by the ordinary globular cluster close binaries may be formed with the components of the different mirrority, O-M binaries. The existence of nonrelativistic invisible companion is the specific feature of such systems (see below).

In the cases, when the gas clouds of ordinary and mirror matter, giving birth to globular clusters , are close to each other in the phase space (see Sec. 3), the mixed globular O-M cluster can be formed.

Besides the evident virial paradox in the compact mixed O-M globular cluster of the radius R , containing N stars, during the time t owing to tidal dissipation binaries can be formed as a result of the large star concentration. For the O-M cluster with $N = 2\ 10^5$ stars, within the radius of $R = 5$ pc during the time 10^{10} years there will be formed

$$N_{bin} \cong 60 \tag{1339}$$

binaries, about half of which will be the O-M binaries.

III. As it was shown by Khlopov et al. (1989; 1991), the effects of mirror matter in the clouds of ordinary molecular gas are strongly enhanced as compared with the general case, considered in Sec. 7.3 .

Giant molecular clouds are the most abundant among the massive unitary objects in galaxies. Having the mass of the order of the one of the globular clusters, i.e.

$$M = 10^4 \div 10^6 M_{\odot} \ , \tag{1340}$$

these clouds are by the order of magnitude more abundant than globular clusters in the Galaxy (Solomon, Edmunds,

1980). They also have the size by the order of magnitude higher, than the one of the globular clusters.

Hence, in the galaxy, containing comparable amounts of ordinary and mirror matter, cross-penetration of O- and M-clouds through each other should be rather frequent.

The molecular clouds contain large amount of internal inhomogeneities, which are the regions of the enhanced density close to the development of the gravitational instability.

The structure of the regions of star formation shows that even relatively small perturbation can initiate the formation of stars within the molecular clouds.

Shock waves are the usual triggers for star formation, so that the youngest objects are found in a thin outer layer of molecular clouds (Thaddeus, 1977).

Penetration of a massive body through molecular cloud may also initiate star formation, so that the region of star formation will be determined by the spatial distribution of gravitational perturbation and may take place in a significant part of the cloud volume.

In the case of the comparable amounts of O- and M-matter in the galaxy, what is considered here, such perturbation may be induced first of all by M-molecular cloud and with the probability by an order of the magnitude smaller by M-globular cluster.

Since molecular clouds intersect with a small relative velocity $v \sim 10$ km/s and are the strong dissipating objects, the intersection of O- and M-clouds may form a giant molecular cloud with the mixed mirrority, inside which the probability to form the mixed O-M stars and O-M binaries is enhanced.

Multiple gravitational interaction of stars with the inhomogeneities within the molecular clouds may lead to capture of some stars by molecular clouds, strongly enhancing the effects of interstellar gas accretion on such stars. It gives rise to higher accretion rate of gas on the star of the other mirrority, i.e. of the O-gas on M-star or M-gas on O-star.

4.3. The Effects of Mirror Matter at Stellar Scales

We have already mentioned in Sec. 3 some effects of the existence of the mirror matter at the stellar scales. Let us give here more general and systematic treatment of such effects, following Khlopov et al. (1989, 1991).

I. The accretion of interstellar gas by the star of the other mirrority results in the admixture of the matter of the other mirrority in stars.

The estimations based on the analysis of the process of accretion, presented in Sec. 3 give for the admixture of O-matter in M-stars and of M-matter in O-stars the magnitude of the order of $\Delta M \sim 10^{-7} \div 10^{-6}\, M_7$.

If such admixture of the mirror matter in the Sun forms a mirror planet near the solar surface, it may give, according to Blinnikov and Khlopov (1980, 1983) the explanation of the source of non-radial solar oscillations with the period $T = 160$ min.

V.F. Shwartsman pointed out the possibility to observe the accretion of interstellar O-gas on the single M-neutron star by the effects of supershort variations of its luminosity (see Sec. 3.3 for more details)

II. The rotation of an ordinary star and of M-matter captured by it around their common center of mass leads to periodical variations of pulsar period due to Doppler effect.

Contrary to the similar effect of hardly observed ordinary-matter planets, the variations, induced by the M-matter, can have the period, corresponding to such close orbits, for which the formation of ordinary planets or their survival after Supernova explosion are impossible.

The searches for these effects need highly precisional timing of pulsars on time intervals, smaller than few hours.

III. As a result of the accretion of ordinary gas and dust on a mirror star, the disc of protoplanet type without young star in its center will be formed in the interstellar medium.

Such discs may be observed by radio line of the molecule CO – 2.6mm.

The mass of the central configuration can be determined by the Doppler effect, and it will be in strong contradiction to the low luminosity of the central body.

The accretion in the regions of higher density of the gas with other mirrority and the formation of O-M binaries give new possibilities of searches for the mirror matter.

As it was pointed out in Sec.4.2 , O-M binaries may be formed inside the globular cluster and at the star formation during the mutual penetration of giant O- and M-molecular clouds.

IV. The star formation or the evolution of O-M binary can lead to a mixed O-M star, containing the comparable amounts of the ordinary and mirror matter.

For the ordinary matter in the mixed O-M star the relation between the main stellar parameters (mass, radius, luminosity, colours, effective temperature etc.) can be strongly violated.

In particular, such star must occupy an unusual position on the Herzsprung-Russel diagram.

After Supernova explosion in the mirror component of the mixed O-M star, the rearrangement of its gravitational field takes place. It is inevitably followed by the rearrangement of the stellar structure and by a one-time change of the properties of the optically visible object. The star should seem to enlarge its size and to decrease its surface temperature.

V. In the close O-M binaries accretion of the ordinary matter on the potential well, induced by the mirror companion star, must result in the formation of accretion discs without visible centres of accretion.

In the case of nonrelativistic mirror star, when the potential well has rather "flat' bottom, the thick accretion disc or the spheroidal configuration of the accreted O-matter should be formed around the invisible M-companion. For the rates of mass exchange typical for close binaries, the configuration typical for highly non-spherical star arises.

One of perspective methods to study such objects is the analysis of linear polarisation of their radiation (Khlopov et al., 1991; Dolginov et al., 1979; Bochkarev, Caritskaya, 1988).

For the low mass O-companion of the O-M binary, the O-disc around the M-companion should remain rather cold, looking like low-luminosity infrared source, and with mass being in the sharp contradiction to the value of the "disc mass", determined by the Doppler effect and equal to the mass of M-star.

In the separated O-M binaries, the effects of accretion of the non-degenerated M-star may be absent. Such star can be discovered as the invisible massive companion of the ordinary star.

The estimation of the fraction of O-M systems among the binaries can be obtained on the basis of the existing catalogue of spectral binaries.

In the catalogues (Batlen et al., 1978; Pedoussaut et al., 1984, 1988) the physical parameters and the elements of orbits for about 1500 spectral binaries are given.

Taking into account, that the theoretical upper limit for the mass of the neutron star is

$$M_{ns} < 3M_{\odot} \tag{1341}$$

(Baym, Pethick, 1979) one can put in the analysis of spectral binaries the constraint

$$M_{inv} > 3M_{\odot} \tag{1342}$$

on the mass of the second invisible companion, in order to exclude the systems with faint white dwarfs and neutron stars.

One can find 45 binaries in the catalogue (Batlen et al., 1978; Pedoussaut et al., 1984, 1988) satisfying this constraint, since the function of masses for these binaries are

$$f(M) > 3M_\odot \ . \tag{1343}$$

Among these 45 binaries only 6 have no lines corresponding to the second companion. With the account for the possibility, that the second invisible companion is faint O-star or black hole, Khlopov et al. (1989) have estimated the fraction of O-M binaries as

$$\alpha < \frac{6}{45} = 13\% \ . \tag{1344}$$

VI. Consider now, following Khlopov et al. (1989; 1991) the properties of these 6 binaries and the possibilities to discriminate in the observations a black hole and a massive M-star as the invisible companion.

The parameters of these 6 spectral binaries are given in the catalogue (Batlen et al., 1978; Pedoussaut et al., 1984, 1988).

In these binaries there are no lines observed, which definitely belong to the second companion.

The function of masses for these binaries satisfy the condition (1343).

From the definition of the function of masses

$$f(M) = \frac{M_2 \sin^3 i}{\left(1 + \frac{M_1}{M_2}\right)} \ , \tag{1345}$$

where i is the inclination of the orbit, one obtains that

$$M = f(M)\frac{\left(1+\frac{M_1}{M_2}\right)^2}{\sin^3 i} > f(M) > 3M_{\odot} \ . \qquad (1346)$$

So in both cases the second component can not be the white dwarf or the neutron star.

If the successive detailed study of these systems will definitely exclude the O-star on the stage of ignition of central thermonuclear burning as the second companion, that means that either the black hole or the M-star is present in the system.

The main criterion to choose between these two variants seems to be the total rate of the energy release in the binary.

In the case of black hole, this rate should be by several orders of the magnitude higher, than in the case of mirror star.

With the account for the possible effective shielding of the radiation by the accretion disc one should consider the integral luminosity over the whole spectrum.

Moreover, very fast variations of luminosity with the minimal time scale of the order of

$$t \sim \frac{r_g}{c} \qquad (1347)$$

should arise in the case of accretion on the black hole.

In this context the binary AO620-00, being the X-ray Nova with millisecond bursts, should be related to the systems with the black hole rather with the M-star.

In the binaries W Cru and V600 Her, the lower estimations of the mass of the second invisible companion give respectively 16 and 32 solar masses.

With such mass, the second component, being neither black hole, nor M-star, should have luminosity higher than the visible component.

VII. The inflow of the ordinary matter on the mirror white dwarf or on the mirror neutron star can lead to the formation of a dense region of O-matter in the centre of accretion disc, having the size of the M-star or its central core. The latter is by the order of magnitude smaller than the size of the whole star.

Observational properties of such a dense region will be close to the corresponding properties of ordinary degenerated stars with some quantitative differences, such as possibly smaller size and higher temperature.

Furthermore, in the case of the mirror white dwarf presumably hydrogen O-object will be observed on its place as the result of O-matter accretion.

Detonation in such an object may lead to phenomena looking like Novae explosions with some possible quantitative differences.

In the case of the mirror neutron star phenomena like bursters can arise with quantitative parameters possibly different from the usual ones.

The ordinary neutron star, being the binary for the normal mirror star, will be observed as radio pulsar in a binary with invisible companion.

Such binary can be distinguished from the binary of relativistic O-stars by the rate of change of the binary orbit, which is by many orders of magnitude more rapid than the rate owing to gravitational wave radiation from the O-relativistic binary.

The rapid evolution of the orbit of O-M binary may follow from

a) the motion of the apsides line induced by the final size of the normal M-companion,
b) accretion of the mirror matter from the normal M-companion on the observed O-neutron star,
c) mass loss by the system owing to non-conservative mass exchange etc.

CHAPTER 11

COSMOPARTICLE PHYSICS OF HORIZONTAL UNIFICATION

1. Physical Grounds for Horizontal Unification

1.1. The Symmetry of Fermion Families

The problem of fermion families (generations) remains one of the central problems of the particle physics.

The standard $SU(3)$x$SU(2)$x$U(1)$ gauge model, as well as its possible "vertical" extensions based on the unification gauge groups $SU(5)$, $SO(10)$ etc., is constructed in the framework of one fermion generation. These models do not contain any deep physical grounds for the existence of mass hierarchy between the fermion families and for the observed weak mixing of quarks (and leptons).

In these models, Yukawa couplings are arbitrary, and one has to input by hands their values for each fermion to reproduce the experimental data on their masses and mixings.

The identity of quark and lepton families

$$\begin{pmatrix} u \\ d \\ \nu_e \\ e \end{pmatrix}, \quad \begin{pmatrix} c \\ s \\ \nu_\mu \\ \mu \end{pmatrix}, \quad \begin{pmatrix} t \\ b \\ \nu_\tau \\ \tau \end{pmatrix}$$

relative to strong and electroweak interactions strongly suggests the existence of a "horizontal" symmetry between these generations.

It is reasonable to consider the concept of local horizontal symmetry $SU(3)_H$, first proposed by Chkareuli (1980).

Under the action of this symmetry transformations left-handed quark and lepton components transform as $SU(3)_H$ triplets, and the right-handed ones transform as antitriplets. Their mass terms transform as

$$3 \otimes 3 = 6 \oplus \overline{3} \tag{1348}$$

and consequently can arise only as a result of horizontal symmetry breaking.

This approach may be trivially generalised on the case of n generations, assuming the respective $SU(n)$ symmetry. In the case of three generations, the choice of the group of horizontal symmetry is unique, since the orthogonal and vector-like gauge groups can not provide different representations for left- and right- handed fermion states.

In the considered approach the hypothesis is reasonable, that the structure of the mass matrices is determined by the pattern of the horizontal symmetry breaking, i.e., by the structure of vacuum expectation values (VEV) of horizontal scalars, maintaining $SU(3)_H$ breaking.

The mass hierarchy between generations is related to the hypothesis of a definite hierarchy in this breaking. Berezhiani, Chkareuli (1982, 1983) and Berezhiani (1983) called it the hypothesis of horizontal hierarchy – HHH.

The simplest realisation of the HHH invokes the introduction of additional superheavy fermions, acquiring their masses via direct coupling with horizontal scalars. The ordinary quark and lepton masses are induced by their "see-saw" mixing (Berezhiani, 1983) with these heavy fermions.

The concept of grand unification (GUT) is another argument in favour of chiral horizontal G symmetry. In the GUT models, left-handed quarks and leptons are put together with antiparticles of their right-handed component into the same irreducible representations of GUT group G_{GUT}. In the framework of $G_{\mathrm{GUT}} \mathrm{x} G_H$ symmetry, left-handed and right-handed components must transform as the conjugated representations of the group G_H, i.e. the G_H symmetry must be chiral.

One may hope, that the complete unification of horizontal and vertical symmetries will be achieved on the basis of unifying fundamental symmetry G , including $G_{\mathrm{GUT}} \mathrm{x} G_H$ in the course of superstring theory development. The most elaborated simplest variant of realistic superstring model $E_8 \mathrm{x} E_8'$ (Candelas et al., 1985; Witten, 1985) does not leave any room for inclusion of horizontal symmetry. However, such inclusion is possible in the framework of wider class of superstring models, e.g., in $SO(32)$ or in heterotic string models with direct compactification to 4-dimentional space–time.

In the latter case (Narain, 1986; Kawai et al., 1986) a wide class of GUT groups with the rank r smaller than

$$r \leq 22 \tag{1349}$$

is possible.

The analysis of the horizontal unification as the phenomenology for theories of everything, to be given here on the basis of cosmoparticle physics, may be useful in this aspect for the proper choice of realistic models from the variety of possibilities, existing for the superstring models.

Here we will not consider supersymmetric extensions of the model, in which a number of new particles are predicted, extending the hidden sector of the theory. Properties of such particles depend critically on the details of supersymmetry breaking and need special study.

To build the realistic model of broken horizontal symmetry, rather wide set of parameters is to be introduced.

However:

1. the number of these parameters is smaller than in the realistic models without horizontal symmetry;
2. the bulk of these parameters is fixed by the experimental data on the quark and lepton properties,

and, finally,

3. the set of new nontrivial physical phenomena, predicted by the model, provides in principle complete check of the model and determination of all the parameters.

These new phenomena arise at high-energy scale of horizontal symmetry breaking F , which is of the order of

$$F > 10^5 \div 10^6 \text{ GeV} , \tag{1350}$$

and can not be achieved at accelerators even in the far future.

However, the combination of experimental searches for their indirect effects in the rare processes with known particles with the analysis of their cosmological and astrophysical effects makes it possible to study physics, predicted at this scale, as well as the cosmological scenarios, based on this physics.

The model proposed by Berezhiani, Khlopov (1990a,b,c; 1991), Berezhiani et al. (1990a,b, 1992), Sakharov, Khlopov (1993; 1994a,b; 1995; 1996) satisfies the following naturality conditions.

1) Natural suppression of flavour changing neutral currents (FCNC) (Glashow, Weinberg, 1977).

The light scalar $SU(2)\text{x}U(1)$ doublets, transforming according to nontrivial representations of G_H (vertical-horizontal fields) and leading to unacceptably strong FCNC, are absent.

Yukawa couplings, responsible for quark and lepton mass generation, include the only standard $SU(2)\text{x}U(1)$

Higgs doublet, though left-handed and right-handed fermion components transform as the conjugate representations of horizontal symmetry group.

The price for it is the introduction of additional superheavy fermions, maintaining the hidden sector of the theory. Mixing with these heavy fermions induces quark and lepton masses. The so-called “see-saw” mechanism (Gell-Mann et al., 1979) is realised not only for neutrinos, but also for all the quarks and leptons either.

Besides that, this approach provides natural explanation of the hierarchy of electroweak and GUT scales on base of mechanisms suggested for GUT models with one generation, e.g., by means of supersymmetric extensions of $SU(5)$, $SO(10)$ etc. (Nilles, 1983; Vysotsky, 1985).

Quark and lepton mass generation via the vertical-horizontal fields would have needed unnatural fine tuning of their parameters even in the supersymmetric case.

2) Natural horizontal hierarchy.

The observed mass hierarchies of families, say, the charged lepton mass ratio

$$m_e : m_\mu : m_\tau = 1 : 200 : 4000 \qquad (1351)$$

is explained by much more moderate hierarchy of horizontal symmetry breaking. The parameters of such breaking are proportional to square root of masses of quarks and leptons. So there is no need in special mechanisms to protect the hierarchy from loop corrections.

3) Natural solution for QCD CP-violation problem (Peccei, Quinn, 1977).

Though there is the only one Higgs doublet, the theory provides natural inclusion of $U(1)$ Peccei-Quinn symmetry (Peccei, Quinn, 1977), associated with heavy Higgs fields, which break the horizontal symmetry G_H at the scale F. Breaking of this global $U(1)$ symmetry results in the existence of pseudo-Goldstone boson α of invisible axion type with the interaction scale F (Zhitnicki, 1980; Dine et al., 1981; Wise at al., 1981).

The boson α has both flavour-diagonal and flavour-non-diagonal coupling with quarks and leptons, i.e., is simultaneously familon of the singlet type (Wilczek, 1982; Anselm, Uraltzev, 1983).

Finally, it is related to neutrino Majorana mass generation, being also in fact the Majoron of the singlet type (Chikashige et al., 1980).

The inevitable consequences of this model are:

a) the flavour-changing neutral transitions, related to axion and horizontal gauge bosons interactions;

b) the existence of neutrino Majorana mass and of the neutrino mass hierarchy of different families;

c) the instability of heavier neutrino relative to axion decay on lighter neutrino;

d) the existence of metastable superheavy fermions.

The model can be tested in the combination of laboratory tests, such as:

- the search for neutrino mass,
- the search for neutrino oscillations,
- the search for double neutrinoless beta decay,
- the study of $K^0 \leftrightarrow \bar{K}^0$ and $B^0 \leftrightarrow \bar{B}^0$ transitions,
- the search for axion decays $\mu \rightarrow e\alpha$, $K \rightarrow \pi\alpha$, etc.

together with the analysis of its predicted cosmological and astrophysical effects.

The latter includes the study of axion emission effects on stellar evolution, investigation of primordial axion field and massive unstable neutrino effects on the dynamics of cosmological large scale structure formation, as well as the analysis of the mechanisms of inflation and baryogenesis, based on the hidden sector of the model.

1.2. The Gauge Model of Family Symmetry Breaking

Following Sakharov, Khlopov (1994a,b) and Khlopov, Sakharov (1994), consider the $SU(2)\mathrm{x}U(1)$ model of electroweak interactions supplemented by the local chiral horizontal symmetry $SU(3)_H$ between the fermion families (2) (Chkareuli, 1980; Berezhiani, Khlopov, 1990a,b,c, 1991; Berezhiani et al., 1990a,b).

Quarks and leptons are joined in the following representations of the group $SU(2)\mathrm{x}U(1)\mathrm{x}SU(3)_H$

$$f_{L\alpha}: \begin{pmatrix} u \\ d \end{pmatrix}_{L\alpha} \left(2, \frac{1}{3}, 3\right); \quad \begin{pmatrix} \nu \\ l \end{pmatrix}_{L\alpha} \quad (2, -1, 3)$$

$$f_R^\alpha: \; u_R^\alpha \left(1, \frac{4}{3}, \overline{3}\right) \; ; \quad d_R^\alpha \left(1, -\frac{2}{3}, \overline{3}\right) \; ;$$

$$e_R^\alpha \; \left(1, -2, \overline{3}\right) , \tag{1352}$$

where we retain the family ($SU(3)_H$) index $\alpha = 1, 2, 3$.

We can choose scalars, breaking the horizontal symmetry, as $SU(3)_H$ sextets and triplets. All of them should be $SU(2)\mathrm{x}U(1)$ singlets, to prevent the electroweak symmetry breaking at the $SU(3)_H$ symmetry breaking scale.

To generate realistic quark and lepton mass matrices, at least three such "horizontal" scalars are needed.

At least one of them with the largest vacuum expectation value (VEV) should be a sextet:

$$\xi^{(0)}_{\{\alpha,\beta\}} \quad ; \quad \alpha,\beta = 1,2,3 \; , \tag{1353}$$

where the brackets {...} denote symmetrisation on the group indices. Otherwise, triplet fields only can not generate realistic mass matrices.

For the two other scalars

$$\xi^{(1)}_{\alpha,\beta} \quad ; \quad \alpha,\beta = 1,2,3 \tag{1354}$$

and

$$\xi^{(2)}_{\alpha,\beta} \; ; \; \alpha, \beta = 1,2,3 \tag{1355}$$

we do not specify further their $SU(3)_H$ content, mentioning only those cases, when sextet and triplet representations result in the different consequences.

Let us introduce additional fermions in the form

$$F_L^\alpha: \; U_L^\alpha \left(1, \frac{4}{3}, \bar{3}\right) \; ; \; D_L^\alpha \left(1, -\frac{2}{3}, \bar{3}\right) \; ; \; E_L^\alpha \; (1, -2, \bar{3})$$

$$F_{R\alpha}: \; U_{R\alpha} \left(1, \frac{4}{3}, 3\right) \; ; \; D_{R\alpha} \left(1, -\frac{2}{3}, 3\right) \; ;$$

$$E_{R\alpha} \; (1, -2, 3) \; ; \; N_{R\alpha} \; (1, 0, 3) \tag{1356}$$

(Berezhiani, 1983; Berezhiani, Khlopov, 1990a,b,c, 1991; Berezhiani et al., 1990a,b).

These fermions cancel the $SU(3)_H$ anomaly of the ordinary quarks and leptons (1352), what implies the existence of the partner for neutrino, neutral lepton N.

The most general Yukawa couplings for quarks and leptons allowed by the symmetry are given by Lagrangian

$$L_{Yuk} = g_f \bar{f}_{L\alpha} F_{R\alpha} \phi^0 + G_F^{(n)} \bar{F}_L^\alpha F_{R\beta} \bar{\xi}^{(n)\alpha\beta} + G_\eta \bar{F}_L^\alpha f_R^\alpha \eta + h.c. \; , \tag{1357}$$

where $n = 0, 1, 2$; $f = u, d, e$; $F = U, D, E$; (Berezhiani, 1983; Berezhiani, Khlopov, 1990a,b,c, 1991; Berezhiani et al., 1990a,b; Khlopov, Sakharov, 1994a,b) and for neutrinos such couplings have the form

$$L_{Yuk,\nu} = g_\nu \bar{\nu}_{L\alpha} N_{R\alpha} \phi^0 + G_N^{(n)} N_{R\alpha} C N_{R\beta} \bar{\xi}^{(n)\alpha\beta} + h.c. \tag{1358}$$

(Berezhiani, Chkareuli 1983; Berezhiani, Khlopov, 1990a,b,c; 1991; Berezhiani et al., 1990a,b; Khlopov, Sakharov, 1994a,b).

Here ϕ^0 is the neutral component of the standard $SU(2)\mathrm{x}U(1)$ Higgs doublet. It corresponds to the representation of the considered $SU(2)\mathrm{x}U(1)\mathrm{x}SU(3)_H$ group, given by

$$\phi \quad (2, \quad -1, \quad 1) \tag{1359}$$

and has the vacuum expectation value

$$\langle\phi^0\rangle = v = \left(\sqrt{8}G_F\right)^{-1/2} = 250 \text{ GeV} . \tag{1360}$$

In the Eq. (1357) the Higgs field η is the real $SU(2)\mathrm{x}U(1)\mathrm{x}SU(3)_H$ singlet scalar, having the vacuum expectation value

$$\langle\eta\rangle = \frac{\mu}{G_\eta} . \tag{1361}$$

Yukawa couplings, given by Eqs. (1357)–(1358) are invariant under the global axial $U(1)_H$ transformations:

$$f_L \to f_L \exp(i\omega) \; ; \quad f_R \to f_R \exp(-i\omega) \; ;$$
$$F_L \to F_L \exp(-i\omega) \; ; \quad F_R \to F_R \exp(i\omega) \; ; \quad \phi \to \phi \; ;$$
$$\xi^{(n)} \to \xi^{(n)} \exp(2i\omega) , \tag{1362}$$

where $n = 0, 1, 2$.

The Higgs potential also possesses this $U(1)$ symmetry, provided that there are no trilinear couplings such as

$$L_{tril} = \Lambda\xi^{(0)}_{\alpha\beta}\xi^{(1)\alpha}\xi^{(1)\beta} + h.c. \; . \tag{1363}$$

These couplings are not induced by any other (gauge or Yukawa) interactions. So their absence in the Lagrangian is natural (Berezhiani, Chkareuli, 1983; Berezhiani,

Khlopov, 1990a,b,c, 1991; Berezhiani et al., 1990a,b; Khlopov, Sakharov, 1994a,b)

The analysis of the Higgs potential (see Berezhiani, Chkareuli, 1983; Berezhiani, Khlopov, 1990a,b,c, 1991; Berezhiani et al., 1990a,b; Khlopov, Sakharov, 1994a,b for details) shows, that the VEV matrix can obtain the form:

$$V_H = \left\langle \xi^{(0)} + \xi^{(1)} + \xi^{(2)} \right\rangle = \begin{pmatrix} r_1 & p_1 & p_3 \\ \pm p_1 & r_2 & p_2 \\ \pm p_3 & \pm p_2 & r_3 \end{pmatrix}, \quad (1364)$$

where "+" and "–" signs correspond to sextet and triplet representations for the scalars $\xi^{(1)}$ and $\xi^{(2)}$, respectively. The vacuum expectation values have natural 5–10 fold hierarchy

$$r_1 > p_1 > r_2 > p_2 > p_3 > r_3 \, . \quad (1365)$$

Inserting the VEV of scalar fields into Yukawa couplings, given by Eqs. (1357)–(1358), one obtains full 6 x 6 mass matrix for charged fermions:

$$\begin{matrix} & f_R & F_R \\ \overline{f}_L & & \\ \overline{F}_L & & \end{matrix} \begin{pmatrix} 0 & g_f v \\ \mu & M_F \end{pmatrix} \quad (1366)$$

and the corresponding mass matrix for neutrinos

$$\begin{matrix} & \nu_R & N_R \\ \overline{\nu}_L & & \\ \overline{N}_L & & \end{matrix} \begin{pmatrix} 0 & g_\nu v \\ g_\nu v & M_N \end{pmatrix} . \quad (1367)$$

Here the 3 x 3 mass matrices are given by

$$M_F = \sum \left\langle \xi^{(n)} \right\rangle G_F^{(n)} \tag{1368}$$

with $F = U, D, E, N$; and the neutrino states are defined as

$$N_L = C\overline{N}_R \ ; \quad \nu_R = C\overline{\nu}_L \ .$$

Note, that only sextet scalars contribute into Majorana mass matrix M_N .

So, one has Dirac "see-saw" mechanism of the quark and lepton mass generation and ordinary Majorana "see-saw" mechanism for neutrino mass term. In the latter case the neutral lepton states N_R play the role of right-handed neutrino.

The quark and charged lepton mass matrices obtained from the block-diagonalisation of the matrices, given by the Eq. (1366) have the form:

$$m_f = \frac{\mu}{M_F} g_f v \ , \tag{1369}$$

where $f = u, d, e$; and the mass matrix for neutrinos after the block-diagonalisation of the matrix (1367) is given by

$$m_\nu = \frac{(g_\nu v)^2}{M_N} \ . \tag{1370}$$

The mass hierarchy between the families appears to be inverted with respect to the hierarchy of the $SU(3)_H \mathrm{x} U(1)_H$ symmetry breaking:

$$SU(3)_H \otimes U(1)_H \Rightarrow SU(2)_H \otimes U'(1)_H \Rightarrow U''(1)_H \Rightarrow I \ , \tag{1371}$$

where I is the trivial group of identical transformations.

Here the intermediate $SU(2)_H$x$U'(1)_H$ horizontal symmetry is maintained between the second and the third generations of quarks and leptons. The global $U''(1)_H$ symmetry remaining unbroken after the second step of symmetry breaking is appropriate to the third generation only.

The considered case is called the inverse hierarchy model, in contrast with the direct hierarchy model, where the hierarchy of quark and lepton masses is parallel with the hierarchy of $SU(3)_H$x$U(1)_H$ symmetry breaking (Berezhiani, 1985; Berezhiani, Khlopov, 1990a,b,c, 1991; Berezhiani et al., 1990a,b; Khlopov, Sakharov, 1994a,b).

The global $U(1)_H$ ($U''(1)_H$) symmetry breaking results in the existence of a massless Goldstone boson, α, named archion, having both flavour diagonal and flavour non-diagonal couplings with quarks and leptons and thus being familon of the singlet type (Anselm, Uraltzev, 1983; Anselm, Berezhiani, 1985). We have discussed the model of archion in the Chapter 2.

In the minimal $SU(2)$x$U(1)$x$SU(3)_H$ model the archion α is the particle of the arion type (Anselm, 1982; Anselm, Uraltzev, 1982). It has vanishing couplings with photons and with gluons due to strict compensation of respective triangle diagrams.

The archion interactions with the ordinary matter are suppressed sufficiently to remove the strong astrophysical restrictions (see review in Kim, 1987; Cheng, 1988) on the scale v'_H.

At the lowest possible scale of family symmetry breaking, the flavour changing decays can go with noticeable probability such as

$$\mu \to e\alpha\,, \quad \tau \to \mu\alpha\,, \quad K \to \pi\alpha\,, \quad B \to K\left(K^*\right)\alpha\,,$$
$$D \to \pi(\rho)\alpha\,. \qquad (1372)$$

The search of such decays could provide the valuable information about the structure of fermion mass matrices.

In any realistic extension of the scheme of horizontal unification, e.g. in the framework of the gauge model based on the symmetry $SU(5)\mathrm{x}SU(3)_H$ triangle diagrams owing to the inevitable presence in GUTs of additional heavy fermions induce archion couplings with photons and gluons, so that $U''(1)_H$ symmetry may be identified with the Peccei-Quinn symmetry (Peccei, Quinn, 1977). Archion represents the specific type of invisible axion (see Chapter 2).

The scale

$$v''_H = v_{PQ} \tag{1373}$$

is then restricted from below by astrophysical estimations of stellar energy losses due to archion emission:

$$v_{PQ} > 10^6 \text{ GeV} . \tag{1374}$$

The constraint (1374) is based on the observational data on the Sun and red giants and follows from the theory of stellar evolution.

At each stage of stellar evolution the time scale t_{ev} is determined by the rate of energy losses L for the fixed energy Q released in thermonuclear reactions at the considered stage and is equal to

$$t_{ev} \sim \frac{Q}{L} . \tag{1375}$$

Additional energy losses due to axion emission L_a lead to smaller evolution time scale t_{ev}'

$$t'_{ev} \sim \frac{Q}{L + L_a} \ . \tag{1376}$$

Observational data on the Sun, stars of main sequence and red giant provide lower limit on this time scale, putting upper limit on the rate of axion emission and thus the lower limit on the scale of axion (archion) interaction.

On the other hand, the maximal allowed effect of archion emission accelerates the evolution of denser and hotter central part of stars what may be important in the interpretation of data on presupernova SN 1987A. The predecessor of this supernova did not exhibit the properties expected for stars in the end of their evolution. It may be explained by higher rate of evolution of the central core, which is not observed directly.

Archion emission from the collapsing star can reduce the energy and time scale of neutrino radiation. The claims on the observation of the neutrino signal, preceding the burst of Supernova SN1987A are used to constrain the possible axion luminosity and thus the scale of axion interaction

$$v_{PQ} > 10^{10} \text{ GeV} \ . \tag{1377}$$

In view of the complications related to calculations of neutrino transport in the collapsing star opaque for neutrino the restriction (1377) should be taken with caution.

This restriction assumes free propagation of archion radiation in the collapsing star what is true when the archion interaction is as weak as it is at the scales of the order of (1377). At smaller scales of archion interaction, collapsing star turns out to be opaque for archions reducing the effect of archion radiation. One needs very accurate analysis of archion transport in the dense matter to get reliable restrictions on the allowed values of the scale v_{PQ}. In any case at the scales

$$v_{PQ} \sim 10^6 \text{ GeV} \tag{1378}$$

energy losses due to archion emission are reduced to the extent compatible with the neutrino signal from SN1987A.

2. The Early Universe from Horizontal Unification

2.1. The Inflationary Dynamics and Energy Scale.

The inflation scenario can find its physical grounds in the framework of the MHU, since the Higgs field η (see Eq. (1361)), determining the flavour independent mass term, may play the role of inflaton.

In general, the vacuum expectation value $\langle\eta\rangle$ defines in MHU the suppression of the masses of neutral fermions (light neutrinos) relative to charged fermions (see Sec.1)

$$\frac{m_\nu}{m_f} \propto \frac{g_f\langle\phi^0\rangle}{G_\eta\langle\eta\rangle} . \tag{1379}$$

It is assumed that the global symmetry is spontaneously broken at the energy scale f and that the real field η has the potential

$$V(\eta) = \lambda\left(\eta^2 - \frac{f^2}{2}\right)^2 . \tag{1380}$$

Thus, the discussion here is reduced to the inflationary model with a single rolling-down scalar field with the potential (1380). It means that the simplest realisation of MHU corresponds to the simplest chaotic inflation scenario.

To fix the parameters of the inflaton potential, we can put observational constraints on the inflaton energy density at the time when the fluctuations at the scales observable in the microwave background were generated.

It has taken place as these scales crossed the Hubble radius during inflation, typically when the scale factor was of the order of exp(−60) of its size at the end of inflation. This quantity is normally referred to as 60 *e*-folding from the end of inflation.

With the use of the limit at 60 *e*-folding from the end, we can analyse the evolution of the system and constrain the energy density at the end of inflation.

In such approach to reconstruct the inflaton potential the equations of motion must be written in the Hamilton–Jacobi form (see Khlopov, Sakharov, 1998 and Refs. therein)

$$\left(\frac{d^2H}{d\eta^2}\right)^2 - \frac{12\pi}{m_{Pl}^2}H^2 = -\frac{32\pi^2}{m_{Pl}^4}V(\eta)$$

$$\dot{\eta} = -\frac{m_{Pl}^2}{4\pi}\frac{dH}{d\eta} \; . \qquad (1381)$$

In principle, the Hamilton–Jacobi formalism enables us to treat the dynamical evolution of the scalar field exactly, at least at the classical level (1381).

Using the inflaton potential, one can calculate the amplitude of density perturbation $\delta_H(k)$. One can then satisfy the COBE result (Smoot, 1996), most conveniently taken to be evaluated at 60 *e*-folding from the end of inflation.

For the models with sufficiently flat scalar spectra with negligible gravitational waves such as obtained in the simplest models of MHU (Sakharov, Khlopov, 1998), the appropriate value of δ_H reproducing the COBE results is given by

$$\delta_H \cong 1.7 \cdot 10^{-5} \ . \tag{1382}$$

One can also estimate the value of the Hubble constant in the end of inflation $H_{\rm end}$ being for the wide range of vacuum expectation values f equal to

$$H_{\rm end} \cong 1.8 \cdot 10^{-7} m_{Pl} \tag{1383}$$

(Khlopov, Sakharov, 1998).

2.2. The Formation of PBH in the MHU.

The ultraviolet behaviour of density perturbation spectrum may be effectively constrained by the analysis of primordial black hole (PBH) formation in the early Universe.

Following the discussion in the Chapter 4 consider spherically symmetrical Gaussian fluctuations with a rms. amplitude $\delta(M)$ and a background equation of state $p = \gamma\varepsilon$ with $0 < \gamma < 1$.

The probability for a region of mass M to form PBH is given by the Gaussian tail of the distribution of density fluctuation (compare with Eq. (405))

$$\beta_0(M) \approx \delta(M) \exp\left(-\frac{\gamma^2}{2\delta^2(M)}\right) . \tag{1384}$$

This probability defines the fraction of the total density expected to transform into PBH with the mass M. The mass of a primordial black hole, being formed at time t, must be at least $\gamma^{3/2}$ times the horizon mass, so

$$M \approx \gamma^{3/2} \frac{t}{t_{Pl}} m_{Pl} \ . \tag{1385}$$

Usually the value of γ in the very early Universe is taken to be equal to $\gamma = 1/3$ what corresponds to a radiation dominated equation of state, postulated for the early Universe in the old Big Bang model.

In the case $\gamma = 0$ both the Eqs. (1384) and (1385) are formally inapplicable (see Chapter 4).

During a dust stage, to which formally corresponds the equation of state $p = 0$, density fluctuations grow and form gravitationally bound objects.

The fraction of the total density going into PBH depends on the probability for these objects to collapse within their Schwarzschild radius. The minimal probability corresponds to direct collapse into PBHs just in the period of object formation, given by (Polnarev, Khlopov, 1982; compare with Eq. (441))

$$\beta(M) \approx 2 \cdot 10^{-2} \delta(M)^{13/2} \ . \tag{1386}$$

If the dust-like stage takes place in the period

$$t_1 < t < t_2 \ , \tag{1387}$$

then the probability given by Eq. (1386) is valid for formation of PBHs in the mass range (compare with Eq. (438))

$$M_1 \leq M \leq M_{\max} \ , \tag{1388}$$

where M_1 is the mass within the cosmological horizon at t_1 and $M_{\max}$ is the mass of objects just separated from the expansion at t_2. The latter is given implicitly by (see Eq. (440))

$$M_{\max} = \left[\delta\left(M_{\max}\right)\right]^{3/2} \frac{t_2}{t_{Pl}} m_{Pl} . \qquad (1389)$$

The product of the last two factors in the right hand side of Eq. (1389) is the mass within the cosmological horizon at t_2. In order to define $M_{\max}$ explicitly, one should know the form of $\delta(M)$.

The possibility of a soft equation of state may occur during the reheating phase at the end of chaotic inflation (Khlopov et al., 1985) that is realised in the inflation model based on MHU. In this model the inflaton field oscillates around the potential minimum from the time

$$t_1 = H_{\text{end}}^{-1} \qquad (1390)$$

until the reheating occurs.

The friction generated by the coupling of the scalar inflaton field to other matter fields turns the kinetic energy of motion into the background radiation. The reheating is completed on a time scale determined by the decay width Γ_η. In the considered case Γ_η is given by

$$\Gamma_\eta\left(\eta \to \bar{F} f\right) = \frac{G_\eta^2 m_\eta}{8\pi} . \qquad (1391)$$

We can calculate G_η from the potential, given by Eq. (1380) using the expressions for the constant of self–interaction of the field η and the mass of η. Provided that the minimal self–interaction of inflaton is generated by the fermion loop corrections according to Eq. (1357), the constant λ_η is

$$\lambda_\eta \cong \frac{G_\eta^4}{8\pi^2} . \qquad (1392)$$

On the other hand, one can estimate Γ_η from the reconstruction of inflaton potential and to obtain for the wide range of the scale f the value

$$\Gamma_\eta \cong 10^{-14} m_{Pl} \ . \tag{1393}$$

It is clear from Eqs. (1383) and (1393) that

$$\Gamma_\eta^{-1} >> H_{\rm end}^{-1} \tag{1394}$$

and the sufficiently long stage of coherent scalar field oscillations with the dust-like equation of state should take place.

The dust-like stage, starting at the end of inflation at t_1, lasts until reheating is completed at the time

$$t_2 = \Gamma_\eta^{-1} \ , \tag{1395}$$

when the scalar field decays rapidly into relativistic particles. However, in the case of chosen numerical parameters, the PBH do not have an extended mass spectrum, peaking around M_1, which coincides with the mass given by Eq.(439).

In the more general case of non-minimal inflaton self-coupling its constant λ is fixed by the condition (1382), but the smaller coupling to the other fields leads to an extended PBH mass spectrum.

Substituting the value (1382) into Eq. (1386) one finds the probability of PBH formation leading to insignificant effect of their existence and evaporation. But the necessity in the internal consistency of MHU scenario makes the picture more complicated and the constraints on PBH formation play important role in its analysis.

Indeed, in the MHU it is necessary to impose the condition

$$G_\eta f \leq G_F^{(n)} \left\langle \xi^{(n)} \right\rangle , \tag{1396}$$

which ensures the right structure of fermion mass matrix, generated by Dirac see-saw mechanism (1369). This condition requires that

$$f \leq 10^{-6} m_{Pl} \tag{1397}$$

for the minimal inflaton self-coupling (1392).

However, the inevitable consequence of such low value of f is the domain-wall problem, induced by quantum fluctuations of real inflaton field on the post-inflational dust-like stage (see Khlopov, Sakharov, 1998 and Refs. therein).

To remove this problem, one should either take

$$f \cong m_{Pl} \tag{1398}$$

and remove condition (1392) of minimal inflaton coupling, or to introduce complex inflaton field. In the first case, the duration of the post-inflational dust-like stage is enormously long, and in the second case the non-minimal axion model is realised.

So even at this level of study we find that the solution should lead to much more complicated multi-parameter cosmological scenario.

2.3. Early «horizontal» phase transitions.

The interaction of horizontal scalar fields with inflaton may result in phase transitions on inflational stage. The reason is that due to such interaction Higgs fields acquire positive mass. When in the course of slow rolling the amplitude of inflaton field falls below some critical value η_c, the mass term in the Higgs potential changes its sign and then the phase transition takes place.

As pointed out in (Sakharov, Khlopov, 1993), the «horizontal» phase transitions lead to the appearance of a characteristic spikes in the spectrum of density perturbations. Kofman and Linde (1986) first found the existence of such spikes in chaotic inflation scenario.

These spike-like perturbations, on scales that left the horizon $(40\div1)e$- folds before the end of inflation reenter the horizon during the radiation or dust-like era and could in principle collapse to form primordial black holes.

Let us now evaluate following Khlopov, Sakharov (1997) the typical size and mass of the black holes produced by these perturbations.

Suppose that after the phase transition the Universe inflated by the $\exp(N_e)$ times. Then at the end of inflation the physical scale that left the horizon during the phase transition is

$$l_e \sim H_0^{-1} \exp(N_e) \ , \tag{1399}$$

where H_0 is the Hubble constant during inflation.

If the equation of state is $p = \varepsilon/3$, the scale factor of the Universe grows after inflation as

$$a(t) \propto \sqrt{tH_0} \ . \tag{1400}$$

The scale

$$l = H_0 \sqrt{tH_0} \exp(N_e) \tag{1401}$$

turns out to be equal to the scale of the cosmological horizon in the period

$$t_k = H_0^{-1} \exp(2N_e) \ . \tag{1402}$$

At that time perturbations with the density contrast $\delta \sim 1$ form black holes of the size

$$r_{bh} \sim H_0^{-1}\exp(2N_e) \tag{1403}$$

and of the mass

$$M \cong \frac{m_{Pl}^2}{H_0}\exp(2N_e)\ . \tag{1404}$$

If PBH are formed at the dust-like stage, the time of its formation is

$$t_k = H_0^{-1}\exp(3N_e) \tag{1405}$$

and the PBH mass is equal to

$$M \cong \frac{m_{Pl}^2}{H_0}\exp(3N_e)\ . \tag{1406}$$

We can include the horizontal scalars $\xi^{(0)}$, $\xi^{(1)}$ and $\xi^{(2)}$ in the effective inflationary potential which we take here in the form

$$V\left(\eta,\xi^{(0)},\xi^{(1)},\xi^{(2)}\right) = -\frac{m_\eta^2}{2}\eta^2 + \frac{\lambda_\eta}{4}\eta^4 - \sum_{i=0}^{2}\frac{m_i^2}{2}\left(\xi^{(i)}\right)^2 + \\ + \sum_{i=0}^{2}\frac{\lambda_i}{4}\left(\xi^{(i)}\right)^4 + \sum_{i=0}^{2}\frac{\nu_i}{2}\eta^2\left(\xi^{(i)}\right)^2\ . \tag{1407}$$

In MHU Higgs boson masses m_i are connected with scales of horizontal symmetry breaking, which can be given by the expression

$$m_i^2 \approx 10^{-3}V_i^2\ , \tag{1408}$$

where vacuum expectation values of Higgs fields

$$V_i = \left\langle \xi^{(i)} \right\rangle , \quad i = 0,1,2 \tag{1409}$$

follow the hierarchy

$$V_2 : V_1 : V_0 \cong 1 : 30 : 200 \tag{1410}$$

and

$$\lambda_i = \lambda_\xi \cong 10^{-3} . \tag{1411}$$

The minimal interaction of inflaton with horizontal scalars is generated by fermion loop corrections, induced by Yukawa interactions (1357) and is given by

$$\nu_i = \nu_\xi \cong \frac{G_\eta^2 \left(G_F^{(i)} \right)^2}{8\pi^2} . \tag{1412}$$

Due to the interaction (1407) between the inflaton η and horizontal scalars phase transitions occur, when the amplitude of the field η falls down to

$$\eta_{ci} = \frac{m_i}{\sqrt{\nu_i}} . \tag{1413}$$

According to (Sakharov, Khlopov, 1993) such phase transitions lead to the appearance of characteristic spikes on the spectrum of adiabatic perturbations with density contrast

$$\delta \cong \frac{4}{9s} ; \quad s = \left(\sqrt{\frac{9}{4} + \kappa \cdot 10^5 \left(\frac{V_i}{m_{Pl}} \right)^2} - \frac{3}{2} \right) , \tag{1414}$$

where $\kappa \sim 1$.

One can relate the scale of horizontal symmetry breaking with the *e*-folding and mass of PBH formed by the spikes. From the condition that all three horizontal phase transitions do not lead to formation of PBH with the mass exceeding 1000 g, one obtains the constraint

$$V_2 \leq 10^{13} \text{ GeV} . \qquad (1415)$$

2.4. The Large-Scale Modulation in the Distribution of Primordial Axion Field.

Recent analysis (Sakharov, Khlopov, 1994; Sakharov et al., 1996) of the invisible axion model of cold dark matter found large scale inhomogeneity in the energy density distribution of coherent oscillations of the primordial axionic field. These inhomogeneities, to be shared by all the models of axionic cold dark matter, have been called *archioles* (Sakharov, Khlopov, 1994).

Archioles are the formation that represents a reflection of the percolation Brownian vacuum structure of axionic walls bounded by strings and which is fixed in the strongly inhomogeneous primary distribution of cold dark matter. They seem to appear in all the cosmological models of invisible axion as their necessary element.

In the general case, the axion field appears in the early Universe after the phase transition with Peccei-Quinn (PQ) symmetry breaking. Axion corresponds to the phase θ of the complex Higgs boson, breaking PQ symmetry, and remains practically massless until the QCD phase transition takes place in the early Universe. The true vacuum state is degenerated for all the values of θ.

Due to instanton effects PQ symmetry is not free from colour anomaly, what is effectively described by the potential

$$V(\theta) = \Lambda_1^4\left(1 - \cos(\theta N)\right) . \tag{1416}$$

It removes the degeneracy of the true vacuum state and results in the nonzero axion mass given by

$$m_a = A_c \frac{m_\pi f_\pi}{F_a} , \tag{1417}$$

where F_a is the scale of PQ symmetry breaking and the constant $A_c \sim 1$ depends on the choice of the axion model (see Chapter 2).

In the period of the QCD phase transition, the axion mass depends on the temperature (Ipser, Sikivie, 1982) until the Compton wavelength of the axion becomes comparable with the cosmological horizon and the condition

$$m(\tilde{t}) \cdot \tilde{t} \approx 0.75 \tag{1418}$$

is fulfilled. Then the axion mass is "switched on" and coherent oscillations of the axion field are turned on. Coherent axion field oscillations correspond to condensate with zero momentum and energy density (see for review Berezhiani et al, 1992)

$$\rho_a(T) = \left(\frac{39.14}{2}\right)\left(\frac{T_1^2 m_a}{m_{Pl}}\right)\left(\frac{T}{T_1}\right)^3 \theta^2 F_a^2 , \tag{1419}$$

where T is the temperature of the thermal background radiation and $T_1 = \Lambda_1$.

In the cosmology of an invisible axion, it is usually assumed that the mean energy density of coherent oscillations is distributed uniformly and that it corresponds to the averaged phase value of axion field amplitude

$$\overline{\theta} = 1 .$$

However, the local value of the energy density of coherent oscillations of the axion field depends on the local phase θ that determines the local amplitude of coherent oscillations.

It was shown in (Sakharov, Khlopov, 1994) that the initial large-scale inhomogeneity of the distribution of θ must be reflected in the distribution of the energy density of coherent oscillations of the axion field. The large-scale modulation of the distribution of the phase θ is due to the formation of topological defects in the two (PQ and QCD) phase transitions.

As soon as the temperature of the Universe falls down F_a , the formation of axion strings takes place in PQ phase transition (Harari, Sikivie, 1982). These strings can be infinite or closed. A numerical simulation of string formation has found (Vilenkin, 1982) that about 80% of the total length of strings corresponds to infinite Brownian lines. It means that 80% of the total length of strings within any finite region corresponds to strings spreading out of this region. The remaining 20% of this length are contributed by closed loops. Infinite strings form a random Brownian network with the step

$$L(t) = \lambda t , \tag{1420}$$

where $\lambda \cong 1$.

When the temperature of the Universe becomes as low as $T \cong T_1$, the potential (1416) contributes significantly into the total potential. It removes the degeneracy of the vacuum on θ so that the minimum of energy at the number of families $N = 1$ corresponds to vacuum with

$$\theta = 2\pi k , \tag{1421}$$

where k is an integer, for example, equal to zero.

The vacuum value of the phase θ cannot be zero everywhere, since the phase must change by $\Delta\theta = 2\pi$ along the closed path around string. Hence, we come from the vacuum with $\theta = 0$ to the vacuum with $\theta = 2\pi$ along such path.

The vacuum value of θ is fixed in all points in space around the string. The exception is the sheet along the direction $\theta = \pi$ from the string. Across this sheet, a transition from the one vacuum to another occurs, and the vacuum axion wall is formed along the sheet. The width of such a wall bounded by strings is of the order of

$$\delta \cong m_a^{-1} . \tag{1422}$$

The initial value of the phase θ must be close to π near the wall, and the amplitude of coherent oscillations in Eq. (1419) is determined by the difference of the initial local phase θ and the vacuum value. The maximal energy of the axion field oscillations is thus given by

$$\rho_a(\theta \sim \pi) = \pi^2 \rho_a(\overline{\theta} \sim 1) . \tag{1423}$$

It means that the distribution of coherent oscillations of the axion field is modulated by non-linear inhomogeneities in which the relative variations of density are

$$\frac{\delta\rho}{\rho} > 1 .$$

Sakharov and Khlopov (1994) called such inhomogeneities archioles.

The scale of this modulation in the density distribution exceeds the cosmological horizon because of the presence of 80%-of length infinite string component in the structure of axionic strings.

In other words, the large-scale structure of archioles replicates the vacuum structure that axionic walls bounded by strings had by the moment, when the axion mass is "switched on". But it is frozen at the radiation-dominated stage and retains after the vacuum structure itself decays and disappears.

The spectrum of inhomogeneities developed by the density in response to the large-scale Brownian modulation of the distribution of coherent oscillations of axion field has been studied in (Sakharov et al., 1996).

Density perturbations, associated with the Brownian net of archioles, were described in the terms of a two-point auto-correlation function, which has the form

$$\left\langle \frac{\delta\rho}{\rho}(\mathbf{k})\frac{\delta\rho}{\rho}(\mathbf{k}')\right\rangle = 12\rho_a\mu_a k^{-2}\delta^3(\mathbf{k}+\mathbf{k}')\tilde{t}^{-1}f^{-2}G^2t^4 \ , \qquad (1424)$$

where ρ is the background density, G is the gravitational constant,

$$f = \begin{cases} \dfrac{1}{6\pi} & \text{for } p = \frac{1}{3}\varepsilon \text{ stage} \\ \dfrac{3}{32\pi} & \text{for } p = 0 \text{ stage} \ , \end{cases}$$

ρ_a is the total energy density of the Brownian lines, μ_a is the linear density in the net. The mean square fluctuation of the mass in archioles is given by

$$\left(\frac{\delta M}{M}\right)^2(k,t) = 12\rho_a\mu_a\tilde{t}^{-1}f^{-2}G^2kt^4 \ . \qquad (1425)$$

Using Eqs. (1424)–(1425), one can calculate quadrupole anisotropy of relic radiation induced by archioles structure and given by

$$\frac{\delta T}{T} \cong 2.3 \cdot 10^{-6} \left(\frac{F_a}{10^{10} \text{ GeV}} \right)^2 . \tag{1426}$$

It leads us to a new model-independent constraint on the scale of PQ symmetry breaking in the model of invisible axion

$$F_a \leq 1.5 \cdot 10^{10} \text{ GeV} . \tag{1427}$$

This upper limit on F_a has an essentially different character than the upper limit quantitatively close to the value (1427), but obtained by Davies (1987) in the comparison of the density of axions from decays of axionic strings with the critical density.

The point is that the density of axions formed in decays of axionic strings depends critically on the assumption about the spectrum of such axions. The estimation of the density of axions from strings has thus an uncertainty factor ~ 70. The corresponding uncertainty is in the estimated upper limit on F_a. Arguments that lead to the constraint (1427) are free from these ambiguities, since they have a model-independent character.

In addition, the low-scale evolution of archioles can provide physical grounds for rms. fluctuations in the mass distribution within spheres of radius $8h^{-1}$Mpc. Such fluctuations are used for normalisation of the amplitude of the density perturbation spectra in the standard CDM.

At the smallest scales, corresponding to the horizon in the period when the axion mass is "switched on", evolution of archioles just in the beginning of axionic CDM dominance in the Universe should lead to formation of the smallest gravitationally bound axionic objects.

Such objects formed at red shifts $z_{MD} \cong 4 \cdot 10^4$ with the minimal mass

$$M \cong \rho_a \tilde{t}^3 \cong 10^{-6} M_\odot \tag{1428}$$

should have typical minimal size

$$R \cong \tilde{t}\left(\frac{1+z_\Lambda}{1+z_{MD}}\right) \cong 10^{13} \text{ cm} , \tag{1429}$$

where z_Λ is the red shift corresponding to the period, when the axion mass is "switched on".

One can expect the mass distribution of axionic objects at small scale to peak around the minimal mass, so that the existence of halo objects with the mass

$$M \sim \left(10^{-6} \div 10^{-1}\right) M_\odot \tag{1430}$$

and size

$$R \sim 10^{13} \div 10^{15} \text{ cm} \tag{1431}$$

is rather probable, what may have interesting application to the theoretical interpretation of ACHO micro-lensing events.

Even at the presented level of MHU the model provides a mechanism for baryogenesis without GUT-induced baryon nonconservation. The mechanism combines the $(B+L)$ non-perturbative electroweak nonconservation at high temperatures with $\Delta L = 2$ non-equilibrium transitions, induced by the physics of Majorana mass of neutrino.

It is possible that under the right order of the magnitude and sign of the mean baryon asymmetry local spatial variation of baryon charge may take place. Such a variation can be induced by modulation of $\theta(\mathbf{x})$ if the axion-induced CP violation takes part in processes of baryosynthesis. The baryon asymmetry in this case may depend on $\theta(\mathbf{x})$ (see for example Ioshimura (1983)), being the sum of constant and spatial varying contributions

$$\Delta(\mathbf{x}) = \Delta_0 + \Delta_1 \sin\theta(\mathbf{x}) \ . \tag{1432}$$

In the case of the amplitude of spatial varying axion-induced contribution exceeds the constant baryon excess $\Delta_1 > \Delta_0$ antibaryon excess is generated in the regions for which the condition

$$\left|\theta(\mathbf{x}) - \frac{3\pi}{2} + 2\pi k\right| > \arccos\left(\frac{\Delta_0}{\Delta_1}\right) \tag{1433}$$

is valid.

Axion-dominated baryosynthesis can provide non-trivial picture of antimatter domain evolution (Khlopov, 1998). Small domains annihilate in this scenario before 1 s of expansion, thus escaping the severe constraints on antimatter annihilation (see Chapters 6 and 7).

Larger domains, satisfying these constraints, can in the result of their evolution form antimatter objects in the modern Universe. The minimal mass of such objects is determined by the condition of their survival in the annihilation with the surrounding matter. The total mass of such domains is restricted by the observed gamma background.

It was shown recently that the modern data do not exclude the existence of a globular cluster of anti-stars in the halo of our Galaxy (Khlopov, 1998).

Such cluster should be the source of anti-nuclear component in galactic cosmic rays what can be tested in AMS experiment.

So the MHU-based cosmology may realise the ideas of anti-world in baryon-asymmetrical Universe.

3. MHU Dark Matter Cosmology

3.1. Dark Matter Unification in the Hidden Sector of Models of Horizontal Unification

In the case of family symmetry breaking the see-saw mechanism of neutrino mass generation provides inverse hierarchy of neutrino masses with the respect to the hierarchy of horizontal symmetry breaking.

The mass of ordinary (light) neutrino is to be inversely proportional to the Majorana mass of its heavy partner(see Eq.(1370))

$$m_\nu \propto M_N^{-1} \ . \tag{1434}$$

In the case of the model of inverse hierarchy the same inverse proportionality to the masses of their heavy partners takes place for quarks and charged leptons. According to Eqs. (1369) and (1370), the hierarchy of neutrino Majorana masses is similar to the ordinary quark and lepton mass hierarchy:

$$m_{\nu_e} : m_{\nu_\mu} : m_{\nu_\tau} \sim m_e : m_\mu : m_\tau \ . \tag{1435}$$

In the model of horizontal unification the mixing angles, defining the amplitude of neutrino oscillations, are determined by the relative rotation in the horizontal group space of the neutrino and charged lepton mass matrices. Therefore the experimental constraints on the neutrino mixing angles from the searches for neutrino oscillations impose upper limits on deviations of neutrino mass ratio from the strict proportionality to the masses of respective charged leptons. With the account for the existing uncertainties such deviations should be within the factor of 2.

For the sextet representation of horizontal Higgs fields $\xi^{(1)}$ and $\xi^{(2)}$ the neutrino mass matrix, given by Eq.

(1370) is non-diagonal, and decays of heavier neutrino ν_H on lighter neutrino ν_L and archion are possible with the lifetimes

$$\tau\left(\nu_H \to \nu_L + \alpha\right) = \frac{16\pi}{g^2_{HL} m_H} , \tag{1436}$$

where

$$g_{HL} = g_{\nu_H \nu_L} = \frac{m_{HL}}{v_{PQ}} \tag{1437}$$

and m_{HL} is the corresponding non-diagonal element of the neutrino mass matrix.

For the scale of family symmetry breaking being as "small" as

$$v_{PQ} \sim 10^6 \text{ GeV} \tag{1438}$$

the predicted effect of Majorana masses of neutrino is rather close to the modern sensitivity of the searches for double neutrinoless beta decay.
Since the archion couplings to fermions of the lightest family (u, d, e, ν_e) are suppressed, the existing constraints on the respective invisible axion scale are weakened down to

$$v_{PQ} \geq 10^6 \text{ GeV} \tag{1439}$$

making the model of archion rather close to the model of hadronic axion (Kim, 1979; Shifman et al., 1980; see Chapter 2).

However, it turns out (Berezhiani et al., 1992), that the archion model escapes the serious problem of primordial superheavy stable Q quarks, predicted in the model of hadronic axion.

One can estimate the frozen concentration of Q quarks and corresponding Q hadrons in the Universe and find it contradicting the upper limits on such concentration, following from the search for anomalous nuclei (so called "crazy isotopes"). Hence the theory of hadronic axion should introduce the mechanism of superheavy quark instability.

The inclusion of the hadronic axion model into GUT models leads inevitably to the existence of superheavy lepton, coupled to axion. The mixing of superheavy quark Q with the light (ordinary) quarks, inducing the instability of Q, would lead to the existence of the axion coupling to leptons, so that the axion is not hadronic.

In view of these troubles of the model of hadronic axion the model of archion is of special interest, since it naturally provides both superheavy quark instability and the suppression of the axion coupling to leptons.

Since the mass of neutrino depends in MHU on the scale of horizontal symmetry breaking,

$$m_{\nu} \propto v_{PQ}^{-1} , \tag{1440}$$

the neutrino lifetime

$$\tau(\nu_H \to \nu_L a) \propto v_{PQ}^{5} \tag{1441}$$

and the mean density of the primordial axion field

$$\rho_a = \rho_a(\overline{\theta} = 1) \propto v_{PQ} , \tag{1442}$$

changing the parameter v_{PQ} one reproduces all the main types of cosmological models of the formation of the structure of the Universe.

At larger v_{PQ} the axion field should dominate in the Universe. Massive stable neutrino dominance corresponds to smaller v_{PQ}. Finally, the smallest possible values of this parameter correspond to cosmological models with massive unstable neutrino (see Chapter 9).

In the framework of MHU the formal continuous change of the fundamental physical parameter v_{PQ} results in continuous transition from one to another form of dark matter, dominating in the Universe. The model makes definite predictions for each type of dark matter from the combination of cosmological, astrophysical and physical constraints.

The total cosmological density ρ_{tot} and the baryon density ρ_b being fixed, the relationship

$$\rho_a(v_{PQ}) + \sum_i \rho_{\nu i} + \rho_a^{dec}(v_{PQ}) + \sum_i \rho_{\nu i}^{dec}(v_{PQ}) + \rho_b = \rho_{tot} \quad (1443)$$

turns out to be an equation relative to v_{PQ}.

In Eq. (1443) both the contribution of primordial axions and neutrinos with the lifetime, exceeding t_U, and the contribution of axions and neutrino from the decay of unstable neutrino are taken into account.

The solutions of this equation define a discrete set of cosmological models with different types of dark matter, forming the structure of the Universe.

In general, there are six different dark matter scenarios which may be realised in the framework of MHU.

3.2. DM Models from Horizontal Unification

1.Cold dark matter (CDM) scenario. The cosmological evolution of the axion field after Peccei-Quinn symmetry breaking follows in MHU the general features of the "invisible" axion cosmology.

In the result of hierarchy of $SU(3)_H \mathrm{x} U(1)_H$ symmetry breaking, $U(1)''_H$ turns out to be Peccei-Quinn symmetry of only one quark-lepton generation. It resolves automatically the cosmological θ-domain problem in the axion theory (Ipser, Sikivie, 1983).

In the axion theory with $N > 1$ fermion families, the discrete degeneracy of vacuums with different N after

PQ phase transitions leads to overproduction of vacuum walls at the border of θ domains. This problem does not exist for the axion theory with only one generation to which the case of MHU is reduced.

As we discussed in Sec. 2, the stochastic distribution of the axion field

$$\theta = \frac{a}{v_{PQ}} \tag{1444}$$

can lead to its change on 2π along a closed path, resulting in the appearance of string structure, decaying rapidly after the axion mass switches on at

$$T \leq 800 \text{ MeV} . \tag{1445}$$

The axion field oscillates with the local amplitude $\theta(\mathbf{x}) - \theta_{\text{vac}}$. The energy density of the oscillations decreases in the course of expansion as (see Eq. (1419))

$$\rho_a \propto a(t)^{-3} , \tag{1446}$$

where $a(t)$ is the scale factor and the modern axion mass density is equal to

$$\rho_a = \left(\frac{v_{PQ}}{4 \cdot 10^{12} \text{ GeV}} \right) \cdot \rho_{cr} , \tag{1447}$$

where ρ_{cr} is the critical density (Abott, Sikivie, 1983; Preskill et al., 1983; Dine, Fishler, 1983).

According to Davis (1986), the intensive axion emission by decaying axion cosmic string structure results in the growth of the modern cosmological axion density up to

$$\rho_a = \left(\frac{v_{PQ}}{2 \cdot 10^{10} \text{ GeV}} \right) \cdot \rho_{cr} \ . \tag{1448}$$

Comparing these predictions with the total cosmological density, taken to be equal to ρ_{cr}, one obtains the upper limit on v_{PQ}.

However, within the framework of inflational cosmology the situation is possible, when the axion field in the observed part of the Universe has an amplitude

$$|\theta(\mathbf{x}) - \theta_{vac}| << 1 \ . \tag{1449}$$

(Linde, 1988).

Exponential expansion of the region with the axion field given by Eq. (1449) provides also the absence of axion strings in this region, so that there is no increase of the axion density due to axion emission of such strings.

Cosmological upper limits on the scale v_{PQ} seem to be absent in this case, so that this scale can reach even Planck values $v_{PQ} \sim m_{Pl}$. However in MHU the scale v_{PQ} can be constrained even in this case.

In MHU cosmology, both the axion and the inflational cosmology follow from the same physical model. It provides self-consistent treatment of axions in the framework of inflational cosmology.

Following the discussion of the Sec. 2, one can fix the inflaton potential from the data of COBE and take into account the condition on the absence of phase transition at the inflational stage.

Based on the chaotic inflation scenario following from MHU, in order to avoid the peaks in density fluctuations, one should exclude the possibility of horizontal symmetry phase transitions at the inflationary stage, what leads to the restriction

$$v_{PQ} < 2 \cdot 10^{10} \text{ GeV} \tag{1450}$$

(Sakharov, Khlopov, 1993).

Under the condition given by Eq. (1450), horizontal phase transitions can take place at post-inflational dust-like stage or at radiation-dominance stage after reheating.

In both cases after PQ symmetry breaking the axion string structure is formed. When the axion mass switches on, the vacuum wall surrounded by string structure appears and dissipates, replicated in the structure of archioles (see Section 2). From the observational data on the quadrupole anisotropy and the estimated by Eq. (1426) effect induced by archioles one has the model-independent constraint (1427).

According to Berezhiani et al. (1992) at

$$v_{PQ} < 10^{8} \text{ GeV} \tag{1451}$$

coherent axion field oscillations are thermalised after QCD phase transition. The rate of axion interaction with equilibrium nucleon-antinucleon pairs in reactions

$$a + N(\overline{N}) \leftrightarrow \pi + N(\overline{N}) \tag{1452}$$

is under the condition (1451) higher than the rate of expansion. As a result, the axion condensate evaporates, and the archioles structure dissipates.

At small values of v_{PQ} corresponding to Eq. (1451), the evaporation of axion condensate results in thermal axion background. Primordial axion gas represents in this case thermal gas of particles with zero spin and with the mass m_a decoupled from plasma and radiation at the temperature 60–160 MeV, depending on the value of v_{PQ} .

In the period of decoupling, relative concentration of thermal axions and neutrino is given by equilibrium values (see Chapter 3)

$$\frac{n_a}{n_{\nu\bar{\nu}}} = \frac{2}{3} \ . \tag{1453}$$

Depending on the exact value of v_{PQ}, the modern relative concentration of thermal axion gas should be corrected by the effect of neutrino heating by muon pairs and pions, present in the equilibrium in the period of thermal axion decoupling. Say, for $v_{PQ} \sim 10^6$ GeV the modern relative concentration of thermal axion gas is

$$\frac{n_a}{n_{\nu\bar{\nu}}} = 0.57 \ . \tag{1454}$$

The existence of axion field distribution before the QCD phase transition and the condensate evaporation can lead to inhomogeneous baryosynthesis, discussed in Sec.2.

So at "low" values of the scale v_{PQ} (1451) primordial axions are not cold rather hot dark matter. However, this form of hot dark matter is not dominant in MHU cosmology.

2. Hot dark matter (HDM) scenario. The dominance of tau neutrino ν_τ with the standard Big Bang model concentration (285) corresponds to the mass of ν_τ

$$m_{\nu_\tau} = 20 \text{ eV} \tag{1455}$$

and the lifetime exceeding the age of the Universe

$$\tau(\nu_\tau \to \nu_\mu + a) > t_U \ . \tag{1456}$$

The cosmological density of such tau neutrino is given in terms of MHU parameters as

$$\rho_{\nu_\tau} = \frac{6.6 \cdot 10^{13}\,\text{GeV}}{v_{PQ}} \cdot \left(\frac{g_\nu^2}{G_N}\right) \cdot \rho_{cr} \; . \tag{1457}$$

For the combination of Yukawa constants taken in MHU in the interval

$$10^{-6} < \frac{g_\nu^2}{G_N} < 10^{-4} \tag{1458}$$

the HDM scenario of tau neutrino dominance is realised in MHU at the scales in the respective range

$$10^{8} < v_{PQ} < 10^{10} \text{ GeV} \; . \tag{1459}$$

The range given by Eq. (1459) practically coincides with the interval corresponding to axionic CDM and defined by the constraints (1427) and (1451). It makes the mixed CDM+HDM dark matter model rather natural solution of MHU cosmology.

3. Relativistic unstable dark matter (UDM) scenario. This solution corresponds in MHU to the dominance in the Universe of relativistic archions and ν_μ, being the products of the decay of ν_τ with the mass

$$m_{\nu_\tau} \cong 50 \div 100 \;\; \text{eV} \tag{1460}$$

and the lifetime

$$\tau\left(\nu_\tau \to \nu_\mu + a\right) = 4 \cdot 10^{15} \div 10^{16} \text{ s} \; . \tag{1461}$$

Muon neutrino from the decay (1461) are relativistic in the modern Universe provided that their mass does not exceed 5 eV.

The considered solution of Eq. (1443) follows from the expression for the energy density of relativistic products of tau neutrino decay given by

$$\rho_{\nu_\mu}^{dec} + \rho_a^{dec} = \left(\frac{v_{PQ}}{10^{10}\text{GeV}}\right)^{3/2} \cdot \frac{1}{x} \cdot \left(\frac{G_N}{g_\nu^2}\right)^{1/2} \cdot \rho_{cr} , \quad (1462)$$

where $x \sim 1$ is the neutrino mixing parameter and the red shift of the relativistic decay products is taken into account.

4. Nonrelativistic UDM scenario. The dominance of non-relativistic ν_μ with the mass ~10 eV, both primordial and from the decay of ν_τ with the mass ~100 eV and the lifetime ~10^{15} s, is realised provided that

$$\tau(\nu_\mu \to \nu_e + a) > t_U . \quad (1463)$$

The density of non-relativistic muon neutrino dominating in the Universe in the considered solution of Eq. (1443) is given in terms of MHU parameters as

$$\rho_{\nu_\mu} = \frac{6.6 \cdot 10^{12}\text{GeV}}{v_{PQ}} \cdot \left(\frac{g_\nu^2}{G_N}\right) \cdot \rho_{cr} . \quad (1464)$$

5. Relativistic hierarchical neutrino decay (HND) scenario. In the modern Universe the dominance of relativistic archions and electron neutrino from the decay of muon neutrino with the mass

$$m_{\nu_\mu} \cong 50 \div 100 \text{ eV} \quad (1465)$$

and the lifetime

$$\tau(\nu_\mu \to \nu_e + a) = 4 \cdot 10^{15} \div 10^{16} \text{ s} \quad (1466)$$

is realised under the condition of rapid decay of tau neutrino with the mass

$$m_{\nu_\tau} \cong 1 \div 10 \text{ keV} \tag{1467}$$

and the lifetime

$$\tau\left(\nu_\tau \to \nu_\mu + a\right) \leq 10^8 \div 10^{10} \text{ s} . \tag{1468}$$

The modern density is given in this case by

$$\rho_{\nu_e}^{dec} + \rho_a^{dec} = \left(\frac{v_{PQ}}{10^8 \text{GeV}}\right)^{3/2} \cdot \frac{1}{x} \cdot \left(\frac{G_N}{g_\nu^2}\right)^{1/2} \cdot \rho_{cr} . \tag{1469}$$

6. Non-relativistic HND scenario. The dominance of non-relativistic or semi-relativistic archions, originated from both early decay of ν_τ , satisfying Eqs. (1467) and (1468), and successive decays of ν_μ, satisfying Eqs. (1465) and (1466), is realised in the case

$$m_a > m_{\nu_e} . \tag{1470}$$

The main contribution into the inhomogeneous dark matter (in rich galaxy clusters and halos of galaxies) is maintained by both primordial thermal archion background and non-relativistic archions from early ν_τ decays. The total density accounting also for the dominant contribution of the homogeneous non-relativistic archion background from recent ν_μ decays is given by

$$\rho_a = \frac{5 \cdot 10^5 \text{GeV}}{v_{PQ}} \cdot \rho_{cr} . \tag{1471}$$

In the opposite case,

$$m_a < m_{\nu_e} \,, \tag{1472}$$

dominate non-relativistic ν_e , both primordial and from ν_τ and ν_μ decays, and their contribution to the modern cosmological density as

$$\rho_{\nu_e} = \frac{3.3 \cdot 10^{10}\,\mathrm{GeV}}{v_{PQ}} \cdot \left(\frac{g_\nu^2}{G_N}\right) \cdot \rho_{cr} \,. \tag{1473}$$

So MHU provides the unique physical basis for CDM (primordial axion condensate), HDM (stable tau neutrino), relativistic and non-relativistic UDM and HND scenarios, being realised as the solutions of Eq. (1443).

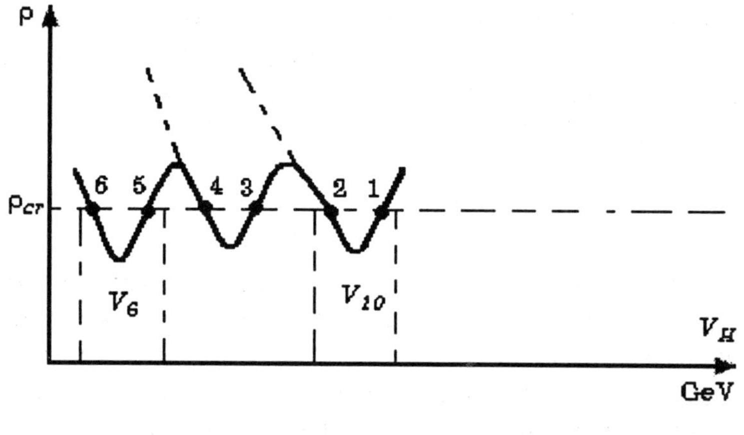

Fig.26. The set of the MHU dark matter scenarios determined by the scale of horizontal symmetry breaking.

The complete set of the solutions of Eq. (1443) can be realised subject to v_{PQ} for the combination of Yukawa couplings in the interval

$$1.5 \cdot 10^{-6} < \frac{g_\nu^2}{G_N} < 4 \cdot 10^{-5} \ . \tag{1474}$$

The graphical solution of Eq. (1443) is shown in the Fig.26.

On the other hand, the set of cosmological and astrophysical constraints considered in Sec. 2 leaves only two small intervals near

$$v_{PQ} \sim 10^{10} \text{ GeV} \ , \tag{1475}$$

where CDM and HDM scenarios or their mixture are possible with the admixture of archioles; and near

$$v_{PQ} \sim 10^{6} \text{ GeV} \ , \tag{1476}$$

in which HND scenarios 5 and 6 can be realised.

The UDM scenarios 3 and 4 are excluded from SN1987A restrictions on v_{PQ} .

The first possibility given by Eq. (1475) corresponds to more conservative "standard" CDM model of large-scale structure formation with the natural extension to the mixed CDM + HDM scenario. The nontrivial effect predicted by MHU is the percolation structure of archioles, which may be desirable for CDM model for better correspondence to the observations of the large scale structure.

The second solution (1476) offers more nontrivial possibility for large sale structure formation. HND scenarios 5 and 6 combine the attractive features of HDM, CDM and UDM models. It makes HND scenarios appealing physically relevant theoretical basis for detailed models of the cosmological large scale structure formation and for comparison of the predictions of such models with the astronomical data.

Indeed in the HND scenarios the dominance of ν_τ with the mass (1÷10) keV in the period (10^8÷10^{10}) s induces short wave fluctuations in the spectrum of density perturbations of ν_μ with the mass $(50 \div 100)$ eV. These short-wave fluctuations correspond to the scales of galaxies since they induce the growth of baryon inhomogeneities with the masses

$$10^7 \leq M \leq 10^{10} M_\odot \,. \tag{1477}$$

Muon neutrino from decays of ν_τ are relativistic in the period of decay t_d but then at

$$t_{dr} \sim t_d \cdot \left(\frac{m_{\nu_\tau}}{2m_{\nu_\mu}} \right)^2 \tag{1478}$$

they are red-shifted down to non-relativistic energy and enhance in the spectrum of fluctuations by the factor of 2 the long-wave component, inherent to HDM models and stimulating the formation of more clear cell structure of voids and superclusters.

Though only one species of dark matter particles – muon neutrino – induce the formation of the structure, the model is effectively two-component. The structure is formed by the mixture of cool component of primordial ν_μ, having the spectrum typical for the thermal gas of particles with mass $(1 \div 10)$ keV, and of hot component of ν_μ from ν_τ, behaving in respect to gravitational instability as the thermal gas of particles with a few eV mass.

Finally, decays of the both types of ν_μ at $(10^{15} \div 10^{16})$ s slow down the rate of the evolution of the structure and provide its survival to the present time. It also removes the contradiction between the total density equal to critical density, predicted by inflational models, and the estimations of the density in inhomogeneities, giving the values less than the critical density.

The primordial thermal archion background, being in these models the coldest component of the modern dark matter play the important role in the evolution of the shortest wavelength part of density perturbations, inducing, in particular, the formation of massive halos outside the visible parts of galaxies. Note that according to (Jensen, 1990) the phase space restrictions on the mass of halo particles (Tremaine, Gunn, 1979) can be weakened or even completely removed in the case of Bose gas.

The both MHU DM solutions, corresponding to the Eqs. (1475) and (1476), provide natural basis for the mixed CDM + HDM model of large scale structure formation. The data of COBE experiment (Smoot et al., 1992) seem to favour such mixed scenario.

3.3. The History of the Universe Based on MHU

To conclude, the cosmological phenomena, discussed above, maintain the MHU-based history of the Universe — the two physically self-consistent inflationary DM scenario of baryon asymmetrical Universe.

In these scenarios, all steps in the cosmological evolution are in quantitative correspondence with the parameters of particle theory.

The physics of inflaton corresponds to the Dirac seesaw mechanism of quark and charged lepton mass generation, lepton number ($\Delta L = 2$) non-conserving decays in combined ($\Delta L = 2$) + EWBNC nucleosynthesis are based on the physics of neutrino Majorana mass generation. The axionic CDM parameters in the high branch solution (1475) and masses and lifetimes of neutrino in HND find physical relevance to the scale and pattern of family symmetry breaking.

Even in the minimal realisation of CDM inflationary scenario, based on the solution (1475), one finds additional nontrivial patterns, such as post-inflational dust-like stage, horizontal phase transitions, axionic string-network, archioles structure, possible PBH formation and evaporation as well as possible baryon charge spatial variation and antibaryon domains formation in the Universe dominated by baryons.

Qualitatively similar picture of the very early Universe on the basis of low-energy branch of MHU is supplemented by the succession of stages of unstable massive neutrino and their decay products dominance, in which the large scale structure of the Universe is formed.

Two-photon decays of archions in HND scenario, causing small ionising effect in the period of recombination, completely ionise baryon matter in the recent period of muon neutrino decays and should lead to the existence of non-thermal electromagnetic background in UV, optical and IR ranges.

From the physical viewpoint, both low- and high-energy branches of MHU are still incomplete versions of phenomenology for Theories of Everything.

One needs supersymmetry to remove the problem of Higgs mass divergence, grand unification — to provide axionic solution of strong CP violation, shadow or mirror matter – to restore the equivalence between left and right co-ordinate systems etc.

Therefore, though the presented cosmological scenarios do not find direct observational contradictions, these scenarios do not and *should not* provide the best fit to the observational data.

For instance, the predicted spectrum of fluctuations grows to the red and not to the blue side as it seem to be favoured by the set of the data on large and medium angle BBBR anisotropy.

Archioles and hot dark matter admixture may be not sufficient to remove the troubles of CDM model of structure formation.

On the contrary, the cool component in HND may be insufficient to explain the effects of dark matter at the scale of small galaxies. The dominance of relativistic products of neutrino decays may find troubles in putting together the observed large scale structure with the observation of the relic radiation anisotropy.

All these disadvantages will be easily removed in more sophisticated scenarios, based on more complete physical grounds. Such scenarios will exhibit nontrivial cosmological patterns, and cosmoparticle physics is sufficiently strong now to face them on and to analyse them in all the physically motivated details, to what the cosmoparticle physics analysis of MHU should serve as a proof.

In any case, MHU provides the illustration for the unified fundamental basis for theoretical description both of the structure of elementary particles and of the structure of the Universe.

Such unified approach to cosmological and particle phenomena is the striking feature of cosmoparticle physics. Unifying the separate results of studies of partial problems of cosmology and particle physics, the model of horizontal unification seems to be the first step on the way towards realistic unified description of unique fundamental grounds of the micro- and macro-world structure.

The way to the highlights of the theory, based on the detailed elaboration of its "low-energy" basis, can give valuable recommendations for the choice of realistic variant of the complete unified "theory of everything" (superstring theory, for example), what is of the evident importance in view of the existing theoretical uncertainties in the searches for the fundamental grounds of physics and cosmology.

The important epistemological aspect of the presented studies is to be pointed out. We have demonstrated the principal possibility of the detailed study of multiparameter "hidden" sector of particle theory.

The example of the model of horizontal unification implies the hope to test any realistic multiparameter model of super-high energy physics. Being elaborated in details, it will lead to indirect effects, accessible to experimental and observational tests. If the amount of such effects exceeds the number of independent model parameters, so the over-determined system of equations relative to these parameters can be deduced.

So the general approach to the experimental test of the theory, based on the over-determined system of equations for unknown parameters, can be realised.

In the framework of cosmoparticle physics the combination of indirect effects, predicted by the theory, provides its detailed analysis in the case, when direct experimental test is impossible.

REFERENCES

Abbot, L., Farhi, E. and Wise, M. (1982) *Phys. Lett.* **117B**, 29.

Abbot, L.F. and Sikivie, P. (1983) *Phys. Lett.* **B120**, 133.

Abbot, L.F. and Wise, M.B. (1984) *Nucl. Phys.* **B244**, 541.

Adler, S.L. (1969) *Phys. Rev.* **177**, 2426.

Affleck, I. and Dine, M. (1985) *Nucl. Phys.* **B249**, 361.

Akhmedov, E.Kh. and Khlopov, M.Yu. (1988a) *Yadernaya Fizika*, **47**, 1079.

Akhmedov, E.Kh. and Khlopov, M.Yu. (1988b) *Mod. Phys. Lett.* **A3**, 451.

Albrecht, A. and Steinhardt, P.J.(1982) *Phys. Rev. Lett.* **48**, 1220.

Albrecht,A. and Turok, N. (1985) *Phys.Rev.Lett.* **54**, 1868.

Albrecht, A. and Turok, N. (1989) **D40**, 973.

Albrecht, A., Steinhardt, P., Turner, M. and Wilczek, F. (1982) *Phys. Rev. Lett.* **48**, 1437.

Alcock, C. and Farhi, E. (1985) *Phys. Rev.* **D32**, 1273.

Alcock, C. and Olinto, A. (1988) *Ann. Rev. Nucl. Part. Phys.* **38**, 161.

Alcock, C.R., Fuller, G.M. and Mathews, C.J. (1987) *Astrophys. J.* **320**, 439.

Alford, M.G. and Wilczek, F. (1989) *Phys. Rev. Lett.* **62**, 1071.

Alichanian, A.I. and Nikitin, S.J.(1938) *Phys. Rev.* **53**, 767.

Alichanian, A.I., Alichanov, A.I. and Dzelepov, B.S. (1938) *Phys. Rev.* **53**, 766.

Alpher, R.A. and Herman, R. (1988) *Phys. Today* **41**, 24.

Anselm, A.A. (1982) *Pisma ZhETF* **36**, 46.

Anselm, A.A. and Berezhiani, Z.G. (1985) *Phys. Lett.* **162B**, 349.

Anselm, A.A. and Uraltzev, N.G. (1982) *Phys. Lett.* **116B**, 161.

Anselm, A.A. and Uraltzev, N.G. (1983) *ZhETF* **84**, 1961.

Anselm, A.A., Uraltzev, N.G. and Khlopov, M.Yu. (1985) *Yadernaya Fizika* **41**, 1678.

Appelquist, T. and Chodos, A. (1983) *Phys. Rev.* **D28**, 772.

Applegate, J.H. and Hogan, C.J. (1985) *Phys. Rev.* **D31**, 3037.

Applegate, J.H., Hogan, C.J. and Scherrer, R.J. (1987) *Phys. Rev.* **D35**, 1151.

Arnold, P. and McLerran, L. (1987) *Phys. Rev.* **D36**, 581.

Aryal, M. and Everett, A.E. (1987) *Phys. Rev.* **D35**, 3105.

Babul, A., Paczynsky, B. and Spergel, D. (1987) *Astrophys. J. Lett.* **316**, L49.

Bahcall, J.N. (1984) *Astrophys. J.* **276**, 156; **276**, 169; **287**, 926.

Bahcall, J.N., Flynn, C. and Gould, A. (1992) *Astrophys. J.* **389**, 234.

Balakirev, A.N. (1976) *Astron. Zh.* **53**, 119.

Balestra, F., Guaraldo, K., Piragino, G., Sapozhnikov, M.G., Falomkin, I.V., Khlopov, M.Yu. et al (1985) *JINR Rapid Comm.* **6**, **11**.

Balestra, F., Piragino, G., Pontecorvo, D.B., Sapozhnikov, M.G., Falomkin, I.V. and Khlopov, M.Yu. (1984) *Yadernaya Fizika* **39**, 990 [*Sov. J. Nucl. Phys.* **39**, 626].

Bardeen, J., Steinhardt P., and Turner, M. (1983) *Phys. Rev.* **D28**, 679.

Bardeen, J.M.et al. (1986) *Astrophys. J.* **304**, 15.

Barnett, R.M. et al. Particle Data Group. (1996) *Review of Particle Physics. Phys. Rev.* **D54**, 1.

Belavin, A.A., Polyakov, A.M., Schwartz, A.S., and Tyupkin, Yu.S. (1975) *Phys. Lett.* **59B**, 85.

Belinsky, V.A., Grishchuk, L.P., Khalatnikov, I.M. and Zeldovich, Ya.B. (1985) *Phys. Lett.* **155B**, 232.

Bell, J.S. and Jackiw, R. (1969) *Nuovo Cim.* **A60**, 47.

Bellido, J.C. a-, Linde, A. and Wands, D. (1996) *Phys. Rev.* **54D**, 6040.

Belotsky, K.M., Khlopov, M.Yu., Sakharov, A.S., Shklyaev, A.A. and Sudarikov, A.L. (1998) *Proc. Cosmion'96* Eds. Khlopov M.Yu. et al (World Scientific) in press.

Belova,T.I. and Kudryavtzev,A.E. (1988) *Physica* **D32**,18.

Bennet,D.P. and Bouchet,F.R.(1988) *Phys.Rev.Lett.***60**,257.

Berestetsky, V.B., Lifshits, E.M. and Pitaevsky, L.P. (1968) *Relativistic Quantum Theory. Part 1* (Moskow, Nauka).

Berezhiani, Z.G. and Chkareuli, J.K. (1982) *Pisma ZhETF* **35**, 494.

Berezhiani, Z.G. and Chkareuli, J.K. (1983) *Pisma ZhETF* **37**, 285.

Berezhiani, Z.G. and Chkareuli, J.K. (1983) *Yadernaya Fizika*, **37**, 1043.

Berezhiani, Z.G. (1983) *Phys. Lett.* **129B**, 99.

Berezhiani, Z.G. (1985) *Phys.Lett.* **150B**, 177.

Berezhiani, Z.G. and Khlopov, M.Yu. (1989) in: *Proc. XXIV Rencontres de Moriond* ed. Tran Thanh Van, J. (Editions Frontieres), p. 179.

Berezhiani, Z.G. and Khlopov, M.Yu. (1990a) *Yadernaya Fizika*, **51**, 1157 [*Sov. J. Nucl. Phys.*, **51**, 739].

Berezhiani, Z.G. and Khlopov, M.Yu. (1990b) *Yadernaya Fizika*, **51**, 1479 [*Sov. J. Nucl. Phys.*, **51**, 935].

Berezhiani, Z.G. and Khlopov, M.Yu. (1990c) *Yadernaya Fizika* **52**, 96 [*Sov. J. Nucl. Phys.*, 52, 60].

Berezhiani, Z.G. and Khlopov, M.Yu. (1991) *Z.Phys.C-Particles and Fields*, **49**, 73.

Berezhiani, Z.G., Khlopov, M.Yu. and Khomeriki, R.R. (1990a) *Yadernaya Fizika* **52**, 104 [*Sov. J. Nucl. Phys.* **52**, 65].

Berezhiani, Z.G., Khlopov, M.Yu. and Khomeriki, R.R. (1990b) *Yadernaya Fizika* **52**, 538 [*Sov. J. Nucl. Phys.* **52**, 344].

Berezhiani, Z.G., Sakharov, A.S. and Khlopov, M.Yu. (1992) *Yadernaya Fizika*, **55**, 1918 [*Sov. J. Nucl. Phys.*, **55**, 1063].

Berezhiani, Z.G., Vysotsky, M.I. and Khlopov, M.Yu. (1987) *Yadernaya Fizika* **45**, 1719.

Berezhiani, Z.G., Vysotsky, M.I., Doroshkevich, A.G., Khlopov, M.Yu. and Yurov, V.P. (1990c) *Yadernaya Fizika* **51**, 1614.

Berkov, A.V., Nikitin, Yu.P. , Sudarikov, A.L. and Khlopov, M.Yu. (1987) *Yadernaya Fizika* **46**, 1729.

Berkov, A.V., Nikitin, Yu.P. , Sudarikov, A.L. and Khlopov, M.Yu. (1988) *Yadernaya Fizika* **48**, 779.

Bernard, C.W. (1974) *Phys. Rev.* **D9**, 3312.

Bisnovatyi–Kogan, G.S. and Novikov, I.D. (1980) *Astron. Zh.* **57**, 899.

Bisnovatyi–Kogan, G.S., Lukash, V.N. and Novikov, I.D. (1980) *Proc IAU Symposium.*

Bjorken, B.J. and Drell, S.D. (1965) *Relativistic Quantum Mechanics* (McGraw-Hill, New York).

Blau, S.K. and Guth, A.H. (1987) in *300 Years of Gravitation*, eds. Hawking, S.H. and Israel, W. (Cambridge University Press).

Blinnikov, S.I. and Khlopov, M.Yu. (1980) *Preprint ITEP N-126*.

Blinnikov, S.I. and Khlopov, M.Yu. (1982) *Yadernaya Fizika* **36**, 809.

Blinnikov, S.I. and Khlopov, M.Yu. (1983) *Astron. Zh.* **60**, 623.

Blinnikov, S.I. and Khlopov, M.Yu. (1983) *Solar physics* **82**, 383.

Bond, J.R., Efstathiou G. ans Silk, J. (1980) *Phys. Rev. Lett.* **45**, 300.

Bond, J.R., and Szalay, A.S. (1983) *Astrophys. J.*. **274**, 443.

Boris, S., Golutvin, A., Laptin, L., Lubimov, V., Nagovizin, V., Nozik, V., Novikov, E., Soloshenko, V., Tihomirov, I. and Tretyakov, E. (1987) *Phys. Rev. Lett.* **58**, 2019.

Bouchet, F., Bennet, D. and Stebbins, A. (1988) *Nature* **335**, 410.

Brandenberger, R.H. (1987) *Nucl. Phys.* **B293**, 812.

Bugrii, A.I. and Trushevsky, A.A. (1977) *Astrophysica* **13**, 361.

Callan, C.G.,Jr. and Coleman, S. (1974) *Phys. Rev.* **D16**, 1762.

Candelas, P., Horovitz, G.T., Strominger, A. and Witten, E. (1985) *Nucl. Phys.* **B258**, 46.

Carr, B.J. (1975) *Astrophys. J.* **205**, 1.

Carr, B.J., Gilbert, J.H. and Lidsey, J.E. (1994) *Phys. Rev.* **50D**, 4853.

Chechetkin, V.M., Khlopov, M.Yu. and Sapozhnikov, M.G. (1982) *Rivista Nouvo Cim.* **5, N-10,** 1.

Chechetkin, V.M., Khlopov, M.Yu., Sapozhnikov, M.G. and Zeldovich, Ya.B. (1982) *Phys. Lett.* **118B**, 329.

Cheng, H.Y. (1988) *Phys. Rep.* **158** 1

Chikashige, Y., Mohapatra, R.N., and Peccei, R.D. (1980) *Phys. Rev. Lett.* **45**,1926.

Chikashige, Y., Mohapatra, R.N., and Peccei, R.D. (1981) *Phys. Lett.* **98B**, 265.

Chiu, H.Y. (1966) *Phys. Rev. Lett.* **17**, 712.

Chkareuli, J.K. (1980) *Pisma ZhETF* **32**, 684.

Chudnovsky, E.M., Field, G.B., Spergel, D.N., and Vilenkin, A. (1986) *Phys. Rev.* **D34**, 944.

Cohen, A. and Kaplan, D. (1987) *Phys. Lett.* **199B**, 251.

Cohen, A. and Kaplan, D.(1988) *Nucl. Phys.* **B308**, 913.

Cohen, A.G., De Rujula, A. and Glashow S.L. (1997) *A Matter-Antimatter Universe? Preprint CERN-TH-97-154*.
Coleman, S. (1974) *Phys. Rev.* **D15**, 2929.
Coleman, S. and Weinberg, E. (1973) *Phys. Rev.* **D7**, 1888.
Colpi, M., Shapiro, S.L., and Wasserman, I. (1986) *Phys. Rev. Lett.* **57**, 2485.
Cowie, L. and Hu, E. (1987) *Astrophys. J.* **318**, L33.
Dannenberg,A. and Hall,L.J. (1988) *Phys.Lett.* **198B**, 411.
Davis, M., Lecar, M., Pryor, C. and Witten, E. (1981) *Astrophys. J.* **250**, 423.
Davis, M. and Peebles, P.J.E. (1983) *Astrophys. J.* **267**, 465.
Davis, R. (1986) *Phys. Lett.* **180B**, 277.
Davis, R.L. (1986) *Phys. Lett.* **180B**, 225.
Deser, S. and Zumino, B. (1976) *Phys. Lett.* **62B**, 335.
De Vaucouleurs, J. (1959) *Astron. Zh.* **36**, 977.
De Vaucouleurs, J. (1976) *Astrophys. J.* **203**, 33.
Dicus, D.S., Kolb, E.W., Teplitz, V.L. and Wagoner, R.V. (1978) *Phys. Rev.* **D18**, 1829.
Dimopoulos, S. and Hall, L.J. (1987) *Phys. Lett.* **196B**, 135.
Dine, M. and Fischler, W. (1983) *Phys. Lett.* **B120**, 137.
Dine, M., Fischler, W. and Srednicky, M. (1981) *Phys. Lett.* **104B**, 199.
Dirac, P.A.M. (1934) *Proc. Phys. Soc.* **A133**, 60.
Dirac, P.A.M. (1948) *Phys. Rev.* **74**, 817.
Dolan, L. and Jackiw, R. (1974) *Phys. Rev.* **D9**, 3320.
Dolgov, A.D. (1980) *Yadernaya Fizika* **32**, 1606.
Dolgov, A.D. (1980) *ZhETF*, **79**, 337.
Dolgov, A.D. (1983) in *The Very Early Universe*, eds. Gibbons, G., Hawking, S.W., and Siklos, S.T. (Cambridge University Press, Cambridge, 1983), p. 449.
Dolgov, A.D. and Linde, A.D. (1982) *Phys. Lett.* **116B**, 329.

Dolgov, A.D. and Zeldovich, Ya.B. (1981) *Rev. Mod. Phys.* **53**, 1.

Dolgov, A.D., Illarionov, A.F., Kardashev, N.S. and Novikov, I.D. (1987) *ZhETF* **94**, 1.

Doroshkevich, A.G. (1980) *Astron. Zh.* **57**, 259.

Doroshkevich, A.G. and Khlopov, M.Yu. (1981) *Astron. Zh.* **58**, 913.

Doroshkevich, A.G. and Khlopov, M.Yu. (1983) *Pisma Astron. Zh.* **9**, 323.

Doroshkevich, A.G. and Khlopov, M.Yu. (1984) *Mon. Not. Roy. Astr. Soc.* **211**, 277.

Doroshkevich, A.G. and Khlopov, M.Yu. (1984) *Yadernaya Fizika* **39**, 869 [*Sov. J. Nucl. Phys.* **39**, 551].

Doroshkevich, A.G. and Khlopov, M.Yu. (1985) *Pisma Astron. Zh.*, **11**, 563.

Doroshkevich, A.G. and Zeldovich, Ya.B. (1981) *ZhETF* **80**, 801.

Doroshkevich, A.G., Klypin, A.A. and Khlopov, M.Yu. (1985) *Pisma Astron. Zh.*, **11**, 483.

Doroshkevich, A.G., Klypin, A.A. and Khlopov, M.Yu. (1988b) in: *Large scale structures of the Universe* eds. Audouze, J. and Szalay, A.S. (Dordrecht, Reidel), p. 293.

Doroshkevich, A.G., Saar, E.M. and Shandarin, S.F. (1978) *Mon. Not. Roy. Astr. Soc.* **184**, 643.

Doroshkevich, A.G., Zeldovich, Ya.B., Sunyaev, R.A. and Khlopov, M.Yu. (1980a) *Pisma Astron. Zh.* **6**, 456.

Doroshkevich, A.G., Zeldovich, Ya.B., Sunyaev, R.A. and Khlopov, M.Yu. (1980b) *Pisma Astron. Zh.* **6**, 465.

Doroshkevich, A.G., Zeldovich, Ya.B., Sunyaev,, R.A., Szalay, A.S. and Khlopov, M.Yu. (1980) *Proc. "Neutrino-80"* (Erice, 1980).

Doroshkevich, A.G., Zeldovich, Ya.B., Sunyaev,, R.A., Szalay, A.S. and Khlopov, M.Yu. (1981) *Ann. N. Y. Acad. Sci.* **375**, 32.

Doroshkevich, A.G.,, Klypin, A.A. and Khlopov, M.Yu. (1988a) *Astron. Zh.*, **65**, 248.
Dubrovich, V.K. and Khlopov, M.Yu. (1989) *Astron. Zh.*, **66**, 232.
Duff, M.J., Nilsson, B.E.W. and Pope, C.N. (1976) *Phys. Rep.* **130**, 1.
Dymnikova, I.G. and Krawczyk, M. (1995) *Mod. Phys. Lett.* **A10**, 3069.
Dymnikova, I.G. and Khlopov, M.Yu. (1998) *Proc. Cosmion'96* Eds. Khlopov M.Yu. et al (World Scientific) in press.
Einasto, J. et al (1980) *Mon. Not. Roy. Astr. Soc.***193**,353.
Ellis, J. and Olive, K.A. (1987) *Phys. Lett.* **193B**, 525.
Ellis, J., Enqvist, K., Nanopoulos, D.V. and Olive, K.A. (1987) *Phys. Lett.* **191B**, 343.
Ellis, J., Enqvist, K., Nanopoulos, D.V. and Zwirner, F. (1986) *Nucl. Phys.* **B276**,14.
Ellis, J., Nanopoulos, D., Olive, K., and Tamvakis, K. (1983) *Nucl. Phys.* **B221**, 524.
Ellis, J., Nanopoulos, D.V. and Quiros, M. (1986) *Phys. Lett.* **174B**, 176.
Englert, F. and Brout, R. (1964) *Phys. Rev. Lett.* **13**,321.
Enrico, N. and Esteban, R. (1990) *Phys. Lett.* **B245**, 109.
Ershov, A.A., Smirnov, G.T., Sorochenko, R.L. and Tsivilev, A.P. (1988) *Pisma Astron. Zh.* **14**, 209.
Everett, E. (1981) *Phys. Rev.* **D24**, 858.
Fabian, A.C., Pringle, J.E. and Rees, M.J. (1975) *Mon. Not. Roy. Astr. Soc.* **172**, 15P.
Fargion, D., Khlopov, M.Yu., Konoplich, R.V. and Mignani, R. (1995) *Phys. Rev.* **D52**, 1828.
Feynman, R.P. and Hibbs, A.R. (1965) *Quantum Mechanics and Path Integrals* (McGraw-Hill, New York).
Field, G.B. (1965) *Astrophys. J..* **142**, 531.
Filippov, S.S., Khlopov, M.Yu., Levitan, Yu.L., Sedelnikov, E.V. and Sobol, I.M. (1996) in: *Proc. 1 Interna-*

tional conference on cosmoparticle physics "Cosmion-94", dedicated to 80 Anniversary of Ya.B.Zeldovich and 5 Memorial of A.D.Sakharov, (Moscow, December 5-14, 1994) eds. Khlopov, M.Yu., Prokhorov, M.E., Starobinsky, A.A. and Tran Thanh Van, J. (Editions Frontieres), p. 93.

Ford, L.H. (1987) *Phys. Rev.* **D35**, 2339.

Freedman, D.Z., van Nieuwenhuizen, P., and Ferrara, S. (1976) *Phys. Rev.* **D13**, 3214.

Freund, P.G.O. and Rubin, M.A. (1980) *Phys. Lett.* **B97**, 233.

Fukugita, M. and Watamura S. (1982) *Phys. Rev. Lett.* **48**, 1522.

Fuller, G.M., Mathews, C.J. and Alcock, C.R. (1988) *Phys. Rev.* **D37**, 1380.

Gamow, G. (1946) *Phys. Rev.* **70**, 572.

Garfinkle, D. and Vachaspati, T. (1987) *Phys. Rev.* **D36**, 2229.

Gasilov, V.A., Maslyankin, V.I. and Khlopov, M.Yu. (1985) *Astrofizika*, 23, 191.

Gell-Mann, M., Ramond, P. and Slansky, R. (1979) in: *Supergravity*, eds Van Nieuwenhuizen, P. and Freedman, D. (North Holland), p.186.

Gelmini, G., Schramm, D.N. and Valle, J.W.F. (1984) *Phys. Lett.* **146B**, 311.

Gelmini, G.B. and Roncadelli, M. (1981) *Phys. Lett.* **99B**, 411.

Georgi, H. and Glashow, S.L. (1974) *Phys. Rev. Lett.* **32**, 438.

Gershtein, S.S. and Zeldovich, Ya.B. (1966) *Pisma ZhETF* **4**, 174.

Gibbons, G.W. and Hawking, S. (1977) *Phys. Rev.* **D15**, 2738.

Glashow, S.L. and Manohar, A. (1985) *Phys. Rev. Lett.* **54**, 2306.

Glashow, S.L. and Weinberg, S. (1977) *Phys. Rev.* **D15**, 1958.

Gliese, W. (1977) *Highlights Astron.* **4**, P.II, 11.
Gliner, E.B. (1965) *ZhETF* **49**, 542.
Gliner, E.B. (1970) *Doklady Akad. Nauk* **192**, 771.
Glinner, E.B. and Dymnikova, I.G. (1975) *Pisma Astron. Zh.* **1**, 7.
Goddard,P. and Mansfield,P.(1986) *Rep.Progr.Phys.***49**,725.
Goldstone, J. (1961) *Nuovo Cim.* **19**, 154.
Goncharov, A.S. and Linde, A.D. (1986) *Fiz. Element. Chastyts i Atomn. Yadra.* **17**, 837.
Goncharov, S.A., Linde, A.D. and Mukhanov, V.F. (1987) *Int. J. Theor. Phys.* **A2**, 561.
Gradschtein, I.S. and Ryzhik, I.M. (1962) *Tables of Integrals, Sums, Series, and Products* (Fizmatgiz, Moscow).
Grawert, G., Luders, G. and Rollnik, H. (1959) *Fortschritte Der Phys.* **7**, 291.
Green, M.B. and Schwarz, J.H. (1984) *Phys. Lett.* **B149**, 117.
Grigoriev, D.Yu. and Rubakov, V.A. (1988) *Nucl. Phys.* **B299**, 67.
Gross, D.J. and Wilczek, F. (1973) *Phys. Rev. Lett.* **30**, 1343.
Gross, D.J., Pisarsky, R.D., and Yaffe, L.G. (1981) *Rev. Mod. Phys.* **53**, 43.
Gross, D.J., Harvey, J.A., Martinec, E. and Rohm, R. (1985) *Phys. Rev. Lett.* **54**, 502.
Gulyaev, S.A., Sorochenko, R.L. and Tsivilev, A.P. (1997) *Pisma Astron. Zh.* **23**, 191.
Guralnik, G.S., Hagen, C.R., and Kibble, T.W.B. (1964) *Phys. Rev. Lett.* **13**,585.
Guth, A. (1981) *Phys. Rev.* **D23**, 347.
Guth, A. and Pi, S.-Y. (1982) *Phys. Rev. Lett.* **49**, 1110.
Guth, A. and Weinberg, E. (1981) *Phys. Rev.* **D23**, 876.
Guth, A. and Weinberg, E. (1983) *Nucl. Phys.* **B212**, 321.
Haber, H.E. and Kane, G.L. (1985) *Phys. Rep.* **117**, 75.

Harari, D. and Sikivie, P. (1987) *Phys. Lett.* **195B**, 361.
Harrari, D. and Sikivie, P. (1987) *Phys. Lett.* **195B**, 361.
Harrison, E.R. (1970) *Phys. Rev.* **D1**, 2726.
Harvey, J.A., Kolb, E.W., Reiss, D.B. and Wolfram, S. (1982) *Nucl. Phys.* **B201**, 16.
Hawking, S.W. (1974) *Nature* **248**, 30.
Hawking, S.W. (1975) *Comm. Math. Phys.* **43**, 199.
Hawking, S., Moss, I.G., and Stewart, J.M. (1982) *Phys. Rev.* **D26**, 2681.
Hawking, S.W. (1982) *Phys. Lett.* **115B**, 295.
Hawking, S.W. and Moss, I.G. (1982) *Phys. Lett.* **110B**, 35.
Higgs, P.W. (1964) *Phys. Rev. Lett.* **13**, 508.
Hill, C.T., Schramm, D.N. and Fry, J.N. (1989) *Comm. Nucl. Part. Phys.* **19**, 25.
Hill, C.T., Schramm, D.N., and Walker. T.P. (1987) *Phys. Rev.* **D36**, 1007.
Hill, G., Hilditch, R.W. and Barnes, J.W. (1979) *Mon. Not. Roy. Astr. Soc.* **186**, 813.
Hogan, C. and Rees, M. (1984) *Nature* 311, 109.
Holtzman, J.A. (1989) *Astrophys. J. Suppl.* **71**, 1.
Hosotani, Y. (1983) *Phys. Lett.* **126B**, 309.
House, F. and Kilkenny, D. (1980) *Astron. and Astrophys.* **81**, 251.
Huang, K. (1982) *Quarks, Leptons, and Gauge Fields* (World Scientific. Singapore).
Hut, P. (1977) *Phys. Lett.* **69B**, 85.
Ignatiev, A,.Yu., Krasnikov, N.V., Kuzmin, V.A. and Tavkhelidze, A.N. (1978) *Phys. Lett.* **76B**, 436.
Ipser, J. and Sikivie, P. (1983) *Phys. Rev. Lett* **50**, 925.
Ipser, J. and Sikivie, P. (1984) *Phys. Rev.* **D30**, 712.
Itoyama, H., McLerran, L., Taylor, T.R. and van der Bij, J.J. (1987) *Nucl. Phys.* **B279**, 380.
Jones, B.J.T. (1976) *Rev. Mod. Phys.* **48**, 107.
Jones, D.H.P. (1972) *Astrophys. J.* **178**, 467.

Kadnikov, A.F., Maslyankin, V.I. and Khlopov, M.Yu. (1989) *Astrofizika*, **31**, 177.

Kajante, K. and Kurkio-Suonio, H. (1986) *Phys. Rev.* **D34**, 1719.

Kalashnikov, O.K. and Khlopov, M.Yu. (1983) *Phys. Lett.* **127B**, 407.

Kaluza, Th. (1921) *Sitzungsber. Preuss. Akad. Wiss. Berlin. Math. Phys.* **K1**, 966.

Kane, G.L., and Kani,I. (1986) *Nucl. Phys.* **B277**, 525.

Kardashev, N.S., Blome, H.J. and Priester, W. (1987) *Com. Astrophys.* **13**, 87.

Kawai, H., Lewellen, D. and Tye, S.H.H. (1986) *Phys. Rev. Lett.* **57**, 1832.

Kazama, Y., Yang, C.N., and Goldhaber, A.S. (1977) *Phys. Rev.* **D15**, 2287.

Kazanas, D. (1980) *Astrophys. J.* **241**, L59.

Khlopov M.Yu. and Sakharov A.S. (1998) *Proc. Cosmion'96* Eds. Khlopov M.Yu. et al (World Scientific) in press.

Khlopov, M.Yu (1990) *Astronuclear Experiment Astrobelix* (M., Znanie) *Ser.Fizika,* **7/90**.

Khlopov, M.Yu. (1979) *Priroda* **12**, 90.

Khlopov, M.Yu. (1980) *Yadernaya Fizika* **32**, 1600.

Khlopov, M.Yu. (1981) *Pisma ZhETF* **33**, 100.

Khlopov, M.Yu. (1985) *The Universe as Laboratory for Elementary Particles. Priroda* **5**, 20.

Khlopov, M.Yu. (1987) *Universe as a Giant Accelerator* (M.,Znanie), *Ser. Fizika,* **1/87**.

Khlopov, M.Yu. (1988) in: *Magnetic stars*, eds. Glagolevsky, Yu.V. and Kopylov, I.M. (Leningrad, Nauka, Leningrad branch), p. 201.

Khlopov, M.Yu. (1989) *Cosmoparticle physics* (M., Znanie) *Ser. Fizika*, **89/3**.

Khlopov, M.Yu. (1992) *Sov. J. Nucl. Phys.*, **55**, 765.

Khlopov, M.Yu. (1995) in. *Dark matter in cosmology, clocks and tests of fundamental laws* eds. Guiderdoni, B., Greene, G., Hinds, D. and Tran Thanh Van, J. (Editions Frontieres), p. 133.

Khlopov, M.Yu. (1996a) in: *Dark Matter in Cosmology. Quantum Measurement. Experimental Gravitation* eds. Tran Thanh Van, J. et al. (Editions Frontieres), p. 255.

Khlopov, M.Yu. (1996b) in: *Proc. 1 International conference on cosmoparticle physics "Cosmion-94", dedicated to 80 Anniversary of Ya.B.Zeldovich and 5 Memorial of A.D.Sakharov, (Moscow, December 5-14, 1994*) eds. Khlopov, M.Yu., Prokhorov, M.E., Starobinsky, A.A. and Tran Thanh Van, J. (Editions Frontieres), p. 1.

Khlopov, M.Yu. (1996c) in: *Proc. 1 International conference on cosmoparticle physics "Cosmion-94", dedicated to 80 Anniversary of Ya.B.Zeldovich and 5 Memorial of A.D.Sakharov, (Moscow, December 5-14, 1994*) eds. Khlopov, M.Yu., Prokhorov, M.E., Starobinsky, A.A. and Tran Thanh Van, J. (Editions Frontieres), p. 67.

Khlopov, M.Yu. and Chechetkin, V.M. (1985) *Test of GUTs in DUMAND. KIAM preprint N-41*.

Khlopov, M.Yu. and Chechetkin, V.M. (1987) *Fiz. Elementarn. Chastits i Atomn. Yadra* **18**, 627 [*Sov. J. Part. Nucl.*, **18**, 267].

Khlopov, M.Yu. and Linde, A.D. (1984) *Phys. Lett.* **138B**, 265.

Khlopov, M.Yu. and Petcov, S.T. (1981) *Phys. Lett.* **B99**, 17.

Khlopov, M.Yu. and Polnarev, A.G (1980) *Phys. Lett.* **B97**, 383.

Khlopov, M.Yu. and Polnarev, A.G (1983) in: *Proc. of the Nuffield Workshop "The very early Universe"* eds. Gibbons, G.W., Hawking, S.W. and Siclos, S.T.C. (Cambridge University Press), p. 407.

Khlopov, M.Yu. and Sakharov, A.S. (1994) in: *Particle astrophysics, atomic physics and gravitation* eds. Tran Thanh Van, J., Fontaine, G. and Hinds, E. (Editions Frontieres), p. 197.

Khlopov, M.Yu. and Sakharov, A.S. (1995) *Gravitation & Cosmology* **3**, 219.

Khlopov, M.Yu. and Sakharov, A.S. (1996) in: *Proc. 1 International conference on cosmoparticle physics "Cosmion-94", dedicated to 80 Anniversary of Ya.B.Zeldovich and 5 Memorial of A.D.Sakharov, (Moscow, December 5-14, 1994)* eds. Khlopov, M.Yu., Prokhorov, M.E., Starobinsky, A.A. and Tran Thanh Van, J. (Editions Frontieres), p. 273.

Khlopov, M.Yu. et al. Astrodamus collaboration (1996a) in: *Proc. 1 International conference on cosmoparticle physics "Cosmion-94", dedicated to 80 Anniversary of Ya.B.Zeldovich and 5 Memorial of A.D.Sakharov, (Moscow, December 5-14, 1994)* eds. Khlopov, M.Yu., Prokhorov, M.E., Starobinsky, A.A. and Tran Thanh Van, J. (Editions Frontieres), p. 99.

Khlopov, M.Yu. et al. Astrodamus collaboration (1996b) in: *Proc. 1 International conference on cosmoparticle physics "Cosmion-94", dedicated to 80 Anniversary of Ya.B.Zeldovich and 5 Memorial of A.D.Sakharov, (Moscow, December 5-14, 1994)* eds. Khlopov, M.Yu., Prokhorov, M.E., Starobinsky, A.A. and Tran Thanh Van, J. (Editions Frontieres), p. 107.

Khlopov, M.Yu. et al. Astrodamus collaboration (1996c) in: *The Dark Side of the Universe* eds. Bernabei, R. and Incicchitti, A. (World Scientific), p. 242.

Khlopov, M.Yu., Beskin, G.M., Bochkarev, N.G., Pustilnik, L.A. and Pustilnik, S.A. (1989) *Preprint FERMILAB–Pub–89/193A.*

Khlopov, M.Yu., Beskin, G.M., Bochkarev, N.G., Pustilnik, L.A. and Pustilnik, S.A. (1991) *Astron. Zh.* **69**, 121.

Khlopov, M.Yu., Konoplich, R.V. and Rubin, S.G. (1996) in: *Proc. 1 International conference on cosmoparticle phys-*

ics "Cosmion-94", dedicated to 80 Anniversary of Ya.B.Zeldovich and 5 Memorial of A.D.Sakharov, (Moscow, December 5-14, 1994) eds. Khlopov, M.Yu., Prokhorov, M.E., Starobinsky, A.A. and Tran Thanh Van, J. (Editions Frontieres), p. 307.

Khlopov, M.Yu., Malomed, B.A. and Zeldovich, Ya.B. (1985) *Mon. Not. Roy. Astr. Soc.* **214**, 100.

Khlopov, M.Yu., Naselsky, P. D. and Polnarev, A.G. (1985) *Proc. III Int. Sem. "Quantum gravity"*(Moscow, 1984), p. 306.

Khlopov, M.Yu., Sakharov, A.S., Shklyaev, A.A. and Sudarikov, A.L. (1996) in: *Proc. 1 International conference on cosmoparticle physics "Cosmion-94", dedicated to 80 Anniversary of Ya.B.Zeldovich and 5 Memorial of A.D.Sakharov, (Moscow, December 5-14, 1994)* eds. Khlopov, M.Yu., Prokhorov, M.E., Starobinsky, A.A. and Tran Thanh Van, J. (Editions Frontieres), p. 289.

Kibble, T.W.B. (1980) *Phys. Rep.* **67**, 183.

Kibble, T.W.B. (1985) *Nucl. Phys.* **B252**, 227.

Kibble, T.W.B., Lazarides, G. and Shafi, Q. (1982a) *Phys. Rev. Lett.* **113B**, 237.

Kibble, T.W.B., Lazarides, G. and Shafi, Q. (1982b) *Phys. Rev.* **D26**, 435.

Kim, J.E. (1979) *Phys. Rev. Lett.* **43**, 103.

Kim, J.E. (1987) *Phys. Rep.* **150**, 1.

Kirzhnits, D.A. (1972) *Pisma ZhETF* **15**, 745.

Kirzhnits, D.A. and Linde, A.D. (1972) *Phys. Lett.* **42B**, 471.

Klein, O. (1926) *Z. Phys.* **37**, 895.

Klinkhammer, F.R. and Norman, C.A. (1981) *Astrophys. J.* **243**, L1.

Klypin, A., Holtzman, J.A., Primack, J.R.and Regos, E. (1993) *Astrophys. J.* **416**, 1.

Kofman, L., Linde, A. and Starobinsky, A.A. (1996) *Phys. Rev. Lett.* **76**, 1011.

Kofman, L., Linde, A.D. and Starobinsky, A.A. (1987) *Phys. Lett.* **157B**, 361.

Kofman, L.A. and Linde, A.D. (1987) *Nucl. Phys.* **B282**,555.

Kofman, L.A. and Pogosyan, Yu.D. (1988) *Phys. Lett.* **214B**, 508.

Kofman, L.A., Linde, A.D. and Einasto, J. (1987) *Nature* **326**, 48.

Kogan, Ya.I. and Khlopov, M.Yu. (1986) *Yadernaya Fizika* **44**, 1344.

Kogan, Ya.I. and Khlopov, M.Yu. (1987) *Yadernaya Fizika* **46**, 314.

Kolb, E.W. and Turner, M.S. (1983) *Ann. Rev. Nucl. Part. Sci.* **33**, 645.

Kolb, E.W., Perry, M.J., and Walker, T.P. (1986) *Phys. Rev.* **D33**, 869.

Komachenko, Yu.Ya. and Khlopov, M.Yu. (1990) *Yadernaya Fizika*, **51**, 1081.

Konoplich, R.V., and Khlopov, M.Yu. (1988) *Yadernaya Fizika*, **47**, 891.

Konoplich, R.V. and Khlopov, M.Yu. (1994) *Yadernaya Fizika* **57**, 452 [*Phys. Atomn. Nucl.*, **57**, 425].

Konoplich, R.V., Sorokina, E.V. and Khlopov, M.Yu. (1994) *Yadernaya Fizika* **57**, 1338. [*Phys. Atomn. Nucl.*, **57**, 1268].

Konoplich, R.V., Rubin, S.G., Sakharov, A.S. and Khlopov, M.Yu., (1998) *Pisma Astron. Zh.* **24**, 1.

Kosover, D,A., Hall, L.J. and Krauss, L.M. (1985) *Phys. Lett.* **150B**, 436.

Krauss, L.M. (1983) *Phys. Lett.* **128B**, 37.

Kuijken, K. and Gilmore, G. (1989) *Mon. Not. Roy. Astr. Soc.* **239**, 571; **239**, 605; **239**, 651.

Kuijken, K. and Gilmore, G. (1991) *Astrophys. J. Lett.* **367**, L9.

Kurki-Suonio, H., Matzner, R.A., Centrello, J.M., Rothman, T. and Wilson, J.R. (1988) *Phys. Rev.* **D38**, 1091.

Kuzmin, V.A. (1970) *Pisma ZhETF* **12**, 335.

Kuzmin, V.A., Rubakov, V.A. and Shaposhnikov, M.E. (1985) *Phys. Lett.* **B155**, 36.

Kuzmin, V.A., Shaposhnikov, M.E. and Tkachev, I.I. (1981) *INR Preprints* P-0209; P-0219; *Phys. Lett.* **105B**, 159.

Kuzmin, V.A., Shaposhnikov, M.E. and Tkachev, I.I. (1992) *Phys. Rev.* **45D**, 466.

Landau, L.D. and Lifshitz, E.M. (1964) *Statistical Physics* (Moscow, Nauka).

Landau, L.D. and Lifshitz, E.M. (1973) *Field Theory* (Moscow, Nauka).

Landsman, N.P. and van Weert, Ch.G. (1987) *Phys. Rep.* **45**, 141.

Langacker, P. and Pi, S.-Y. (1980) *Phys. Rev. Lett.* **45**, 1.

Lazarides, G. and Shafi, Q. (1982) *Phys. Lett.* **115B**, 21.

Lazarides, G., Panagiotakopoulos, C. and Shafi, Q. (1986) *Phys. Rev. Lett.* **56**, 557.

Lazarides, G., Panagiotakopoulos, C., and Shafi, Q. (1987) *Phys. Lett.* **192B**, 323.

Lazarides, G., Shafi, Q., and Walsh, T. (1982) *Nucl. Phys.* **B195**, 157.

Lee, T.D. (1973) *Phys. Rev.* **D8**, 1226.

Lee, T.D. and Yang, C.N. (1955) *Phys. Rev.* **98**, 1501.

Lee, W. and Weinberg, S. (1977) *Phys. Rev. Lett.* **39**, 165.

Levitan, Yu.L., Sedelnikov, E.V., Sobol, I.M. and Khlopov, M.Yu. (1994) *Yadernaya Fizika* **57**, 1466.

Levitan, Yu.L., Sobol, I.M., Khlopov, M.Yu. and Chechetkin, V.M. (1988) *Yadernaya Fizika* **47**, 168 [*Sov. J. Nuc. Phys.* **47**, 109].

Liddle, A.R. (1993) *Phys. Rev.* **D 49**, 739.

Liddle, A.R. and Lyth, D.H. (1993) *Phys.Rep.* **231**, 1.

Lidsey, J.E. at al. (1995) *Preprint FERMILAB-Pub-95/280-A/astro-ph/9508078.*

Linde, A.D. (1979) *Repts. on Progress in Physics* **42**, 389.

Linde, A.D. (1982a) *Phys. Lett.* **108B**, 389.

Linde, A.D. (1982b) *Phys. Lett.* **116B**, 335.

Linde, A.D. (1983a) *Pisma ZhETF*. **38**, 149.

Linde, A.D. (1983b) *Phys. Lett.* **129B**, 177.

Linde, A.D. (1984) *Uspekhi Fiz. Nauk* **114**, 177.

Linde, A.D. (1985) *Phys. Lett.* **160B**, 243.

Linde, A.D. (1986) *Mod. Phys. Lett.* **A2**, 81.

Linde, A.D. (1988) *Phys. Lett.* **200B**, 272.

Linde, A.D. (1988) *Phys. Lett.* **B201**, 437.

Linde, A.D. (1990) *Particle Physics and Inflationary Cosmology* (Harwood, Chur, Switzerland).

Lubimov, V.A., Novikov, E.G., Nozik, V.Z., Tret'yakov, E.F. and Kozik, V.S. (1980) *Yadernaya Fizika* **32**, 310 [*Sov. J. Nuc. Phys.* **32**, 280].

Luty, M.A. (1992) *Phys. Rev.* **D45**, 455.

Ma, C.P. (1996) *Astrophys. J.* **471**, 13.

Madsen, J. (1990) in: *Particle Astrophysics. The Early Universe and Cosmic Structures* eds. Alimi, J.M., Blanchard, A., Bouquet, A., Martin de Volnay, F. and Tran Thanh Van, J. (Editions Frontieres), p.205.

Mao, S. and Paczynski, B. (1996), *Astrophys. J.* **473**, 57.

Martemyanov, B.V., Khlopov, M.Yu. and Schepkin M.G. (1980) *Pisma ZhETF* **32**, 484.

Matzner, M. (1988) *Computer in Physics* **2**, 51.

Mohapatra, C.N. (1987) *Phys. Lett.* **186B**, 303.

Morgan, D.L. and Hudges, V.W. (1970) *Phys. Rev.* **D2**, 1389.

Nambu, Y. (1970) *Proc. Int. Conf. on Symmetries and Quark Models* p.269.

Narain, K.S. (1986) *Phys. Lett.* **169B**, 41.

Naselsky, D. and Polnarev, A.G. (1986) *Astron. Zh.* **62**, 833.

Nielsen, H.B. and Olesen, P. (1973) *Nucl. Phys.* **B61**, 45.

Nilles, H.P. (1983) *Phys. Rep.* **86**, 200.

Nilles, H.P. (1984) *Phys. Rep.* **110**, 1.

Novikov, I.D., Polnarev, A.G., Starobinsky, A.A. and Zeldovich, Ya.B. (1979) *Astron.&Astrophys.* **80**, 104.

Novikov, V.A., Shifman, M.A., Vainshtein, A.I. and Zakharov, V.I. (1981) *Nucl. Phys.* **B191**, 301.

Okun, L.B. and Zeldovich, Ya.B. (1976) *Comm. Nucl. and Part. Phys.* **6**, 69.

Okun, L.B. (1981) *Leptons and Quarks* (Moscow, Nauka).

Okun, L.B. (1984) *Elementary Particle Physics* (Moscow, Nauka).

Olshanetsky, M.A. (1982) *Uspekhi Fiz. Nauk.* **136**, 421.

Oort, J.H. (1965) in: *Stellar Dynamics. Galactic Structure* (Chicago-London), p. 455.

Ostriker, J., Thompson, C. and Witten, E. (1986) *Phys. Lett.* **B180**, 231.

Ozernoy, L.M. and Chernin, A.D. (1966) *Astron. Zh.* **44**, 1131.

Parker, L. (1968) *Phys. Rev. Lett.* **21**, 562.

Peacock, J.A., and Dodds, S.J. (1994) *Mon. Not. Roy. Astr. Soc.* **267**, 1020.

Peccei, R.D. and Quinn, H.R. (1977a) *Phys. Rev. Lett.* **38**, 1440.

Peccei, R.D. and Quinn, H.R. (1977b) *Phys. Rev.* **D16**, 1791.

Penzias, A.A. and Wilson, R.W. (1965) *Astrophys. J.* **142**, 419.

Pi, S.-Y. (1984) *Phys. Rev. Lett.* **52**, 1725.

Pogosyan, Yu.D., and Starobinsky, A.A. (1995) *Astrophys. J..* **447**, 465.

Politzer, H.D. (1973) *Phys. Rev. Lett.* **30**, 1346.

Polnarev, A.G and Khlopov, M.Yu. (1981) *Astron. Zh.* **58**, 706.

Polnarev, A.G and Khlopov, M.Yu. (1982a) *Astron. Zh.* **59**, 15.

Polnarev, A.G and Khlopov, M.Yu. (1982b) *Astron. Zh.* **59**, 639.

Polnarev, A.G. and Khlopov, M.Yu. (1985) *Uspekhi Fiz. Nauk* **145**, 369 [*Sov. Phys. Uspekhi* **28**, 213].

Polyakov, A.M. (1974) *Pisma ZhETF*. **20**, 430.

Preskill, J., Wise, M., and Wilczek, F. (1983) *Phys. Lett.* **B120**,127.

Preskill, J.R. (1979) *Phys. Rev. Lett.* **43**, 1365.

Press, W.H., and Schechter, P. (1974) *Astrophys. J.* **187**, 425.

Press, W.H., and Teukolsky, S. (1977) *Astrophys. J.* **213**, 183.

Raffelt, G. and Seckel, D. (1988) *Phys. Rev. Lett* **60**, 1793.

Rajaraman, R. (1982) *An Introduction to Solitons and Instantons in Quantum Field Theory* (North-Holland Publishing Company. Amsterdam-New York-Oxford).

Reiss, D. (1982) *Phys. Lett.* **115B**, 217.

Rodionov, B.U. (1996) in: *Proc. 1 International conference on cosmoparticle physics "Cosmion-94", dedicated to 80 Anniversary of Ya.B.Zeldovich and 5 Memorial of A.D.Sakharov, (Moscow, December 5-14, 1994*) eds. Khlopov, M.Yu., Prokhorov, M.E., Starobinsky, A.A. and Tran Thanh Van, J. (Editions Frontieres), p. 473.

Rubakov, V.A. (1981) *Pisma ZhETF*. **33**, 658.

Rubakov, V.A. (1982) *Nucl. Phys.* **B203**, 311.

Rubakov, V.A., Sazhin, M.V., and Veryaskin, A.V.(1982) *Phys. Lett.* **115B**, 189.

Ruffini, R. and Bonazzola, S. (1969) *Phys. Rev.* **187**, 1767.

Ruiz, F.R. and Alvarez-Estrado, R.F. (1987) *Z. Phys.* **C34**, 131.

Sachs, R.K. and Wolf, A.M. (1967) *Astrophys. J.* **147**, 73.

Sakharov, A.D. (1967) *Pisma ZhETF* **5**, 32.

Sakharov, A.D. (1979) *ZhETF* **76**, 1172.

Sakharov, A.D. (1989) *Vestnik AN SSSR* **4**, 39.

Sakharov, A.S. and Khlopov, M.Yu. (1993) *Yadernaya Fizika*, **56**, 220 [*Sov. J. Nucl. Phys.*, **56**, 412].

Sakharov, A.S. and Khlopov, M.Yu. (1994) *Yadernaya Fizika* **57**, 514 [*Phys. Atomn. Nucl.*, **57**, 485].

Sakharov, A.S. and Khlopov, M.Yu. (1994) *Yadernaya Fizika* **57**, 690 [*Phys. Atomn. Nucl.*, **57**, 651].

Sakharov, A.S., Sokolov, D.D. and Khlopov, M.Yu. (1996) *Yadernaya Fizika* **59**, 1050.

Sakharov, A.S., Sudarikov, A.L., Khlopov, M.Yu. and Shklyaev, A.A. (1996a) *Yadernaya Fizika*, **59**, 295.

Sakharov, A.S., Sudarikov, A.L., Khlopov, M.Yu. and Shklyaev, A.A. (1996b) *Yadernaya Fizika*, **59**, .

Salopek, D.S. and Bond, J.R. (1990) *Phys. Rev.* **D 42**, 3936.

Sato, H. and Takahara, F. (1980) *Progr. Theor. Phys.* **54**, 2029.

Sato, K. (1981a) *Phys. Lett.* **99B**, 66.

Sato, K. (1981b) *Mon. Not. Roy. Astr. Soc.* **195**, 467.

Sato, K. and Kobayashi, H. (1987) *Progr. Theor. Phys.* **58**, 1775.

Sato, N., Kodama, H. and Sato, K. (1985) *Progr. Theor. Phys.* **74**, 405.

Sazhin, M.V. and Khlopov, M.Yu. (1989) *Astron. Zh.* **66**, 191.

Schechter, J. and Valle, J.W.F. (1982) *Phys. Rev.* **D25**, 774.

Scherk, J. (1979) *Phys. Lett.* **88B**, 265.

Schmidt, M., Schneider, D.P., and Gunn, J.E. (1986) *Astrophys. J.* **306**, 411.

Schramm, D.N. and Steigman, G. (1981) *Astrophys. J.* 151, 459.

Schramm, D.N. and Copi, C. (1996) in: *Proc. 1 International conference on cosmoparticle physics "Cosmion-94"*, *dedicated to 80 Anniversary of Ya.B.Zeldovich and 5 Memorial of A.D.Sakharov, (Moscow, December 5-14, 1994)* eds. Khlopov, M.Yu., Prokhorov, M.E., Starobinsky, A.A. and Tran Thanh Van, J. (Editions Frontieres), p. 77.

Schwartz, A.S. (1976) *Nucl. Phys.* **B112**, 358.

Schwartz, A.S. (1982) *Nucl. Phys.* **B208**, 141.

Schwartz, A.S. and Tyupkin, Yu.A. (1982) *Nucl. Phys.* **B209**, 427.

Schwinger, J. (1966) *Phys. Rev.* **144**, 1087.

Sedelnikov, E.V., Filippov, S.S. and Khlopov, M.Yu. (1995) *Yadernaya Fizika* **58**, 280.

Sedelnikov, E.V. and Khlopov, M.Yu. (1996) *Yadernaya Fizika* **59**, 1045.

Severny, A.B., Kotov, V.A. and Tsap, T.T. (1976) *Nature* **259**, 87.

Severny, A.B., Kotov, V.A. and Tsap, T.T. (1979) *Astron. Zh.* **56**, 1137.

Shafi, Q. and Wetterich, C. (1983) *Phys. Lett.* **129B**, 387.

Shaposhnikov, M.E. (1987) *Nucl. Phys.* **B287**, 757.

Shaposhnikov, M.E. (1988) *Nucl. Phys.* **B299**, 797.

Shifman, A.I., Vainshtein, A.I., and Zakharov, V.I. (1979) *Nucl. Phys.* **B147**, 385.

Shifman, M.A., Vainshtein, A.I., and Zakharov, V.I. (1980) *Nucl. Phys.* **B166**, 493.

Shuryak, E.V. (1980) *Phys. Rep.* **61**, 71.

Shvartsman, V.F. (1969) *Pisma ZhETF*. **9**, 315.

Shvartsman, V.F. (1977) *Comm. of SAO*. **19**, 1.

Shvartsman,V.F., Braginsky,V.B., Gershtein,S.S., Zeldovich,Ya.B. and Khlopov,M.Yu.(1982)*PismaZhETF* **36**,224.

Sikivie, P. (1982) *Phys. Rev. Lett.* **48**, 1156.

Silk, J. (1968) *Astrophys. J.* **151**, 459.

Silk, J. and Turner, M. (1987) *Phys. Rev.* **D35**, 419.

Smoot, G.F. et al. (1992) *Astrophys. J. Lett.* **300**, 90.

Sohnius, M.F. (1985) *Phys. Rep.* **128**, 39.

Starobinsky, A.A. (1979) *Pisma ZhETF*. **30**, 719.

Starobinsky, A.A. (1980) *Phys. Lett.* **91**, 99.
Starobinsky, A.A. (1982) *Phys. Lett.* **117B**, 175.
Starobinsky, A.A. (1983) *Pisma Astron Zh.* **9**, 1983.
Stauffer, D. (1979) *Phys.Rep.* **54**, 1.
Stecker, F.W. and Shafi, Q. (1983) *Phys.Rev.Lett.* **50**, 928.
Stecker, F.W. (1985) *Nucl. Phys.* **B252**, 25.
Sunyaev, R.A. and Zeldovich, Ya.B. (1969) *Astrophys. Space Sci.* **4**, 285.
Sunyaev, R.A. and Zeldovich, Ya.B. (1970) *Astrophys. Space Sci.* **7**, 20.
Terasawa, N. and Sato, K. (1989) *Phys. Rev.* **D39**, 2893.
t'Hooft, G. (1974) *Nucl. Phys.* **B79**, 276.
t'Hooft, G. (1976) *Phys. Rev. Lett.* **37**, 8.
Toussaint, D., Treiman, S.B., Wilczek, F., and Zee, A. (1979) *Phys. Rev.* **D19**, 1036.
Tremaine, S. and Gunn, J.E. (1979) *Phys. Rev. Lett* **42**, 407.
Tryon, E.P. (1973) *Nature.* **246**, 396.
Tsivilev, A.P. (1991) *Pisma Astron. Zh.* **17**, 3.
Tully, R.B. (1982) *Astrophys. J.* **257**, 389.
Turner, M.S. (1990) *Phys. Rep.* **197**, 67.
Turner, M.S. and Widrow, L. (1986) *Phys. Rev. Lett.* **57**, 2237.
Turner, M.S. and Widrow, L.M. (1988) *Phys. Rev.* **D37**, 2743.
Turner, M.S., Cohen, A.G. and Kaplan, D.B. (1988) *Phys. Lett.* **B216**, 20.
Turner, M.S., Steigman, G. and Krauss, L.M. (1984) *Phys. Rev. Lett* **92**, 2090.
Turok, N. and Bhattacharjee, P. (1984) *Phys. Rev.* **D29**, 1557.
Vachaspati, T. and Vilenkin, A. (1984) *Phys. Rev.* **D30**, 2936.
Vachaspati, T. and Vilenkin, A. (1985) *Phys. Rev.* **D31**, 3052.

Vainshtein, A.I., Zakharov, V.I., Novikov, V.I. and Shifman, A,I. (1982) *Uspekhi Fiz. Nauk.* **136**, 553.
Valle, J.W.F. (1983) *Phys. Lett.* **131B**, 87.
Vilenkin, A. (1981a) *Phys. Rev. Lett.* **46**, 1169; 1496 (E)
Vilenkin, A. (1981b) *Phys. Rev.* **D24**, 2082.
Vilenkin, A. (1981c) *Phys. Lett.* **B107**, 47.
Vilenkin, A. (1981d) *Phys. Rev.* **D23**, 852.
Vilenkin, A. (1982) *Nucl. Phys.* **B196**, 240.
Vilenkin, A. (1983a) *Phys. Rev.* **D27**, 2848.
Vilenkin, A. (1985) *Phys. Rep.* **121**, 263.
Vilenkin, A. and Everett, A.E. (1982) *Phys. Rev. Lett.* **48**, 1867.
Vishniac, E.T., Olive, K. and Seckel, D. (1987) *Nucl. Phys.* **B289**,717.
Voloshin, M.B. and Dolgov, A.D. (1982) *Yadernaya Fizika* **35**, 213.
Voloshin, M.B. and Okun, L.B. (1978) *Pisma ZhETF* **28**, 156.
Voloshin, M.B. and Ter-Martirosyan, K.A. (1984) *Theory of Gauge Interactions of Elementary Particles* (Moscow, Energoatomizdat).
Vysotsky, M.I. (1985) *Uspekhi Fiz. Nauk* **147**, 100;
Vysotsky, M.I., Dolgov, A.D., and Zeldovich, Ya.B. (1987) *Pisma ZhETF.* **26**, 200.
Vysotsky, M.I., Zeldovich, Ya.B., Khlopov, M.Yu. and Chechetkin, V.M. (1978) *Pisma ZhETF*, **27**, 533.
Vysotsky, M.V. (1985) *Uspekhi Fiz. Nauk* **146**, 591.
Wagoner, R.V., Fowler, W.A. and Hoyle, F. (1967) *Astrophys. J.* **148**, 3.
Warren, S.J., Hewlett, P.C., Irwin, M.J., McMahou, R.J., Bridgeland, M.T., Bunclark, P.S. and Kibblewhite, E.J. (1987) *Nature* **325**, 131.
Watkins, R. and Widrow, L.M. (1992) *Nucl. Phys.* **B374**, 446.

Weinberg, S. (1972) *Gravitation and Cosmology* (John Wiley and Sons, Inc. New York-London-Sidney-Toronto).
Weinberg, S. (1974) *Phys. Rev.* **D9**, 3357.
Weinberg, S. (1978) *Phys. Rev. Lett.* **40**, 223.
Weinberg, S. (1979) *Phys. Rev. Lett.* **42**, 850.
Widrow, L.M. (1988) *Phys. Rev.* **D39**, 3571.
Wilczek, F. (1978) *Phys. Rev. Lett.* **40**, 279.
Wilczek, F. (1982) *Phys. Rev. Lett.* **49**, 1549.
Wilczek, F. (1983) in *The Very Early Universe* eds. Gibbons, G., Hawking, S.W., and Siklos, S.T. (Cambridge University Press, Cambridge) p. 9.
Wise, M.B., Georgi, H., and Glashow, S. L. (1981) *Phys. Rev. Lett.* **47**, 402.
Witten, E. (1981) *Nucl. Phys.* **B177**, 477.
Witten, E. (1984) *Phys. Rev.* **D30**, 272.
Witten, E. (1985) *Nucl. Phys.* **B258**, 75.
Wright, E.L. et al. (1992) *Astrophys. J. Lett.* **300**, 100.
Yahil A. and Beaudet, G. (1976)) *Astrophys. J.* **206**, 26.
Yamamoto, K. (1986) *Phys. Lett.* **168B**, 341.
Yamamoto, K. (1987) *Phys. Lett.* **194B**, 390.
Yang, C. N. and Mills, R.L. (1954) *Phys. Rev.* **96**, 191.
Yokoyama, J., Kodama, H., Sato, K. and Sato, N. (1987) *Int. J. Mod. Phys.* **A2**, 1808.
Yoshimura, M. (1978) *Phys. Rev. Lett.* **41**, 281; **42**, 746 (Erratum).
Zavelski, A.S. (1939) *Phys. Rev.* **55**, 317.
Zeldovich, Ya.B. (1961) *ZhETF.* **41**, 1609.
Zeldovich, Ya.B., Okun, L.B. and Pikelner, S.B. (1965) *Uspekhi Fiz. Nauk* **84**, 113.
Zeldovich, Ya.B. and Poduretz, M.A. (1965) *Astron. Zh.* **42**, 963.
Zeldovich, Ya.B. (1968) *Uspechi Fiz. Nauk* **95**, 209.
Zeldovich, Ya.B. (1972) *Mon. Not. Roy. Astr. Soc.* **160**, 1.
Zeldovich, Ya.B., Illarionov,A.F. and Sunyaev,R.A. (1972) *ZhETF.* **62**, 1217.

Zeldovich, Ya.B., Kobzarev, I.Yu. and Okun, L.B. (1974) *ZhETF* 67, 3.

Zeldovich, Ya.B. and Novikov, I.D. (1975) *Structure and Evolution of the Universe* (Moscow, Nauka).

Zeldovich, Ya.B. (1976) *Pisma ZhETF* **24**, 29.

Zeldovich, Ya.B., Starobinsky, A.A. , Khlopov, M.Yu. and Chechetkin, V.M. (1977) *Pisma Astron. Zh.* **3**, 208.

Zeldovich, Ya.B. and Khlopov, M.Yu. (1978) *Phys. Lett.* **79B**, 239.

Zeldovich, Ya.B. (1980) *Mon. Not. Roy. Astr. Soc.* **192**, 63.

Zeldovich, Ya.B. and Sunyaev, R.A. (1980) *Pisma Astron. Zh.* **6**, 451.

Zeldovich, Ya.B., Klypin, A.A., Khlopov, M.Yu. and Chechetkin, V.M. (1980) *Yadernaya Fizika* **31**, 1286. [*Sov. J. Nucl. Phys.* **31**, 664].

Zeldovich, Ya.B. and Khlopov, M.Yu. (1981) *Pisma ZhETF*, **34**, 148.

Zeldovich, Ya.B. and Khlopov, M.Yu. (1981) *Uspekhi Fiz. Nauk* **135**, 45.

Zeldovich, Ya.B. and Khlopov, M.Yu. (1984) *Astrophysics and Elementary Particles. MEPI lectures* (MEPI, Moscow).

Zeldovich, Ya.B. and Khlopov, M.Yu. (1988) *Drama of Ideas in the Knowledge of Nature* (M.,Nauka).

Zhitnitcki, A.R. (1980) *Yadernaya Fizika.* **31**, 497.

Zumino, B. (1975) *Nucl. Phys.* **B89**, 535.

SUBJECT INDEX

A

age of the Universe, 35, 94, 392, 400
Alice strings, 56, 440–444
antibaryon domains, 158–160, 238, 240, 260, 298–299, 524
antinucleon annihilation, 274–276, 299
antiproton sources, 274, 279–281, 282, 301
antimatter objects, 524
archioles, 521, 523
archion, 87, 504–507
arion, 87, 504
asymptotic freedom, 42
axion, 48, 78–89
axion decay, 78
axion mass, 78

B

background radiation, 6, 7, 100, 103, 104, 107, 128, 149, 245, 247, 274, 303, 304, 311, 333, 386, 394, 396, 401, 412, 447, 508, 521, 540
baryon asymmetry, 103, 127, 155, 158, 159, 225, 226, 259, 286, 287, 299, 326, 361, 448, 524
baryon charge nonconservation, 47, 49, 137, 159, 325, 523
baryon excess, 155, 156, 158, 159, 298, 524
baryon, 34
baryosynthesis, 127, 128, 129, 155, 156, 158, 159, 524
Big Bang nucleosynthesis, 99, 125, 161, 205, 207, 210, 211, 223, 224, 240, 248, 255, 257, 260, 450
Big Bang scenario, 6– 10, 91-129, 166, 539–541
black hole evaporation, 181–183, 260–273, 282–290, 302, 326, 539
black hole formation, 169, 171, 177, 192, 200
Bose-Einstein spectrum, 244

bubble nucleation, 151, 152, 193, 199, 201
bubble walls, 169, 185, 188, 200

C

chaotic inflation, 152, 154, 507–508, 513, 531
chemical potential, 224, 225, 243
closed world, 95
cold dark matter, 161, 162, 359, 404, 517, 528, 537, 539
collision terms, 305, 306, 314,
colour anomaly, 77, 518
colour gauge group, 40
colour interaction, 39
colour symmetry, 38, 438
comoving volume, 96
compactification, 57
confinement, 42, 264, 438–439
cosmic string, 8, 52, 219, 349, 442–443, 519–520
cosmoarcheology, 14, 165, 319–320, 324, 420
cosmological density, 94–99, 160, 161, 163, 174, 176, 200, 204, 207, 210, 216, 247, 255, 267, 268, 283, 285, 286, 289, 295, 350, 355, 356, 391, 400, 402–404, 411, 414, 426, 442, 451–452, 528, 530, 533–537
cosmological time, 107, 112, 113, 114, 115, 117, 123, 155, 173, 174, 175, 190, 200, 201, 219, 220, 221, 234–240, 242, 248–250, 253–258, 264–265, 268, 271–272, 276, 284–286, 288–289, 291, 293, 298–299, 301, 303–304, 311–312, 324, 325, 333, 335–337, 347, 380–385, 392, 398, 511, 514, 537, 538
CP violation, 47, 156, 158, 159, 163, 260, 299, 325, 361, 437, 448–449, 497, 524, 540
critical density, 94, 95, 97, 98, 99, 102, 148, 160–162, 197, 201, 356, 391, 396, 402, 403, 406, 411, 412, 414, 426, 530, 533, 534, 535, 536, 537

D

dark matter, 8, 10, 11, 15, 16, 17, 18, 19, 48, 97, 127, 128, 129, 155, 160–163, 271, 281, 357–361, 386–388, 401–408, 409–416, 419, 431, 451–453, 461, 464–466, 480, 522, 528–541

decoupling, 114, 118, 121, 122, 123, 227, 229, 292, 293, 532
density perturbations, 126, 150, 152, 153, 168, 170–180, 286–291, 351–356, 360–362, 369–387, 389–390, 392–400, 401, 403–409, 412, 413, 509, 511, 516, 537, 538, 540
deuterium abundance, 212–213, 221, 255, 274–277, 281, 282, 317–319, 321–324, 326, 447
deuterium photodesintegration, 124, 257–258, 316–318
dissipation of antimatter domains, 160, 238–239, 260, 298–299, 524
domain walls, 52, 521, 529
double beta decay, 72–73, 497
dust–like stage, 18, 105–106, 152–153, 169–171, 174–176, 200, 202, 219–221, 247–249, 254–257, 284–288, 289–290, 367–368, 375, 379, 382, 397, 405, 510, 511, 513, 515, 539

E

e-folding, 147–149, 154, 508, 513
Einstein equations, 91, 96, 97, 105, 144
electric charge, 1, 19, 21, 23-29, 31–32, 37, 41, 62, 130, 436–439
electroweak interaction, 3, 10, 19, 39, 43–47, 78, 438, 494, 497, 498
electroweak scale, 52, 70, 78
elementary particle, 1–6. 16–17, 19, 22–23, 27–29, 32–39, 43–46, 48, 49–57, 58, 61–66, 69–71, 77–79, 80–82, 84, 86–87, 107, 112–115, 123–124, 130–134, 137–138, 155–158, 163, 165, 184, 202, 204, 208–211, 214, 216–219, 226–227, 237, 245–246, 248, 252–256, 272–273, 283–288, 291–296, 302–304, 309–310, 315–316, 318, 326, 330–331, 332, 334–336, 345–347, 356–359, 390–394, 396–401, 403, 412–415, 416, 421–423, 424–426, 429–431, 433–434, 436–440, 493–501, 503–505, 517–518, 525–527, 532–538
entropy of the Universe, 101, 103, 121–122, 203, 204, 206, 286, 353–354

equation of state, 104, 105, 106, 115, 144–146, 151–153, 168, 169, 171, 177, 200, 207, 215, 219, 221–223, 249, 254, 285, 288, 368, 371, 381, 509–513, 514, 521

F

false vacuum, 149, 169, 185–188, 190–195
false vacuum bag, 187–190, 193–195, 199
families, 17, 69, 71, 83, 87, 162, 208, 417, 423, 493, 497, 499, 503–504, 519, 526, 539
familon, 71, 162, 498, 504
family symmetry, 69, 71, 87, 162, 417, 499–505, 525, 526, 539
fermion mass matrix, 68, 70, 250, 502–503, 525
flavour changing decays, 4, 67–72, 496, 498, 504
flavour symmetry, 35, 39
flavour, 35, 38, 39, 496, 498, 504, 507
fractional electric charge, 37, 439–440
fracton, 439–440
free quark abundance, 37–38
freezing out, 114, 123, 133, 166, 209, 215–219, 221–226, 229, 231, 233, 237, 424, 439, 450
freezing temperature, 133, 220, 221, 424
frozen concentration, 123, 133, 210, 213, 215, 216, 219–226, 229, 231–239, 248–249, 254–256, 260, 316, 334, 391–395, 397–401, 412–414, 424, 439, 450

G

global symmetry, 71, 76, 82, 85, 86, 87, 497, 501, 504
gluino, 53, 158, 295–296, 326, 331
gluon, 38, 41–43, 49, 74, 77, 86, 262, 263, 266, 296, 302, 309, 310, 326, 331, 504, 505
Goldstone boson, 3, 55, 69, 71, 76, 77, 79, 86, 162, 497, 504
grand unification, 4, 9, 17, 48–56, 127, 131, 134, 137, 146, 152, 158, 159, 225, 254, 297, 299, 326, 420, 440, 493, 495, 497, 505, 523, 527, 540
gravitational instability, 126, 153, 170, 247, 369–386, 389–400, 405–410, 413–415, 419, 452, 453, 538

gravitino, 3, 54, 291–297, 304, 326, 330–331
GUT models, 4, 17, 49–54, 127, 131, 137, 152, 158, 159, 225, 254, 297, 299, 326, 420, 440, 449, 493, 495, 497, 505, 523, 527
GUT phase transition, 51, 134, 137, 146, 152, 299, 441
GUT symmetry breaking, 50, 52, 53, 131, 134, 137, 146, 440, 497

H

hadronic axion, 84–86, 89, 526–527
hadrons, 32, 34–36, 38, 43, 145, 302, 310, 326, 437, 439, 527
Hagedorn temperature, 115
heavy stable neutral leptons, 208–209, 359, 421–431
heterotic string, 5, 57, 163, 210, 218, 495
hidden sector, 4, 5, 8, 16, 18, 47–89, 154, 157, 162, 166, 207–210, 216–223, 226–227, 237–238, 244–246, 254, 283–284, 287, 291–296, 315–319, 330–331, 345–349, 358–362, 391–394, 396–401, 412–416, 419–420, 429–431, 437–443, 449–454, 494–498, 500–507, 517, 525–528, 536–541
Higgs boson, 3, 10, 44–45, 48, 64, 79, 80, 84, 87, 146, 425, 429, 430, 440, 494, 497, 499–503, 507, 515–517, 525
Higgs mechanism, 44–46, 64, 497
Higgs potential, 44–46, 80, 82, 84, 146, 149–150, 501–502, 507–508, 513–517
horizon problem, 149, 150
horizontal symmetry, 17–18, 86–87, 162, 417, 494–505, 515–517, 525–527, 539
horizontal unification, 17–18, 86–87, 159, 417, 493–508, %11–513, 515–517, 523–541
hot dark matter, 8, 163, 358, 373–376, 379–380, 383, 386–388, 394–395, 404, 405, 406–410, 532–533, 536, 537, 538, 539
Hubble constant, 92–93, 97–98, 102, 145, 197, 199, 202, 203, 305, 307, 335, 391, 424, 508–509, 511–512, 514–515, 522
Hubble law, 92

I

inflational cosmology, 9, 10, 128, 148, 152, 161, 162, 164, 444, 507–509, 513–517
inflaton, 154, 155, 157, 293, 507–509
inhomogeneous baryosynthesis, 158, 238, 524
invisible axion, 53, 76–89, 158, 497–498, 504–507, 517–524
island Universe, 161, 450

L

large-scale structure, 8, 152, 161, 162, 164, 351-361, 382–387, 389–415, 444–449, 451–453, 528–538
lepton number, 62, 160
lepton, 33, 69, 423–426, 498–500, 502–504
lepton asymmetry, 225, 226
light element abundance, 99, 125, 127, 162, 205, 212, 241, 279, 282, 283, 302
local gauge invariance, 31
local gauge transformation, 30
local phase transformation, 30
long derivative, 31
LSSP, 54, 293, 429

M

magnetic monopole, 51, 131, 133, 138, 145, 146, 148
Majorana mass, 62, 160
Majoron, 70, 163, 497
matter dominance, 96, 105, 172, 175, 205
maximon, 350
metastable massive particles, 258
mirror particles, 56, 210, 433–440
mirrority, 56, 442
multicomponent dark matter, 164

N

neutralino, 54, 429
neutrino decays, 69, 251–255, 557
neutrino dominance, 256, 362–388, 527, 532–538
neutrino magnetic moment, 65
neutrino mass, 8, 33, 34, 59–64, 127, 160, 163, 362–388, 497, 502–504, 525, 527, 532–538
neutrino mixing, 66,
neutrino oscillations, 66, 510
neutrino species, 208, 213, 214, 217, 230, 231, 232, 253
neutron capture, 278
neutron lifetime, 211, 278

non-equilibrium cosmological nucleosynthesis, 305, 306, 318, 323
non-equilibrium neutrino, 332–348,
non-equilibrium particles, 235, 241, 247, 248, 260, 262, 282–284, 302–307, 309–314, 316, 314, 320, 325, 328, 342
non-equilibrium processes, 241
nuclear cosmoarcheology, 323, 328, 331

O

old Big Bang scenario, 110, 115, 116
open Universe, 95

P

parity nonconservation, 43, 434–436
PBH evaporation, 185, 204, 206, 233, 236, 241, 259, 263, 270, 287, 292, 304, 324, 331
PBH formation, 154, 166–180, 188–195, 509–512, 539,
PBH mass spectrum, 201, 526
PBH relics, 202–204
PBH, 167–170, 175, 176, 178–182, 184, 185, 186, 200–204, 262–274, 282, 287, 288, 290–294, 300, 301, 526
PBH formation, 292
Peccei-Quinn symmetry, 55, 77, 497, 505, 517–519, 528
period of recombination, 126
photino, 53, 298
Planck length, 108
Planck mass, 109, 167, 184
Planck scale, 55
Planck spectrum, 241, 242, 244, 249, 256, 306
Planck temperature, 109, 149
Planck time, 108, 109, 149, 150, 151
primordial monopole abundance, 135
proton decay, 35, 50, 128, 138

Q

QCD confinement, 42
QCD phase transition, 43, 548
QCD vacuum, 78
QCD, 38
quantum electrodynamics, 40
quantum field theory, 29
quark mass matrix, 76, 502
quark, 38, 42, 71, 264, 265, 287, 517

R

radiation dominance, 104, 105, 147, 155, 156, 534, 535,
RD stage, 108, 240, 241, 247, 261, 262, 267, 269, 271, 274–276, 280, 281–286, 288, 292, 297, 301–304, 311, 318, 321, 324, 330, 331–333, 337, 341, 342
red shift, 92, 103, 104, 164, 233, 242, 244, 274, 275, 309, 340, 343, 348, 363, 364, 402, 406, 408, 448, 523
reheating, 148, 151, 155, 293, 300, 512
R-parity, 54, 293

S

scale factor, 95, 96, 103, 105, 147, 148, 149, 150, 151, 202, 207, 285, 365, 374, 375, 514
see- saw mechanism, 55, 503, 525
shadow particles, 58, 413, 450
shadow world, 58
slow rolling, 152, 156
specific entropy, 206, 258, 289
spontaneous CP violation, 158, 160, 238, 262, 299, 300
spontaneous symmetry breaking, 45
sterile neutrino states, 211, 227, 230, 231
strong CP violation, 59, 163
supergravity, 54, 298
superheavy fermions, 510
superheavy particles, 288, 290, 320, 339, 341–344
superstring theory, 5, 16, 57, 163, 210, 218, 495
supersymmetric models, 291–293, 298
supersymmetric particles, 53, 54
supersymmetry, 48, 54, 158

T

theories of everything, 5, 16, 57, 495
thermal neutrino background, 118, 338, 343, 346, 349
thin wall approximation, 191
topological defects, 51, 533
tritium, 280, 281

U

unstable dark matter, 163, 164, 410–415, 533–539
unstable neutrino, 66–71, 255, 257, 526, 527, 533–539

V

very early Universe, 110, 115, 116, 129, 147, 157, 165, 166, 175, 185, 198, 204, 507–517, 523
voids, 198

W

W boson, 43, 425, 426
wall-surrounded-by-strings, 53, 521
weak isospin group, 43
weak source, 304, 309, 311, 313, 341
WIMP, 420–430

Y

Yukawa interaction, 81, 500, 502

Z

Z boson, 43, 209, 253, 254, 421, 423, 425
Z boson width, 209, 253, 421